The name of M Nelkon is associated with physics throughout the English-speaking world. His classic *Advanced Level Physics* is in its fifth edition. Now a full-time author, he was previously Head of Science at William Ellis School, London and an examiner for the London, Cambridge and Welsh Boards.

M V Detheridge is the co-author with M Nelkon in the Pan Study Aids GCSE series. He has wide teaching experience, and is currently Deputy Head Teacher at William Ellis School, London. He has been an examiner for the Oxford and Cambridge Board.

Both authors have first-class degrees in physics.

Pan Study Aids for A level include:

Advanced Biology

Advanced Chemistry

Advanced Computing Science

Advanced Economics

Advanced Mathematics

Advanced Physics

Advanced Sociology

ADVANCED PHYSICS

M. Nelkon and M. V. Detheridge

Pan Books London and Sydney

First published in 1982 by Pan Books Ltd,
in association with Heinemann Educational Books Ltd

This edition published 1987 by Pan Books Ltd,
Cavaye Place, London SW10 9PG

9 8 7 6 5 4 3 2 1

© M Nelkon & M V Detheridge 1987

ISBN 0 330 29431 8

Text design by Peter Ward
Text illustrations by ML Design
Photoset by Parker Typesetting Service, Leicester
Printed and bound in Spain by
Mateu Cromo Artes Gráficas, Madrid

CONTENTS

Contents

Contents

ACKNOWLEDGEMENTS

The publishers are grateful to the following Examination Boards, whose addresses are listed below, for permission to reproduce questions from examination papers:

Associated Examination Board, University of Cambridge Local Examinations Syndicate, Joint Metriculation Board, University of London Schools Examination Board, Northern Ireland Schools Examinations Council, Oxford Delegacy of Local Examinations, Oxford and Cambridge Schools Examination Board, Southern Universities Joint Board, Welsh Joint Education Committee.

THE EXAMINATION BOARDS

The addresses below are those from which copies of syllabuses and past examination papers may be ordered. The abbreviations (AEB etc) are those used in the text to identify actual questions.

Associated Examining
Board (AEB),
Stag Hill House, Guildford,
Surrey GU2 5WJ

University of Cambridge Local
Examinations Syndicate (C),
Syndicate Buildings,
1 Hills Road,
Cambridge CB1 2EU

Joint Matriculation Board (JMB),
Manchester MI5 6EU

University of London School
Examinations Department
(publications office) (L),
52 Gordon Square,
London WC1H 0PJ

Northern Ireland Schools
Examinations Council (NI),
Examinations Office,
Beechill House, Beechill Road,
Belfast BT8 4RS

Oxford Delegacy of Local
Examinations (O),
Ewert Place, Summertown,
Oxford OX2 7BZ

Oxford and Cambridge Schools
Examination Board (O & C),
10 Trumpington Street,
Cambridge CB2 1QB

Southern Universities Joint
Board (SUJB)
Cotham Road,
Bristol BS6 6DD

Welsh Joint Education
Committee (WEL),
245 Western Avenue,
Cardiff CF5 2YX

TO THE READER

No book can hope to replace two years of steady and conscientious study. There is, however, a lot which every student can do to help make the revision period more effective. This book is designed to give you a brief presentation of the main ideas in A level physics and so aid the revision process.

Each chapter is laid out in a similar form. First there is the *basic text* giving a brief account of the main topics. Then there is a section of *worked examples* which follows the relevant text. Each example has a title where necessary to make it easier for you to find the ones you want to study. Where appropriate, an overview of the question is given showing the main strategy for solution. Next in each chapter are *two summaries*. The first is a verbal summary which reminds you of the main topics covered and some of the important formulae. The second summary is in diagram form and shows the relationships between the various ideas covered. Finally there are some *questions* from recent papers for you to try yourself. *Answers* to all the numerical questions are given in the back of the book.

Some of the things you can do to help yourself are listed below. It is always necessary to find the best way of working to suit each individual, but most teachers agree that students would do better if they followed these suggested guidelines.

REVISION

1 MAKE A TIMETABLE

First decide the number of hours in the week you can spend on revision. Don't forget that you need time to sleep and time for relaxation if the process is to be effective. Decide how to space your work – you might choose to do this evenly, or work a bit harder on some days and so get a day off. Work out the priorities in your revision and concentrate on timetabling those activities first. Above all be realistic – you will soon get very miserable if you set yourself impossible targets and then fail to meet them.

2 DECIDE WHERE YOU WILL STUDY

Always try to study in the same place; then you have everything ready. *Sit* at a table or desk. Don't sit in an armchair, or worse, lie on a bed – and can you *really* work hard while watching television or listening to music?!

3 SET OUT OBJECTIVES

At the start of each study period take two or three minutes to write down what you expect to achieve in that period. Again be realistic, but expect to work hard. It is sometimes helpful to relax for 5 minutes or so in the middle of any period of work, but *beware* of telephoning your boy/girl friend where 5 minutes easily becomes an hour; and cups of tea/coffee can easily turn into a feast when there is the prospect of a return to the study table.

4 MAKE BRIEF NOTES

This helps to keep your mind on the job and gives you a written summary for later reference. Often notes in diagrammatic form, like the diagram summaries, are very helpful. Don't let the note taking become an end in itself – it can easily take too much time if you do.

5 REVIEW YOUR WORK

In a period of study, try to spend some time going over again what you learned earlier in that session, and what you studied the day before. It has been shown that if you can go over work at intervals, your memory is very much helped. It is certainly much better to do this than to spend a lot of time on one topic and then never go back to it. In this way you forget a topic very easily.

EXAMINATIONS

Most physics exams have a variety of elements.

1 MULTIPLE CHOICE

By now you are probably familiar with this type of exam. If you are faced with a question which you do not know how to do, it is often helpful to eliminate some of the options. If need be you can then guess between those remaining!

2 SHORT ANSWERS

The tendency of candidates in these papers is to get too long winded and run out of time. Discuss with your teacher what is a sensible time to spend on each part of a paper, and then stick to that time allocation.

3   LONG ANSWERS

Here you may be asked, for example, for a description of some experiment or an explanation. Decide quickly what needs to be put in the answer. Use diagrams to communicate information – often this is a lot quicker than words. Be concise and to the point.

In numerical questions (in this paper or short answer papers) make sure you explain the physics. Always write down clearly any formulae you use. Put in numerical values in the same order as the symbols appear in the equation – the examiner can then see what you are doing. Don't forget to put in the correct units.

4   USE OF DATA

Often you may need to plot a graph. Make this of reasonable size covering most of the graph paper. If a straight line relationship of the form $y=mx+c$ is plotted, then m is the *gradient* and c is the *intercept on the y-axis* (where $x=0$). Always find the gradient of a line (i) by drawing the line as long as possible, and (ii) using the numerical values on the x- and y-axes to work out the gradient. If an intercept on the y-axis is required then you must plot the x-axis starting at $x=0$.

5   COMPREHENSION PAPER

This tests your ability to come quickly to terms with new information about physics. Don't expect *all* the information to be in the article – the examiner is assuming that you have studied A level physics and will expect you to use that knowledge where relevant. It is probably helpful to 'scan' the article very quickly to get the gist of it before reading it more carefully. If necessary read the article *twice* to get the sense of it.

6   PRACTICAL PAPER

Be careful to do what you are told and *don't* do things unless asked. For example, if you are told to make an observation and record your answer, then don't write lengthy explanations unless specifically asked to do so. Keep apparatus on the bench – particularly circuits – neatly laid out. This can save error and confusion.

MECHANICS: MOTION, FORCE, MOMENTUM, ENERGY

CONTENTS

EQUATIONS OF LINEAR MOTION

In the *Pan Study Aids: Physics* (London, Pan Books by M. Nelkon and M. V. Detheridge) we discussed the topics of speed, velocity and acceleration. If an object has a uniform acceleration a while it changes from an initial velocity u to a final velocity v in a time t, then, if s is the displacement or distance travelled

$$v = u + at \tag{1}$$
$$s = ut = \tfrac{1}{2}at^2 \tag{2}$$
$$v^2 = u^2 + 2as \tag{3}$$

Velocity is the rate of change of displacement, or ds/dt, and has unit m s^{-1}. **Acceleration** is the rate of change of velocity, or dv/dt, and has unit m s^{-2}. The acceleration of free-fall, or acceleration due to gravity, g, is about 9.8m s^{-2} or 10m s^{-2} in round figures.

MOTION AND DIRECTION

A car moving in a certain direction with a speed of 20m s^{-1} may be said to have a velocity of '+20m s^{-1}' taking this direction as positive. Another car moving with a speed of 20m s^{-1} in the opposite direction then has a velocity of '−20m s^{-1}'.

Similarly, a ball thrown vertically upwards with an initial velocity 20m s^{-1} has a velocity measured *upwards* of +20m s^{-1}. Under the downward pull of gravity the ball decelerates at say 10m s^{-2}(g), that is, its upward acceleration can be written as $a = -g = -10$m s^{-2}. The correct sign must always be used in the equations of motion. This is illustrated in Example 1.1 on motion under gravity on page 26.

DISTANCE (*s*)–TIME (*t*) GRAPHS

If we plot a *distance* (*s*)–*time* (*t*) graph for an object moving in a certain direction, the velocity v at any instant is found from the *gradient* to the graph at that instant since $v = ds/dt =$ rate of change of s with time t. Figure 1.1(*a*) shows the graph of distance s against time t when a ball is thrown vertically upwards. At the start O, the velocity, v, is the gradient, a/b, of the *tangent* to the curve at O. At the greatest height A the tangent to the curve is horizontal and the gradient is zero so $v = O$. At B, when the ball is falling, the gradient is downwards, or *negative*.

Velocity = GRADIENT of s–t graph

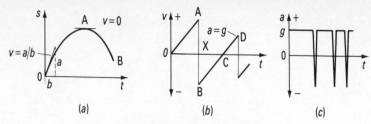

Fig 1.1 Graphs of distance, velocity and acceleration against time

VELOCITY (v)–TIME (t) GRAPHS

Figure 1.1(*b*) shows the *velocity (v)–time (t)* graph of a ball released from a height to the ground and then bouncing upwards at the ground. Generally, the gradient of the graph at any instant is the *acceleration a* at this instant since $a=dv/dt$.

Acceleration=GRADIENT of *v–t* graph

In this case the acceleration of the falling ball is constant and equal to *g*. The graph is therefore a straight line OA until it hits the ground. At this instant the ball bounces upwards and so its velocity is negative as shown by B. As it moves upwards the velocity decreases along BC, becomes zero at C and then increases along CD as it falls again. The gradient of the lines OA and BC is the same. The gradient is equal to *g*, the acceleration due to gravity, which is in a downward direction whether the ball is rising or falling. As shown, the ball bounces up with a slightly smaller velocity than the velocity with which it hits the ground.

DISTANCE FROM VELOCITY–TIME GRAPHS

The velocity–time graph is useful because the *distance* travelled can be found from it, in addition to the acceleration. Whatever the shape of the *v–t* graph, the distance travelled is given by the *area between the graph and the time axis*, since distance=velocity×time and we can divide any area into small strips where the velocity is constant over very small time intervals. So in Fig 1.1(*b*), the distance travelled by the ball just before it hits the ground is equal to the area of the triangle OAX. The distance travelled to its greatest height after it rebounds is equal to the area of the triangle BXC.

Distance=AREA between *v–t* graph and time-axis

Figure 1.1(*c*) shows how the *acceleration a* of the bouncing ball varies with time *t*. The value is *g* except for the short time the ball hits the ground and is given an opposite acceleration. Since the ball rebounds with smaller velocity, the time during which the acceleration is *g* decreases, as shown.

VECTORS AND SCALARS; RESULTANT

Many quantities in physics have direction as well as magnitude (size). They are called **vectors**. Velocity, acceleration and force are examples of vectors. Quantities such as speed, mass and energy have magnitude but no direction. They are called **scalars**.

We can find the sum or **resultant** of two vectors in one of two ways. In the 'parallelogram method', the two vectors P and Q are represented in magnitude and direction by OA and OC respectively (Fig 1.2(a)). When the parallelogram of OABC is completed, the *diagonal* OB represents the resultant R of P and Q in magnitude and direction. In the 'triangle method', one vector P is represented by OA and this is followed by drawing AB to represent the vector Q (Fig 1.2(b). Note that the arrows of P and Q follow each other. Then OB represents the resultant R of P and Q, or $\vec{R}=\vec{P}+\vec{Q}$. The resultant R is the same as in Fig 1.2(a).

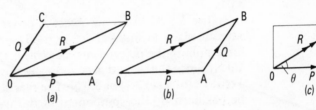

Fig 1.2 Vector addition

The vector sum or resultant of two *perpendicular* vectors is often needed. In this case, from the parallelogram or triangle method the resultant R is given from Pythagoras' theorem for a right-angle tri-angle (Fig 1.2(c)), by

$$R=\sqrt{P^2+Q^2}$$

Also, the angle θ which R makes with P is given by

$$\tan \theta = \frac{BC}{OC}=\frac{Q}{P}$$

RESOLVED COMPONENT OF FORCE

In physics we often need to find the effect of a force in a direction different from the force direction. This is called the resolved **component** of the force in the direction concerned.

Figure 1.3(a) shows a force $\vec{F}$ acting in the direction OA on an object at O, which can only move along rails in the direction OB at an angle θ to OA. From the parallelogram of vectors, we see that F is the resultant or sum of two forces represented by the sides OC and OD of the rectangle OCAD. Now the force represented by OD has no effect along OB since it acts perpendicular to OB. So the force OC represents the resolved component of F in the direction OB.

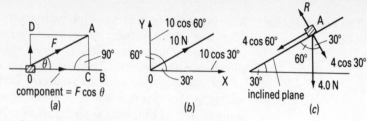

Fig 1.3 Components of forces

Since OC/OA=OC/F=cos θ, then OC=F cos θ. *Remember* that

component of F at angle θ = F cos θ

So in the direction OX at 30° to a force of 10 N, Fig 1.3(b), the component is 10 cos 30° or 8.7 N. In the perpendicular direction to OX, along OY, the component is 10 cos 60° or 10 sin 30°, which is 5.0 N.

Figure 1.3(c) shows an object A of weight 4.0 N supported on a smooth plane inclined at 30° to the horizontal. If A is released, the force pulling the object *down the plane* is the component of its weight 4.0 N. Since the weight acts vertically downwards, the component is 4 cos 60° (or 4 sin 30°) or 2.0 N. The reaction R of the plane on A must be balanced by the component of the weight at right angles to the plane since A never moves at right angles to the plane. So R=4 cos 30° or 3.5 N.

Example 1.2 on page 27 gives more examples of components.

FORCE, ACCELERATION AND MOMENTUM

Newton defined force as 'the rate of change of momentum' it produces in the direction the force acts. 'Momentum' is the product 'mass times velocity' of a moving object.

Momentum = mass × velocity (mv)

$$\text{Force } F=\frac{\text{momentum CHANGE}}{\text{time } t}$$

So if the velocity of an object of mass m is changed from u to v in a time t by a force F, then

$$F \propto \frac{mv-mu}{t}$$

If m remains constant during the time, then, since $(v-u)/t=a$, acceleration,

$$F \propto ma \text{ or } F=kma$$

where k is a constant. The **newton** (N) is defined as the force which gives a mass of 1kg an acceleration of 1m s^{-1}. So with $F=1$, $m=1$ and

$a=1$, then $k=1$. We can therefore find the force F acting on an object from the relation.

$F=ma$

where F is in newtons, m in kilograms and a in metre second^{-2}.

Note carefully that F is the *resultant* force. In Example 1.3 later, you will see that all the forces which act on an object are added vectorially to find F. On the other hand m is a scalar quantity – it has no direction and is the mass of the object in kilograms.

A mass of 1kg falling freely under gravity has an acceleration g of 9.8m s^{-2} or 10m s^{-2} in round figures. So the gravitational pull or **weight** of the mass of 1kg is 9.8 N or 10 N in round figures. Generally

weight$=mg$

where m is the mass.

GRAVITATIONAL FIELD STRENGTH (INTENSITY)

Since $g=$weight/mass, we can write $g=9.8$ or 10 N kg^{-1}. The force per unit mass in a gravitational field is called the **gravitational field strength** or **intensity**. You will often see $g=9.8$m s^{-2} or 9.8 N kg^{-1} in examination papers.

In space well away from the Earth, the mass of an object remains the same but the gravitational field strength diminishes to some value g'. The weight of the object then falls to a value mg'. On the moon's surface the gravitational field strength is about 1.6 N kg^{-1}. So the weight of mass of 1 kg there is 1.6 N.

Electric fields occur inside television tubes or near aerials and electric field strength or intensity is discussed later. A comparison between gravitational fields and electric fields is given on page 84.

HOW TO APPLY $F=ma$ IN PROBLEMS

Problems on acceleration in a lift or on an inclined plane or of a trailer pulled by a truck, for example, can be done by remembering the following rules:

1 In $F=ma$, F is the *resultant* force of all the forces acting. So put in the directions of all the forces in your diagram and work out the resultant.
2 The acceleration a (a vector) is in the direction of the resultant force F. So check that F and a are in the same direction.
3 The mass m is a scalar. You do not have to worry about its direction.

These rules are illustrated in Example 1.3 on page 27.

FORCE AND MOMENTUM CHANGE

Suppose a snooker ball of mass 0.4kg strikes a cushion with a velocity of 5m s^{-1} and rebounds from the cushion in $\frac{1}{10}$ second with the same speed. Then

$$\text{Force on cushion, } F=\frac{\text{momentum change}}{\text{time}}=\frac{0.4\times[5-(-5)]}{1/10}$$
$$=4\times10=40\text{N}$$

Here the mass of the ball stayed constant and the velocity changed with time.

Forces are also produced by momentum change when the mass varies with time, for example, when fast-moving water from a hose-pipe strikes a wall. In this case

$$\begin{aligned}\text{force}\quad F&=\frac{\text{momentum change of water}}{\text{time}}\\&=\frac{\text{mass}\times\text{velocity change}}{\text{time}}\\&=\frac{\text{mass of water}}{\text{time}}\times\text{velocity change}\end{aligned}$$

So $\quad\quad F=$mass of water per second$\times$velocity change

If the mass of water per second coming from a hosepipe is 50kg s^{-1} and the speed is 10m s^{-1} before it hits the wall and zero afterwards, then

$$F=50\times(10-0)=500\text{ N}$$

This is illustrated in Example 1.4 on page 29.

ACTION AND REACTION

In his third law, Newton stated that:

Action and reaction are always equal and opposite.

If you lean on the table with your elbow, the downward force (action) on the table is equal and opposite to the upward force (reaction) of the table on your elbow. When a ball is kicked, the force on the ball by the foot is equal to the force on the foot by the ball at the instant of contact. The effect on the ball by the action force is considerable but the reaction force on the foot is hardly noticeable.

Forces of reaction, however, are used in a lawn sprinkler. The water is accelerated as it passes round the bend towards the nozzle and the force on the water produces an equal and opposite force on the sprinkler, which then rotates backwards. In the jet aeroplane, hot gases are accelerated from the rear with considerable force and the equal force of reaction pushes the aeroplane forward.

Action and reaction are forces which may also act over a distance. For example, the gravitational pull on the Earth as it rotates round the sun is equal and opposite to the gravitational pull of the Earth on the sun. Similarly, when a ball X falls to the ground, the attraction of the Earth on X is equal and opposite to the attraction of X on the Earth. The downward pull on the ball is considerable because its mass is small. The effect of the upward pull on the Earth is not noticeable as the Earth has an enormous mass, about 10^{24}kg.

Helicopters have an uplift force on them due to a reaction force when the blades hit the air downwards as they rotate. This is illustrated by Example 1.5 on page 29.

CONSERVATION OF LINEAR MOMENTUM. EXPLOSIVE FORCES

When a bullet is fired from a rifle. the explosive force F acts for the same time t on the bullet and the rifle. The bullet moves forward under the force of action and the rifle moves *backwards* under the force of reaction. Now from $F \times t = mv - mu$ on page 000, or from Newton's laws we see that

$$F \times t = \text{momentum change}$$

$F \times t$ is called the **impulse** (time-effect)of the force. Since F and t are the same numerically for the bullet and rifle, if the forward momentum change of the bullet is +100 units, for example, the backward momentum change of the rifle is −100 units. So the total momentum of the bullet and rifle $= +100 - 100 = 0$. This is the same as the total momentum *before* the explosion as the bullet and rifle were stationary. We see that the total momentum is 'conserved' when an explosion occurs.

The conservation of momentum helps to calculate velocities. Suppose m and M are the respective masses of the bullet and rifle, and v and V their respective velocities after the explosion (Fig 1.4(a)). Then from the conservation of momentum,

$$mv + MV = 0$$

or
$$V = -\frac{m}{M}v$$

So if the mass m of the bullet is 0.06kg, the mass M of the rifle is 1kg and the velocity of the bullet is 200m s^{-1}, the backward velocity V of the rifle is

$$V = -\frac{0.06 \times 200}{2} = -6\text{m s}^{-1}$$

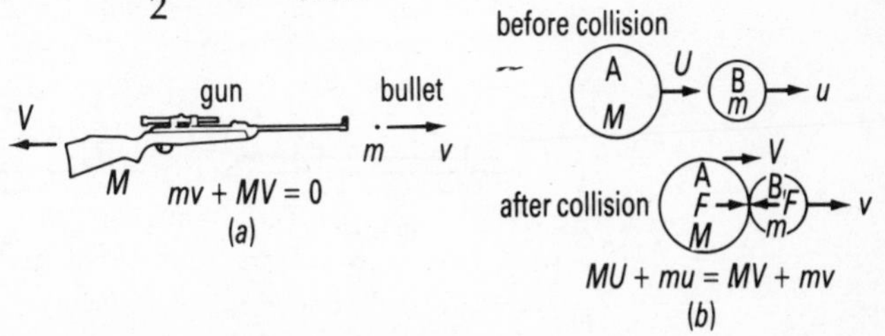

Fig 1.4 Conservation of linear momentum

**MOMENTUM IN
COLLISIONS**

If two moving objects A and B collide, then, from the law of action and reaction, A exerts a force F on B equal and opposite to that exerted by B on A (Fig 1.4(b)). Also, the forces F act for the same time t on A and B respectively. Taking into account the opposite directions of the forces,

> momentum change of A$=+F.t$, momentum change of B$=-F.t$

So the *total* momentum change $=+F.t-F.t=0$. This means that the total momentum of A and B is conserved, or the same before and after the collision.

The conservation of the linear momentum can be expressed as follows:

> **If no external (outside) forces act on two colliding objects, the total momentum in a given direction before collision is equal to the total momentum in the same direction after collision.**

Examples 1.6 and 1.7 on page 30 illustrate how the momentum conservation is applied. Since momentum is a vector we need plus and minus signs in front of values of momentum where their directions are opposite.

WORK, ENERGY AND POWER

WORK

When a force F pulls an object through a distance s, the force is said to do **work**. If s is in the same direction as F (as in Fig 1.5(a)), then, by definition

> work done, $W=F\times s$

1 **joule** (J) is defined as the work done when a force of 1 N moves through a distance of 1m in the direction of the force. So $1\,J=1\,N\times1m$. If the force F is inclined at an angle of θ to the direction of s, we must use the component of F in the direction of s, which is $F\cos\theta$ (Fig 1.5 (b)). In this case $W=F\cos\theta\times s$.

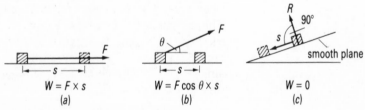

Fig 1.5   Work

We see that a force at 90° to a moving object does no work. In Fig 1.5(c), for example, the normal reaction R of a plane on an object sliding down an inclined plane does no work on the object. Work would be done against the frictional force F acting on the object if the plane were rough, and in this case the work done is $F \times s$ where s is the distance along the plane moved by the object.

KINETIC ENERGY

A shot putter has to provide a considerable muscular force F over a short distance s of his or her arm to throw the shot a long way. The putter does work equal to $F \times s$ and this transfers **energy**. The amount of energy transferred to the shot is equal to the work done. Energy, like work, is measured in joules. Note that the energy transferred to the shot comes from the chemical energy of the shot putter.

A moving object is said to have **kinetic energy** (k.e.). Suppose the shot has a mass m and is thrown with velocity v. Then,

$$\text{kinetic energy} = \text{work done} = F \times s = ma \times s,$$

where a is the acceleration of the shot starting from rest and reaching a velocity v. If F is constant and a is uniform, then, from $v^2 = u^2 + 2as$, we have $v^2 = 2as$ since $u = 0$. So $as = v^2/2$. Hence from above,

$$\text{kinetic energy} = mas = \tfrac{1}{2}mv^2 \qquad (1)$$
$$\text{So} \quad \text{kinetic energy} = \tfrac{1}{2}\,\text{mass} \times \text{velocity}^2$$

The kinetic energy is in J when m is in kg and v is in m s^{-1}.

When it is thrown, a ball has kinetic energy of *translation*. The molecules of a solid have kinetic energy of *vibration*. A diatomic gas molecule has mainly kinetic energy of translation and *rotation* (or spin). During play, a tennis ball or cricket ball may have both translational and rotational energy, that is, it may spin as well as move forward.

GRAVITATIONAL POTENTIAL ENERGY. MOLECULAR POTENTIAL ENERGY

A ball raised from the ground to some height h has **gravitational potential energy** (p.e.) when it is stationary at this level, that is, it is capable of doing work. For example, a pile-driver, a massive weight held stationary above a wooden stake, can drive the stake into the sea-bed when it is allowed to fall on it.

The gravitational potential energy of an object of mass m is the work done in raising it through a height h, assuming the potential energy has a 'zero' value at the ground, say. Now the force exerted in raising the object steadily is mg, its weight. Since work done = force × distance,

$$\text{potential energy} = mgh \qquad (2)$$

The potential energy is in J when m is in kg, $g = 9.8$ N kg^{-1} and h is in m.

Note that h is a small change in height above the ground, in which case the gravitational force mg is constant for the distance h. For small changes in height well above the earth the potential energy change can still be calculated from $mg'h$, where g' is the gravitational field strength at this height. But if the changes in height are big, we can *not* use $mg'h$ because g' will vary appreciably from one height to the other.

Gravitational potential energy, then, is due to position or level. Molecules inside a metal spring are particles in the *molecular field* of the metal, which is a completely different kind of field from the gravitational field. If the spring is wound more tightly, the molecules come closer together and their molecular potential energy has increased. The person winding the spring has transferred some of his or her chemical energy to molecular energy. Also, when a tennis ball is hit, some of the molecular potential energy of the tight racquet strings has been transferred to mechanical kinetic energy of the ball.

ENERGY CHANGES; ELASTIC AND INELASTIC COLLISIONS

Suppose a ball is thrown vertically upwards with an initial kinetic energy of 20 J. As it rises, some of this energy is transferred to potential energy but the total energy (potential plus kinetic) is always constant. So if the kinetic energy falls to 5 J at some height, the potential energy is then (20−5) or 15 J. At the top of the motion the kinetic energy becomes zero and the potential energy is now 20 J.

In this case we see that the total energy remains constant. In collisions, however, some of the kinetic energy may be transferred to energy of a different kind. When two toys cars, for example, collide head-on, most of their kinetic energy before collision is changed to heat and sound after collision. If the metal of the car is compressed, some change in the molecular potential energy of the metal may also occur. This type of collision, in which the total kinetic energy is less after collision, is called an **inelastic collision**. In a container of gas, the molecules are constantly colliding with each other, but their kinetic energy is unchanged. This is called an **elastic collision**. Electrons colliding with a gas atom may make an elastic or an inelastic collision depending on their energy, as we see later.

In elastic collisions, total kinetic energy is conserved. In inelastic collisions, total kinetic energy is not conserved.

gas molecules
(elastic collisions)

inelastic collision

(b)

(a)

Fig 1.6 Elastic and inelastic collisions

A football which is kicked is an example of an inelastic collision (Fig 1.6(*b*)). A lump of putty thrown on to the ground is an inelastic collision. As we stated before, decrease in mechanical energy produces an equal amount of energy in the form of heat and sound. The Principle of Conservation of Energy states: *In a closed system, energy can be transferred from one form to another but the total amount of energy remains constant.*

> *Question*: Is total momentum conserved in an inelastic collision?
> *Answer*: Yes. The conservation of momentum is true for *all* collisions
> *Question*: What is conserved in an elastic collision?
> *Answer*: Total momentum and total kinetic energy are conserved

Examples 8 and 9 on momentum and energy are given on page 32.

KINETIC ENERGY AND MOMENTUM RELATION

The momentum of a moving object, $p=mv$, where m is the mass and v is the velocity.

$$\text{kinetic energy} = \frac{1}{2}mv^2 = \frac{m^2v^2}{2m} = \frac{p^2}{2m}$$

So objects with the *same* momentum p have kinetic energy proportional to $1/m$, that is, the smallest mass has the greatest kinetic energy. This is due to the factor v^2 which occurs in the kinetic energy. The object with the smallest mass has the greatest velocity for a given momentum.

Consider a stationary radioactive nucleus which disintegrates with the emission of an alpha-particle X of mass 4 units, leaving a heavy mass Y of 204 units which recoils backwards. This is similar to the recoil of a rifle after a bullet is fired. From the conservation of momentum, the momentum of the alpha-particle X is equal to the momentum of the heavy mass Y. Suppose the total energy produced by the emission is W joules. Then, since the momentum p of X and Y is the same, from the above relation

$$\frac{\text{kinetic energy of X}}{\text{kinetic energy of Y}} = \frac{1/4}{1/204} = \frac{204}{4}$$

So the kinetic energy of X=204/(204+4) of W joules=204W/208= 51W/52. So the alpha-particle has the major share of the total energy.

POWER

The **power** P of a machine is the *rate of doing work* or the work done per second. Power is measured in **watts**, symbol W. 1W is defined as the rate of working at 1 J s^{-1}. Larger units of power are the kilowatt, kW, and the megawatt, MW. 1kW=1000W and 1MW=10^{6}W.

Consider a car travelling at a constant velocity v. Its acceleration is then zero, so the resultant force on it is zero. This means that the forward force F on the car due to the engine is exactly balanced by the resistance to the motion, which acts in the opposite direction to F. Suppose the total resistance is 200N. Then F=200N. If the car is moving at a constant velocity v of 20m s^{-1}, then

engine power P=work done per second=$F\times$distance per second

So $P=F\times v$ (1)
              $=200\times20=4000W=4kW$

You should remember that

Power=force$\times$velocity

Example 1.10 on page 33 shows how to calculate the power of a car engine when the car moves up an inclined plane.

Example 1.10 on page 33

WORKED EXAMPLES ON MOTION, FORCE, MOMENTUM AND ENERGY

1.1 VERTICAL MOTION UNDER GRAVITY

A tennis ball is thrown vertically upwards with an initial velocity of 20m s^{-1}.
(a)   How long does it take to reach the top of its motion?
(b)   How high did it go?

(Overview
(a)   We need the time. So we use $v=u+at$.
(b)   We need the distance. So we use $s=ut+\frac{1}{2}at^2$.)

(a)   At the top of its motion, the final velocity v=0. So from $v=u+at$,

$v=0=20-10t$, or $=t=2$ s

(b)   The distance s travelled to the top, or greatest height of the ball, is found from $s=ut+\frac{1}{2}at^2$. In this case $u=+20$m s^{-1}, $a=-10$m s^{-2} and $t=2$ s. So

$s=(20\times2)-\frac{1}{2}\times10\times2^2=20$m

1.2 COMPONENT VALUES

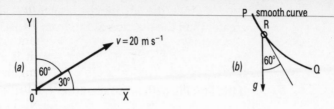

Fig 1.7 Component values

(*a*)   A javelin is thrown initially from O with a velocity *v* of 20m s^{-1} at 30° to the horizontal (Fig 1.7(*a*)).
With what velocity is the javelin moving initially in:
    (i) a horizontal direction, OX; and
    (ii) a vertical direction, OY?

(**Overview** Component of a vector quantity *F* in a direction at an angle to *F* is *F* cos *θ*.)

(i)   In direction OX, component=*v* cos 30°=20 cos 30°=17m s^{-1}
(ii)   In direction OY, component=*v* cos 60° (or *v* sin 30°)=10m s^{-1}

(*b*)   A smooth ring R moves down a smooth wire, PQ (Fig 1.7(*b*)).
What is the acceleration of the ring when the tangent at R to PQ makes 60° with the vertical (assume *g*=9.8m s^{-2})?
Without the wire, the ring R would fall vertically downwards with an acceleration *g* of 9.8m s^{-2}.
Sliding on the wire, R has an acceleration tangentially to the wire at an angle of 60° to the vertical at this instant. So

$$\text{acceleration=component of } g \text{ along tangent}$$
$$=g \cos 60°=9.8 \cos 60°=4.9\text{m s}^{-2}$$

1.3 HOW TO APPLY F=ma   **1 Lift. Weightless phenomenon.** A person of mass 50kg moves
(*a*)   upwards
(*b*)   downwards
in a lift with an acceleration of 2m s^{-2}. Calculate the reaction force at the floor of the lift in each case.
    What is the reaction force if the light falls with an acceleration equal to *g*?
(*a*)   The forces on the person are the reaction *R* acting upwards and the weight *mg* acting downwards (Fig 1.8(*a*)). Since the lift moves upwards, the resultant upward force *F* on the person is (*R*−*mg*). From *F*=*ma* where *F* is in newtons,

$$R-(50 \times 10)=50\times 2$$
So $$R=100+500=600\text{N}$$

(*b*)   Since the lift moves *downwards*, the weight *mg* or 500N is greater than the reaction *R*. So, from *F=ma*,

$$500-R=50\times2=100$$

Hence $R=500-100=400N$

If the lift falls with an acceleration *g*, then

$$mg-R=ma=mg$$

So $R=0$. In this case, the person feels no reaction at his or her feet and this is the sensation of being 'weightless'. So, as we see later, an astronaut inside a spacecraft, falling with the spacecraft, experiences no reaction force and therefore feels 'weightless'.

2 Inclined plane and resistance A vehicle X of mass 1000kg is driven up an incline at 30° to the horizontal with an acceleration of 2m s⁻². The resistance to the motion is 1000N. Calculate the forward force *P* on the car due to the engine (Fig 1.8(*b*)).

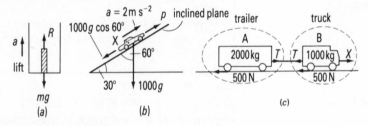

Fig 1.8 Problems on acceleration

The forces on X are
1   the engine force *P* acting *up* the plane,
2   its weight *mg* or 10000N acting *vertically downwards*,
3   the resistance of 1000N acting *down* the plane.
       The component of the weight of 1000*g* or 10000N acts down the plane in opposition to the forward force *P*. Now
       Component=10000 cos 60° (or 10000 sin 30°)
                    =5000N

So for motion up the inclined plane, resultant force *F* on vehicle= $P-5000-1000=P-6000N$

       From *F=ma*,

       $P-6000=1000\times2=2000$
       So $P=6000+2000=8000N$

3 Trailer and truck A trailer A of mass 2000kg is attached by a chain to a truck B of mass 1000kg (Fig 1.8(*c*)). Assuming the chain is taut, calculate
(*a*)   the force X on the truck caused by the engine and

(b) the tension in the chain, when the truck and trailer are moving horizontally with an acceleration of 3m s^{-2}. The resistance to the motion of the truck and trailer is 500N each.

Here we consider *one vehicle at a time*. For the *truck*, if T is the tension (force) in the chain,

resultant force $F=X-T-500=ma=1000\times3$

So $X-T=3500$ (1)

For the *trailer*,

resultant force $F=T-500=ma=2000\times3$

So $T=6500$N   (2)

From (1), $X=3500+T$

                                  $=3500+6500=10\,000$N

1.4 FORCE ON WALL DUE TO WATER FROM HOSEPIPE

A hose-pipe of area of cross-section 10^{-2} m^2 ejects water horizontally with a velocity of 20m s^{-1}. Calculate the force exerted by the water on a vertical wall if the water comes to rest on impact with the wall. Density of water$=1000$kg m^{-3} (Fig 1.9).

Force on wall$=$mass of water per second$\times$velocity change

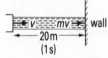

In 1 second the water from the hose would occupy a length of 20m. So

Fig 1.9 Force due to water

mass of water per second$=(20\times10^{-2})\times1000kg=200$kg

since mass$=$volume$\times$density. The velocity of the water changes from 20m s^{-1} to zero on impact with the wall. So

force on wall$=$momentum change per second of water
$=200\times20=4000$N

1.5 REACTION FORCE ON HELICOPTER

A helicopter has blades of radius 2m and hovers in the air when the blades are rotating and hitting the air vertically downwards with a velocity of 30m s^{-1}. Assuming the density of air is 1.2kg m^{-3}, estimate the weight of the helicopter.

(**Overview** Downward force on air$=$mass of moving air per second$\times$velocity change.
Upward reaction force$=$downward force on air.)

Downward force on air$=$mass of moving air$\times$velocity change

In 1 second, volume of air moving down$=$area swept by blades $\times$ velocity of air

$=\pi\times2^2\times30$

So mass of air per second$=$volume per second$\times$density

$$=\pi\times2^2\times30\times1.2=450\text{kg}$$
$$\therefore \quad F=450\times(30-0)=13\ 500\text{N}$$
$$=\text{weight of helicopter}$$

1.6 COLLISIONS IN SAME DIRECTION

Two steel balls A and B of mass 0.1kg and 0.2kg respectively, approach each other along a horizontal groove with velocities of 2m s^{-1} and 5m s^{-1} respectively (Fig 1.10). After the collision, B moves in the same direction with a velocity of 1m s^{-1}. Calculate the velocity of A after collision.

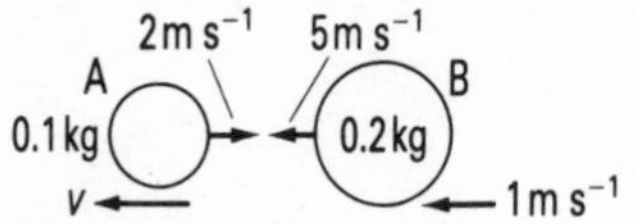

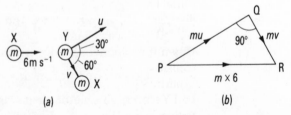

Fig 1.10 Collision calculations

Take the direction of the initial velocity of B, 5m s^{-1}, as a +ve direction. Then

$$\text{initial total momentum}=(0.2\times5)-(0.1\times2)=0.8\text{kg m s}^{-1}$$

and

$$\text{final total momentum}=(0.2\times1)+0.1\ v=0.2+0.1\ v$$

where v is the velocity of A as shown in Fig 1.10. So

$$0.2+0.1\ v=0.8,\ \text{or}\ v=0.6/0.1=6\text{m s}^{-1}$$

1.7 COLLISIONS WITH REBOUND IN DIFFERENT DIRECTIONS

A ball X of mass m, moving with a velocity of 6m s^{-1}, hits a stationary ball Y of similar mass m at a glancing angle (Fig 1.11(a)). Y then moves off at an angle of 30° to the initial direction of X and X moves off at an angle of 60° to its initial direction. Calculate the velocity u of Y and the velocity v of X after the collision.

Fig 1.11 Elastic collision

Since we have two unknowns, u and v, we need to apply conservation of momentum in two different directions.

Direction XY In this direction,

(i)   total momentum of X and Y before collision$=m\times6=6m$ and;

(ii)   total momentum of X and Y after collision$=mv\cos 60°+mu\cos 30°$, using components. $\cos 60°=\frac{1}{2}$, $\cos 30°=\sqrt{\frac{3}{2}}$

So $\dfrac{mv}{2}+\dfrac{\sqrt{3}}{2}mu=6m$ (1)

Direction 90° to XY In this direction, initial momentum of X$=0$ and so total momentum$=0$.

In this direction, total momentum downward of X and Y$=mv\sin 60°-mu\sin 30°=\sqrt{3}\,mv/2-mu/2$. From conservation of momentum,

$$\dfrac{\sqrt{3}}{2}mv-\dfrac{mu}{2}=0$$ (2)

From (1), cancelling m, $v+\sqrt{3}\,u=12$
From (2), cancelling m, $\sqrt{3}v+u=0$

Solving these equations, we find $v=3$m s^{-1} and $u=5.2$m s^{-1}

Figure 1.11(b) shows the initial momentum ($m\times6$) of X represented by a vector PR. This is the total momentum initially of X and Y since Y is stationary. The final momentum of Y is represented by the vector PQ and the final momentum of X by the vector QR. Since u is perpendicular to v, angle PQR is 90°.

From the principle of conservation of momentum, PR is the vector sum of PQ and QR, that is, PQR is a *closed* triangle. Further, angle PQR is 90° and angle QPR is 30°. So

$$\dfrac{PQ}{PR}=\dfrac{mu}{m\times6}=\cos 30°$$

and $u=6\cos 30°=5.2$m s^{-1}

Also $\dfrac{QR}{PR}=\dfrac{mv}{m\times6}=\sin 30°$

So $v=6\sin 30°=3.0$m s^{-1}

This agrees with our results by calculation.

Note

Initial total kinetic energy of X and Y$=\frac{1}{2}mv^2=\frac{1}{2}\times m\times6^2=18m$.

Final total kinetic energy of X and Y$=\frac{1}{2}\times m\times5.2^2+\frac{1}{2}\times m\times3^2=18m$

When the kinetic energy is conserved in addition to momentum in a collision, we call this an *elastic collision* (see page 24). When an α-particle X collides with a helium nucleus Y at rest, the tracks of X and Y after collision are found to be at 90° to each other. This is due to the fact that the masses of X and Y are equal and that an elastic collision was made.

1.8 FORCE IN MOMENTUM AND WORK

Calculate the force which would bring a vehicle of mass 1000kg moving with a velocity of 20m s^{-1} to rest in

(a)   a time of 40 s and

(b)   a distance of 25m.

(a) $F \times t = momentum$ change

So $F \times 40 = mv = 1000 \times 20$, or $F = 500$N

(b) $F \times s = $ work done $= energy$ change

So $F \times 25 = \frac{1}{2}mv^2 = \frac{1}{2} \times 1000 \times 20^2 = 200\,000$

$$F = 200\,000/25 = 8000\text{N}$$

1.9 MOMENTUM AND KINETIC ENERGY IN COLLISIONS

A bullet of mass 50g or 0.05kg is fired with a velocity of 100m s^{-1} into a stationary block of mass 2kg suspended at the end of a long string. Calculate the height the block rises if the bullet becomes embedded in the block (Fig 1.12).

(Overview

(i)   The height h is connected with the gain in potential energy (p.e.) of block and bullet (mgh).

(ii)   gain in p.e. = kinetic energy ($\frac{1}{2}mv^2$) of block and bullet after bullet enters block.

(iii) v can be found from conservation of momentum of block and bullet.)

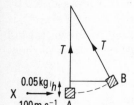

Fig 1.12 Calculation on momentum and energy

Initial momentum of bullet (X) and block (A) $= 0.05 \times 100 = 5$kg m s^{-1} since the block has no momentum

Final momentum of bullet and block $= 2.05v$

where v is the velocity of block and bullet.

So $2.05v = 5$, or $v = 5/2.05 = 2.44$m s^{-1} (step iii)

In Fig 1.12, the block and bullet rise from A and B through a vertical height h. We can apply the conservation of energy under gravity here because the tension T in the string is always at right angles to the displacement of the block as it swings in a circular arc and so does no work. So if m is the total mass of block and bullet

$$mgh = \frac{1}{2}mv^2$$ (step ii)

Then $h = \dfrac{v^2}{2g} = \dfrac{2.44^2}{2 \times 9.8} = 0.30$m   (step i)

1.10 POWER OF CAR MOVING UP INCLINED PLANE

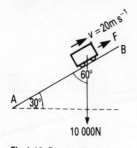

Fig 1.13 Power of moving car

A car is moving at constant velocity of 20m s^{-1} up a road inclined at 20° to the horizontal. The weight of the car is 10 000N and the frictional resistance to the motion is 500N

Calculate the power of the engine (Assume g=10N kg^{-1}).

The car moves with constant velocity along AB (Fig 1.13). Since the acceleration is zero, the resultant force along AB is zero. So

F, force due to engine=downward forces

=500N+10 000 cos 60° (component of weight)

=500+5000=5500N

So power=$F \times v$=5500×20=110 000W=110kW

1 LINEAR MOTION

Uniform acceleration:

$$v=u+at, \quad s=ut+\tfrac{1}{2}at^2, \quad v^2=u^2+2as$$

Vectors must be added (or subtracted) by a parallelogram or triangle method. *Vectors*: velocity, force, momentum. *Scalars*: speed, work, energy.

Resolved component of vector X in direction θ to itself$=X \cos \theta$

2 MOMENTUM AND FORCE

Momentum$=$mass$\times$velocity
Force$=$rate of change of momentum
(Newton's second law)

If mass constant,
then $F=ma$. (F is *resultant* force) Weight$=mg$
If mass varies with time,

$F=$mass per second$\times$velocity change

3 CONSERVATION LAWS

Conservation of momentum applies to *all* types of collisions and explosions.

Conservation of kinetic energy only applies to *elastic* collisions.

4 WORK, POWER AND ENERGY

Work$=$force$\times$distance in direction of force (unit: joule, J)
Power$=$rate of doing work$=F\times v$ (unit: watt, W)

Kinetic energy (k.e.) $=$energy due to motion$=\tfrac{1}{2}mv^2=p^2/2m$ where p is momentum

Potential energy (p.e.) $=$energy due to level or position. Stretched springs or metals have molecular p.e.

Gravitational p.e. $=mgh$

DIAGRAM SUMMARY

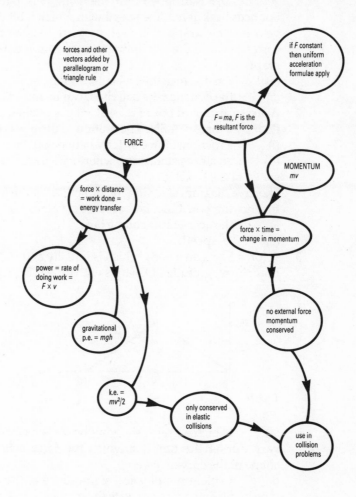

1   A tennis ball is dropped from the hand, falls to the ground and bounces back at half the speed with which it hit the ground. Draw a velocity-time graph of its motion. Mark the point on the graph which corresponds to the ball hitting the ground. Indicate how, from the graph

    (a)   the distance the ball falls, and

    (b)   the distance the ball rises, can be found.   (L)

2   A large cardboard box of mass 0.75kg is pushed across a horizontal floor by a force of 4.5N. The motion of the box is opposed by

    (i)   a frictional force of 1.5N between the box and the floor, and

    (ii)   an air resistance kv^2, where $k=6.0\times10^{-2}$kg m^{-1} and v is the speed of the box in m s^{-1}.

Sketch a diagram showing the directions of the forces which act on the moving box. Calculate maximum values for

    (a)   the acceleration of the box,

    (b)   its speed.   (L)

3   A force F acts on a body which is initially at rest. the graph shows how the magnitude of F varies with time t (Fig 1.14).

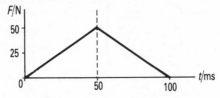

Fig 1.14

    (a)   Sketch a graph showing how the velocity of the body would vary during the time for which the force acts and account for the shape of the curve.

    (b)   Explain the physical significance of the area under the force-time graph and hence calculate the momentum of the body at $t=50$m s.

                                                         (JMB)

4   A horizontal force of 2000N is applied to a vehicle of mass 400kg which is initially at rest on a horizontal surface. If the total force opposing motion is constant at 800N, calculate.

    (i)   the acceleration of the vehicle,

    (ii)   the kinetic energy of the vehicle 5 s after the force is first applied

(iii)   the total power developed 5 s after the force is first applied

(AEB 1985)

5 (a)   State the conditions under which
(i) *linear momentum* and
(ii) *kinetic energy* is conserved in a collision between two masses.

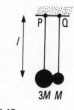

Fig 1.15

(b)   Two small spheres of mass 3M and M hang on strings from two fixed points P and Q at the same level so that their centres of mass are also at the same level and the spheres are in contact (Fig 1.15). It may be assumed that the radii of the spheres are very small indeed compared with the distance *l* of the centre of mass below P and Q. The mass 3M is drawn aside, keeping its string taut until it makes an angle θ with the vertical and then released from rest. Obtain expressions for
(i) the speed with which the mass 3M collides with the stationary mass M, neglecting *air resistance*,
(ii) the speeds which the masses have immediately after collision, assuming it is perfectly elastic.

With the aid of diagrams, discuss the subsequent motion as fully as possible. Explain why it is periodic if θ is small and state the period.

(O&C)

6 A pendulum bob is drawn through an angle of 30° from the vertical. After being released from rest it strikes and adheres to a stationary bob of equal mass at the bottom of its swing.
Calculate the angle through which they swing after impact.

(W)

7 A stone is dropped from a point a few metres above the Earth's surface. Considering the system of stone and Earth, discuss briefly how the principle of conservation of momentum applies *before* the impact of the stone with the Earth.   (C)

8 (a)   State Newton's second law of motion. Explain the steps leading from the second law to the definition of the newton and state this definition.

(b)   If you were asked to investigate in the laboratory the relation between force and *acceleration for a fixed mass*,
(i) outline the procedure you would follow,
(ii) state *the measurements you would make* and
(iii)show how you would establish your conclusion.

(c)   The diagram (Fig 1.16) shows a railway truck of total mass 8000kg travelling at a steady speed of 5.0m s^{-1} towards a fixed buffer B. It strikes the fixed buffer and bounces back to travel at 5.0m s^{-1} the other way.

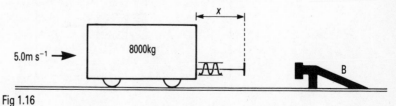

Fig 1.16

The spring in the buffer obeys Hooke's law.

Calculate

(i) the maximum energy stored in the *spring*,

(ii) the maximum force exerted by the truck on the buffer. Sketch graphs showing how the force on the truck varies with

(iii) the distance x (see diagram), and

(iv) the time.

Explain your reasoning.   (L)

9   (a)   A body X of mass m_1 is moving with velocity u towards a stationary body Y of mass m_2 along a line joining their centres of mass. X collides with Y and adheres to it, the two bodies moving with a velocity v after the collision along the same straight line. Write down an equation in terms of m_1, m_2, u and v relating the momentum of the system before the collision with the momentum after the collision.

(b)   Describe, with the aid of a labelled diagram, an experimental arrangement you could use in a school laboratory to test the validity of the equation you have written in (a). Describe the procedures you would adopt to make the relevant measurements and show how you would use them to draw your conclusions.

(c)   One trolley of mass 0.25kg and another of mass 0.50kg, each of length 0.20m are held in contact on a horizontal runway. A spring inside one of the trolleys is released and they are forced apart with a total kinetic energy of 0.75 J.

(i) By considering energy and momentum changes, calculate the velocity of each trolley immediately after separation.

(ii) Each trolley travels a distance of 1.0m after separation, strikes a rigid vertical obstacle and on rebounding loses 19% of its kinetic energy. Ignoring both the effect of friction and the duration of the collisions with the obstacles, calculate the position on the runway at which the trolleys next collide,

(JMB)

10   Define *linear momentum* and state the conditions under which it is conserved. Outline an experimental method of investigating the momentum law. Discuss briefly *two* other examples of conservation laws in physics.

By considering a head-on collision between two isolated bodies, show how the conservation of linear momentum is a consequence of Newton's laws.

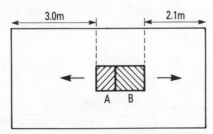

Fig 1.17

The diagram (Fig 1.17) shows a container of mass 45kg floating in deep space. An astronaut, looking into it, observes an object of mass 15kg floating inside the container, explode into two fragments A and B of masses 5kg and 10kg respectively. These move apart in the directions shown in the figure. Initially, the astronaut, container and object have no relative motion. The impulse from the explosion on each fragment is 10kg m s^{-1}.

The fragments adhere to the walls of the container on impact. Describe the motion of the fragments and container. Draw graphs of position and velocity of the container, relative to the astronaut, against time for the first 5 seconds after the explosion (O&C)

CIRCULAR MOTION, GRAVITATION, SIMPLE HARMONIC MOTION AND ROTATIONAL DYNAMICS

CONTENTS

CIRCULAR MOTION

ANGULAR VELOCITY

An object moving from A to B in a circular path is said to have an average **angular velocity** ω defined by θ/t where θ is the angle described about the centre of the circle in a time t.

$$\omega = \frac{\theta \, (\text{rad})}{t \, (\text{s})}$$

Note that θ is normally measured in radians in calculating angular velocity and that $360°$ (1 revolution)$=2\pi$ radians. So an object at the end of a string, whirled round steadily and making 3 revolutions in 2 seconds, has an angular velocity given by

$$\omega = \frac{\text{angle } \theta}{\text{time } t} = \frac{3 \times 2\pi}{2} = 3\pi \text{ rad s}^{-1}$$

A dynamo coil, rotating at 50 revolutions per second, has an angular velocity $\omega = 50 \times 2\pi = 100\pi$ rad s^{-1}. The Earth spins on its axis once per 24 hours approximately. So its average angular velocity is given by

$$\omega = \frac{2\pi \text{rad}}{24 \times 3600 \text{ s}} = 7.3 \times 10^{-5} \text{ rad s}^{-1}$$

An angle θ in radians is defined by the relation $\theta = s/r$, where s is the length of the arc opposite θ and r is the circle radius. So $s = r\theta$. Now the speed v of an object moving from A to B in Fig 2.1 is *distance s/time t*. So

$$v = \frac{s}{t} = r\frac{\theta}{t} = r\omega$$

$$v = r\omega$$

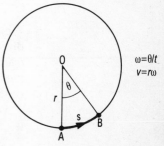

$\omega = \theta/t$
$v = r\omega$

Fig 2.1 Angular velocity

So a runner moving steadily with an angular velocity ω of 0.1 rad s^{-1} round a circular track of radius r of 80m has a speed v given by

$$v=r\omega=80\times0.1=8\text{m s}^{-1}$$

When the speed increased to 10m s^{-1}, the new angular speed ω is

$$\omega=\frac{v}{r}=\frac{10}{80}=0.125\text{ rad s}^{-1}$$

ACCELERATION IN CIRCLE; CENTRIPETAL FORCE

When we whirl a heavy object tied to the end of a rope in a horizontal circle, we can feel that a force is needed to keep the object rotating at a steady speed. This force is exerted by the rope on the object as it moves in its circular path. Now a force produces an **acceleration**. So although the object is moving round the circle with a constant speed, it has an acceleration.

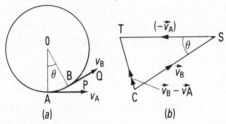

Fig 2.2 Acceleration in circle

Figure 2.2(a) shows how the acceleration is found. At A, the direction of the velocity v is along the tangent AP. In a short time t later, when the object is at B, the velocity has changed in direction to BQ, where BQ is a tangent to the circle at B, although the numerical value is still v. To find the change in velocity, we need a vector subtraction, $\vec{v}_B-\vec{v}_A$, since velocity is a vector.

In Fig 2.2(b), CS represents v_B and ST represents $(-v_A)$. So CT represents $\vec{v}_B-\vec{v}_A$, the velocity change from A to B. See page 17. So

$$\text{acceleration } a=\frac{\text{velocity change}}{\text{time}}=\frac{CT}{t}$$

When B and A in Fig 2.2(a) are close to each other, we see that the direction of the velocity change or acceleration CT in Fig 2.2(b) is *towards the centre* O of the circle. So the force needed to keep an object moving in a circle is directed towards the centre. This is called a **centripetal force**.

When we calculate the acceleration a towards the centre, the result is

$$a=\frac{v^2}{r} \tag{1}$$

where v is the constant speed in the circle of radius r. See *Advanced Level Physics* (Heinemann) by Nelkon and Parker. If ω is the angular speed, then $v=r\omega$. So

$$a= \frac{v^2}{r}=\frac{r^2\omega^2}{r}=r\omega^2$$

Remember that

$$a=r\omega^2 \qquad (2)$$

The centripetal force on the object of mass m is given by $F=ma$. So

$$\text{centripetal force}= \frac{mv^2}{r} \text{ or } mr\omega^2$$

You should remember that to make an object move round a circle, we must always have a force on it *pointing towards the centre* of the circle.

Example 1 on the centripetal force direction is given on page 46.

CONICAL PENDULUM; BANKING OF TRACKS

An example of an object moving in a circle is shown in Fig 2.3(*a*). Here an object A of mass m at the end of a string of length l moves in a horizontal circle of radius r with a constant speed v. This motion is called a 'conical pendulum' because the string SA moves continually round the surface of a cone whose half-angle at the top is the angle θ shown in Fig 2.3(*a*).

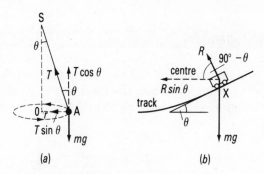

(a)   (b)

Fig 2.3 Examples of circular motion

The forces on A are
(*a*)   its weight mg acting vertically downwards and
(*b*)   the tension T in the inclined string.
The force mg has no component in a horizontal direction. So it cannot provide a centripetal force towards the centre O. This is provided by the tension T. It has a component $T \cos (90°-\theta)$ or $T \sin \theta$ *towards* O. So

$$F=T \sin \theta= \frac{mv^2}{r} \qquad (1)$$

Also, since the mass moves in a *horizontal* circle, the weight mg acting downwards must be exactly counterbalanced by the vertical component of T, which is $T \cos \theta$. So

$$T \cos \theta = mg \tag{2}$$

Dividing (1) by (2), then

$$\frac{\sin \theta}{\cos \theta} = \tan \theta = \frac{v^2}{rg} \tag{3}$$

Knowing the values of v, r and g, (3) can now be used to find the angle θ. If the tension T is required, the relation in (2) can be used, since $T = mg/\cos \theta$.

To avoid *sideslip* or frictional forces, racing tracks are banked as shown roughly in Fig 2.3(b). At X, a car has a normal reaction R and its weight mg acting on it, but no frictional force. In this case, as for the conical pendulum, the centripetal force is provided by the horizontal component of R which is $R \cos (90° - \theta)$, where θ is the angle with the horizontal made by the tangent to the track at X. So $R \sin \theta = mv^2/r$. For vertical equilibrium, $R \cos \theta = mg$. So as in the case of the conical pendulum, $\tan \theta = v^2/rg$. This relation is used to calculate the banking or θ needed for a particular value of the speed v. It can be seen that the angle θ increases when v increases.

Example 2 on page 47 shows how the tension in a string is calculated when an object is whirled in a vertical circle. Example 3 on page 47 shows how the angle of banking for an aeroplane's wings can be found.

WORKED EXAMPLES ON CIRCULAR MOTION

2.1 CENTRIPETAL FORCE DIRECTION

A small mass A is tied to a string OA suspended from the inside of a train (Fig 2.4). When the train moves round a horizontal curve with its centre on the left, which of the diagrams X, Y and Z is correct?

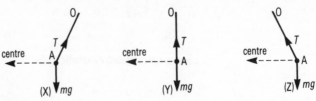

Fig 2.4 Direction of centripetal force

We need a force *towards the centre* of the circle. The weight mg acts vertically and therefore has no component towards the centre. So the tension (force) T in the string must provide the force towards the centre.

In X, T has a component in a direction *away* from the centre. So X is not correct. In Y, T acts vertically and so has no component towards the centre. So Y is not correct. Z is the correct diagram because T has a component towards the centre.

2.2 CIRCULAR MOTION IN A VERTICAL PLANE

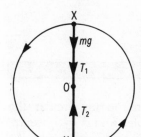

Fig 2.5 Circular motion

A small mass X of 0.2kg is whirled at the end of a string of length 2m in a vertical circle at 3 revolutions per second (Fig 2.5).
  Find the tension in the string when X is just
(a)   at the top of the circle,
(b)   at the bottom of the circle (assume $g=10$N kg^{-1}).

(Overview
(i)   For circular motion, $F=mv^2/r=mr\omega^2$ $F=\frac{mv^2}{r}$ or $F=mr\omega^2$
(ii) F acts *towards the centre*.
(iii) F is the *resultant* force towards the centre.)

(a)   Suppose X is at the top of the circle (Fig 2.5). Then if T_1 is the string tension

$$F=T_1+mg$$

Now $F=mr\omega^2$, where $\omega=2\pi\times3=6\pi$ rad s^{-1}

So $T_1=mr\omega^2-mg=0.2\times2\times(6\pi)^2-0.2\times10$
$$=142-2=140\text{N}$$

(b)   Suppose X is now at the bottom of the circle (Fig 2.5). If T_2 is the tension in the string,

force towards centre $F=T_2-mg$ (T_2 and mg in opposite directions)
So $T_2-mg=mr\omega^2$
and $T_2=mr\omega^2+mg=0.2\times2\times(6\pi)^2+0.2\times10$
$$= 142+2=144\text{N}$$

2.3 BANKING OF AEROPLANE WINGS

An aeroplane A banks its wings to fly in a horizontal circle of radius 10km at 720km h^{-1}.

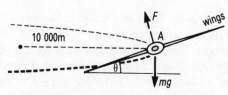

Fig 2.6 Banking of wings

  At what angle to the horizontal are the wings banked (assume $g=10$m s^{-2}).

Suppose the wings are banked at an angle θ to the horizontal (Fig 2.6).

The speed $v=720$km h$^{-1}=200$m s^{-1} and $r=10$km$=10\,000$m.

If $F=$lift on plane due to air pressure, $mg=$weight of plane,

$$F \sin \theta = mv^2/r \text{ and } F \cos \theta = mg \text{ (see page 45)}$$

So, dividing,

$$\tan \theta = v^2/rg = 200^2/(10\,000 \times 10) = 0.4$$
$$\therefore \qquad \theta = 22° \text{ (approx)}$$

GRAVITATION

NEWTON'S DEDUCTION FROM KEPLER'S LAW

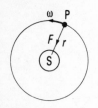

Fig 2.7 Planet motion round Sun

Planets move in orbits round the sun which are roughly circular (the orbits are actually elliptical). About 1619 Kepler found by observations that if r was the average radius of the orbit and T the period of a planet round the sun, then T^2 was proportional to r^3 for the different planets.

Suppose m is the mass of planet P, r the average radius of the circular orbit and ω the angular velocity of the planet (Fig 2.7). The force towards the centre is then $mr\omega^2$, the centripetal force value. Newton *assumed* that the force of attraction of the sun on the planet was inversely proportional to the square of the distance r between them, that is, $F=km/r^2$, where k is the constant. In this case,

$$F=mr\omega^2=\frac{km}{r^2}$$

So $$r^3=\frac{k}{\omega^2}=\frac{kT^2}{4\pi^2},$$

since $\omega=2\pi/T$. Hence $r^3 \propto T^2$, which is Kepler's law.

NEWTON'S GRAVITATIONAL LAW; RELATION BETWEEN G AND g

Newton's law of gravitation can be stated as

$$F=G\frac{m_1m_2}{r^2}$$

where F is the force of attraction between two small masses m_1, m_2 at a distance r apart and G is a constant known as the **gravitational constant**. Unlike g, which varies with its distance from the centre of the Earth G is a *universal* constant. Experiments show that approximately $G=6.7 \times 10^{-11}$ N m^2 kg^{-2} if F is in newtons, m in kilograms and r in metres.

You should note that the force between given small masses varies as $1/r^2$, where r is the distance between them.

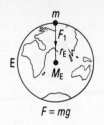

$F = mg$

Fig 2.8 Relation between g and G

Or $F \propto \dfrac{1}{r^2}$ (two given masses)

So if we double the distance r between two masses, the gravitational force F decreases to *one-quarter* of its original value.

For objects on or outside the Earth (but *not* inside), the mass M_E of the Earth can be considered concentrated at its centre. So the gravitational force on a mass m on the Earth's surface is given by

$$F = \frac{GM_E m}{r_E^2}$$

But $F = \text{weight} = mg$, where g is about 9.8 N kg^{-1}. So $mg = GM_E m / r_E^2$, and cancelling m,

$$g = \frac{GM_E}{r_E^2} \text{ or } gr_E^2 = GM_E$$

Using the values $G = 6.7 \times 10^{-11}$ N m^2 kg^{-2}, $g = 9.8$ N kg^{-1} and $r_E = 6.4 \times 10^6$m, the mass M_E of the Earth is

$$M_E = \frac{gr_E^2}{G} = \frac{9.8 \times (6.4 \times 10^6)^2}{6.7 \times 10^{-11}}$$
$$= 6 \times 10^{24} \text{kg}$$

PERIOD OF SATELLITE

Consider a satellite of mass m orbiting the Earth of mass M_E in a circle of radius r round the centre of the Earth. From the gravitational law, the centripetal force $= GM_E m / r^2 = mr\omega^2$, where $\omega = 2\pi/T = $ angular velocity and T is the period.

The mass m of the satellite *cancels* on both sides of the equation. Simplifying,

$$T^2 = \frac{4\pi^2 r^3}{GM_E}$$

If a satellite of greater or lighter mass were in the *same* orbit, the period T would be the same as before because the period is independent of the satellite mass as we have shown.

Now we have shown before that $GM_E = gr_E^2$. So taking the square root of T^2

$$T = 2\pi \sqrt{\frac{r^3}{gr_E^2}} \tag{1}$$

The moon is an Earth satellite. Its distance r from the Earth's centre is about 60.1 r_E, where $r_E = $ Earth radius $= 6.4 \times 10^6$m. Using $g = 9.8$m s^{-2} and the values given for r and r_E, we find from the equation (1) that the period T of the moon round the Earth is about 27.3 days or one month. This agrees with observations of the moon's orbit.

PARKING (CLARKE) ORBITS

The period of the Earth as it rotates on its axis is about 24 hours. If the period of a satellite S in its orbit is exactly equal to the period of the Earth, the satellite will stay over the same place on the Earth as the Earth rotates (Fig 2.9). Such a satellite, in a so-called 'parking orbit', can relay television programmes from one part of the world to another. This is done in relaying the Olympic Games, for example.

Fig 2.9 Parking orbit

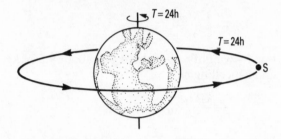

We can find the radius of the required orbit for our relation in (1). The value of r is given by $T^2=4\pi^2r^3/gr_E^2$, from which

$$r=\sqrt[3]{\frac{T^2gr_E^2}{4\pi^2}}=\sqrt[3]{\frac{(24\times3600)^2\times9.8\times(6.4\times10^6)^2}{4\pi^2}}$$
$$=42.4\times10^6\text{m}$$
$$=42\,400\text{km}$$

So the height of the orbit above the Earth's surface = $42\,400-6400=36\,000$km. The satellite is carried to this height by a rocket and over the equator it is given an impulse by firing jets so that it is deflected in a direction parallel to the required orbit with the velocity required for the orbit. If the satellite were sent into an orbit round the Earth other than round the equator, it would not stay over the same place on the Earth while the Earth rotated about its axis.

GRAVITATIONAL FIELD STRENGTH

At a point X in a gravitational field, the *field strength* is defined as the 'force per kg' acting on a mass at X. Since $F=ma=mg$, where g is the acceleration of free-fall on the Earth's surface, the force per kg at the Earth's surface$=F/m=g$. Using the value of g, we can say that $g=9.8$N kg^{-1} at a distance r_E from the Earth's centre, where r_E is the radius of the Earth, as at A (Fig 2.10).

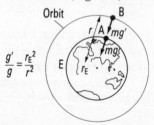

Fig 2.10 Variation of gravitational field strength

At a distance r from the Earth's centre, where r is greater than r_E, the field strength is less than g. Suppose it is g' at a point B, as shown. The gravitational force on the same mass m at B is then mg'. From Newton's law of gravitation, the force between the Earth and the mass m is inversely proportional to the square of the distance between them. So

$$\frac{mg'}{mg} = \frac{1/r^2}{1/r_E^2} = \frac{r_E^2}{r^2} \text{ , or } g' = \frac{r_E^2}{r^2} \times g$$

So $g' \propto 1/r^2$ since r_E and g are constants.

Example 2.4 on page 52 shows how to calculate the force on a mass at a height above the Earth and the speed of a satellite in orbit at this height.

Example 2.5 on page 53 shows how to find the period of two stars rotating about their common centre of mass.

GRAVITATIONAL POTENTIAL; SATELLITE ENERGY

Work or energy is needed to take a satellite from the Earth's surface to a place some distance away. The work is required to overcome the force of attraction of the Earth on the mass of the satellite. The energy of a mass due to its position in a gravitational field is called its *potential energy*. The *potential*, symbol V, at a point in the field is the potential energy per unit mass (kg).

In the Earth's gravitational field, the potential at infinity is taken as the 'zero' of potential. At a point distance r from the centre of the Earth and outside the Earth, work is required to take a mass to infinity against the attractive force of the Earth. So the potential V at a distance r is *less* than at infinity and so has a negative value. Calculation shows that (see Nelkon and Parker *Advanced Level Physics* (London, Heinemann Educational Books)

$$V = -\frac{GM}{r} \tag{1}$$

A satellite in an orbit round the Earth has both potential and kinetic energy. Suppose its mass is m and its orbit is circular and radius r, where r is the distance from the centre of the Earth. Then if M is the mass of the Earth,

$$\text{kinetic energy, k.e.} = \tfrac{1}{2}mv^2 = \frac{GMm}{2r}$$

since $mv^2/r = $ centripetal force $= GMm/r^2$. So

$$\text{total energy} = \text{p.e.} + \text{k.e.} = -\frac{GMm}{r} + \frac{GMm}{2r} = -\frac{GMm}{2r} \tag{2}$$

When a spacecraft re-enters the Earth's atmosphere, the radius r of its orbit decreases. We see from above that its kinetic energy *increases*. But its potential energy decreases twice as much as the increase in kinetic energy. So on the whole there is a loss of mechanical energy, which is transformed to heat.

VELOCITY OF ESCAPE

A rocket used to carry a moon satellite must be fired from the Earth with sufficient energy to escape from the gravitational influence of the Earth. Suppose the rocket has a mass m and is fired with a velocity v to just escape from the Earth's gravitational field. Then the loss in kinetic energy of the rocket=the gain in potential energy from the Earth's surface, radius r_E, to infinity. So

$$\frac{1}{2}mv^2 = m \times \frac{GM}{r_E}$$

Hence $v = \sqrt{\dfrac{2GM}{r_E}} = \sqrt{2gr_E}$, since $GM/r_E^2 = g$

Substituting g=9.8 N kg^{-1} and r_E=6.4×10^{6}m, then v=11km s^{-1} approximately. On the average, molecules in the atmosphere have speeds less than the velocity of escape so they remain in the Earth's neighbourhood. There is little atmosphere round the moon. Here the gravitational attraction is much less than that of the Earth and the velocity of escape is correspondingly low.

Velocity of escape $= \sqrt{2gr_E}$

WORKED EXAMPLES ON GRAVITATION

2.4

Calculate the force on a mass of 10kg at a height of 800km above the Earth, assuming the gravitational field strength is 9.8N kg^{-1} at the Earth's surface, radius 6400km.
What is the speed of a spacecraft in this orbit and its period?

(**Overview** (i) The force on the mass=mg', where g' is the gravitational field strength at the height.
(ii) g' can be found from $g' \propto 1/r^2$, where r is the distance from the Earth's centre.)

At a height of 800km above the Earth, the distance from the Earth's centre is 800+6400=7200km.
From the law of gravitation, the gravitational strength (the force per kg) is inversely proportional to the square of the distance from the centre. So if g' is the field strength at a distance of 7200km and g (9.8N kg^{-1}) is the field strength at a distance of 6400km, then

$$\frac{g'}{g} = \frac{1/7200^2}{1/6400^2} = \frac{6400^2}{7200^2} = \frac{8^2}{9^2} = \frac{64}{81}$$

So   force on 10kg mass$=mg'=10\times\dfrac{64}{81}\times9.8$

$=77.4$N

Speed in orbit. If m is the spacecraft mass, the centripetal force is

$$\frac{mv}{r}=mg'$$

So $v=\sqrt{rg'}=\sqrt{(7200\times10^3)\times(64\times9.8/81)}$

$=7470$m s^{-1}

Period in orbit. The period T=orbit circumference/v. So

$$T=\frac{2\pi r}{v}=\frac{2\pi\times7200\times10^3}{7470}=6056\text{s}$$

2.5   Two stars of masses 10^{20}kg and 3×10^{20}kg rotate about their centre of mass under the gravitational attraction between them. If their separation is 10^4km, calculate the period of rotation. ($G=6.7\times10^{-11}$ N m^2 kg^{-2}.)

(**Overview** (i) Period $T=2\pi/\omega$.
(ii) Gravitational force$=mr\omega^2$, so ω can be found.
(iii) Centre of mass is nearer the larger mass, taking moments.)

The centre of mass divides the distance 10^4km in the ratio 3:1. Its distance r_1 from the *smaller* mass 10^{20}kg is ($3\times10^4/4$)km or $3\times10^7/4$ m.

If m_1 is the smaller mass (10^{20}kg) and m_2 is the larger mass (3×10^{20}kg), the gravitational force of attraction on m_1 is $F=Gm_1m_2/r^2$, where r is the separation 10^4km or 10^7m.

But the centripetal force F on the mass $m_1=m_1r_1\omega^2$. So

$$m_1r_1\omega^2=\frac{Gm_1m_2}{r^2}$$

Cancelling m_1 and simplifying,

$$\omega=\sqrt{\frac{Gm_2}{r_1r^2}}$$

So   period $T=\dfrac{2\pi}{\omega}=2\pi\sqrt{\dfrac{r_1r^2}{Gm_2}}$

$$=\sqrt{\frac{3\times10^7\times10^{14}}{4\times6.7\times10^{-11}\times3\times10^{20}}}$$

$=1.2\times10^6$s

SIMPLE HARMONIC MOTION (S.H.M.)

Oscillations of matter such as air produce sound waves. Oscillations of electric and magnetic forces produce light and radio waves. All oscillations can be built up from those called *simple harmonic*, which we now study.

SIMPLE HARMONIC MOTION

Simple harmonic motion can be defined as the motion of a particle whose acceleration a is proportional to its distance x from a fixed point and is always directed towards that point, that is,

$$\text{acceleration } a = -kx \qquad (1)$$

where k is a constant. The minus shows that, as it moves to and fro, the acceleration a is oppositely directed to the displacement x. For example, if the particle is on the right of O in Fig 2.11(a), then the acceleration is towards the left. This causes the particle to slow down as it moves from O to A and to speed up as it moves back from A to O. It is often useful to think of the motion of the bob of a simple pendulum when considering how the acceleration (or velocity) varies with displacement from O in simple harmonic motion (Fig 2.11(b)).

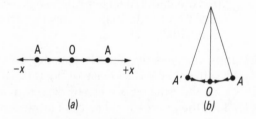

Fig 2.11 Simple harmonic motion

It is useful to write the constant k as ω^2, where ω is a constant related to the frequency or period of the motion, as we show later. So s.h.m. is also represented by

$$\text{acceleration } a = -\omega^2 x \qquad (2)$$

The *amplitude r* of the motion is the maximum displacement. At the end of the oscillation at A in Fig 2.11(a), where $x=r$, the *maximum* acceleration is given by

$$a_{\max} = -\omega^2 r \qquad (3)$$

The *velocity v* at a displacement x can be shown to be

$$v = \omega\sqrt{r^2 - x^2} \qquad (4)$$

See Nelkon and Parker *Advanced Level Physics* (London, Heinemann Educational Books).

So the *maximum* velocity is obtained at the centre O of the oscillation, where $x=0$. Hence, from (4)

$$v_{max}=\omega r \qquad\qquad (5)$$

The *period* T is the time for a complete to and fro movement or cycle. It can be shown that

$$T=\frac{2\pi}{\omega} \qquad\qquad (6)$$

The displacement x itself varies with time t according to a sine relation if t is measured from the *centre* O of the oscillations and

$$x=r\sin\omega t \qquad\qquad (7)$$

If t is measured from the *end* of an oscillation

$$x=r\cos\omega t.$$

Starting with (7), you can find v from $v=dx/dt$. The result is $v=\omega r\cos\omega t$. So v is 90° out of phase with x in their variation with time t. This means that at $t=0$, $x=0$ but $v=\omega r=$maximum velocity. You can see this is true from the gradient of the displacement (x)−time (t) graph at $t=0$ in Fig 2.12(c).

GRAPHS IN S.H.M.

Figure 2.12(a) shows how the acceleration a varies with displacement x.

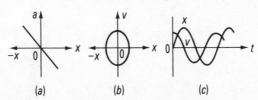

(a) (b) (c)

Fig 2.12 Acceleration, velocity and displacement against time in s.h.m.

It is a straight line sloping downwards and passing through the origin O, since $a=-kx$. Figure 2.12(b) shows how v varies with displacement x. Figure 2.12(c) shows how x and v vary with *time*, t.

Example 2.6 on page 61 shows how to calculate the period and maximum velocity of an oscillating mass.

OSCILLATION OF SPRING AND MASS

The oscillation of a mass attached to a spring is analogous to oscillations in branches of physics other than mechanics. For example, the oscillation of molecules in solids depends on the bond of attraction between the molecules and this bond can be considered similar to (but not the same as) the action of a spring. Similarly, in a transmitter or radio oscillator circuit, the components required act like a 'mass' and a 'spring'.

1   Figure 2.13(*a*) shows a spring S attached at one end to a smooth table with a mass m at the other end O. When the spring is pulled out a little and then released, the mass oscillates about O, which is the centre of the oscillations.

If the extension x of a spring obeys Hooke's law, that is, the spring is not overstretched, then the force or tension T is proportional to x. So we write $T=-kx$, where k is a constant called the *spring constant* and the minus shows that T is pulling in a direction opposite to the way in which x is measured positive. k is numerically the 'force per unit extension' (F/x) of the spring and is therefore measured in N m^{-1}.

At a distance x from O, the force towards O=the tension T in the spring=ma, where a is the acceleration. If the extension x obeys Hooke's law and x is measured positively in the direction OA, then $T=-kx$, where k is the spring constant. So $ma=-kx$, and

$$a=-\frac{k}{m}x=-\omega^2x,$$

where $\omega^2=k/m$. So the motion is simple harmonic. Also, the period T is

$$T=\frac{2\pi}{\omega}=\frac{2\pi}{\sqrt{k/m}}=2\pi\sqrt{\frac{m}{k}} \tag{1}$$

If k is 20N m^{-1} and $m=0.1$kg, then

$$T=2\pi\sqrt{\frac{0.1}{20}}=0.4\text{s}$$

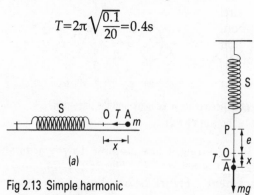

(a)

Fig 2.13 Simple harmonic motion with spring

(b)

2   Suppose the same mass m is attached to the spring S when it is vertical (Fig 2.13(*b*)). Unlike the previous case, *the weight mg extends the spring a distance e to O*. When the mass is pulled down a little and then released, it oscillates about O. So O is the centre of the oscillation.

Suppose at some time that the mass is at A, where OA=x. The upward tension T is then given by $T=k(e+x)$. If the direction OA is the +ve direction of x, the force in this direction is $(mg-T=mg-ke-kx)$. But $mg=ke$, as the weight extended the spring by a length e before the oscillation began. So the resultant force acting down$=-kx$. From $F=ma$, we see that $ma=-kx$. So

$$a=-\frac{k}{m}x=-\omega^2 x$$

and   period $T=\frac{2\pi}{\omega}=2\pi\sqrt{\frac{m}{k}}$ (2)

This is the same result as we obtained in Fig 2.13(a). Further, $m/k=e/g$ as $mg=ke$. So from (2), $T=2\pi\sqrt{e/g}$.

From our two results in (1) and (2), we see that the period is the same whether the spring is horizontal or vertical. When it is vertical and the mass is put on, stretching the spring only alters the centre of oscillation but not the period. If the spring is vertical, it is useful to remember that there are two expressions for the period. They are:

$$T=2\pi\sqrt{\frac{m}{k}}$$

and $T=2\pi\sqrt{\frac{e}{g}}$

where e is the extension due to the weight.

SIMPLE PENDULUM

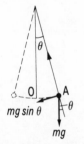

Fig 2.14 Simple pendulum theory

Figure 2.14 shows a simple pendulum of length l oscillating about a point O through a *small* angle. The bob, mass m, moves along the arc of a circle of radius l. Suppose the bob is at A at some instant. The component $mg \sin \theta$ of the weight is the only force which moves the bob along the arc OA. So from $F=ma$, and measuring x, the distance from O, positive in the direction OA, we have

$$ma=-mg \sin \theta$$

Cancelling m and using $\sin \theta=\theta$ in radians when θ is small, then

$$a=-g \sin \theta=-g\theta=-g\frac{x}{l}=-\omega^2 x$$

where $\omega^2=g/l$. So

$$\text{period } T=\frac{2\pi}{\omega}=2\pi\sqrt{\frac{l}{g}}$$

$$T=2\pi\sqrt{\frac{l}{g}}$$

Note that this formula for T is only true when θ is very small, less than about 10°.

KINETIC ENERGY AND POTENTIAL ENERGY IN S.H.M.

When the bob of a simple pendulum oscillates, its kinetic energy (k.e.) and potential energy (p.e.) constantly change from one form to the other. For example, at the end of an oscillation at the top of the swing, the p.e. of the bob is a maximum and its k.e. is zero at this instant. As the bob returns to the centre of oscillation its k.e. increases to an amount x say at a point and its p.e. then decreases by the amount x. At the centre of oscillation its k.e. is a maximum and its p.e. is zero, reckoning the p.e. as zero at the lowest point of the swing. In general, the *total energy* (k.e.+p.e.) is *constant*. Figure 2.15(*a*) illustrates the variation of k.e., p.e. and total energy with *displacement x*, and Fig 2.15(*b*) with *time t*.

The same results are obtained for an oscillating system of a mass M at the end of a spring S on a smooth table. See Fig 2.16(*a*). If the spring has negligible mass, then the spring S can possess potential energy and the mass M can possess kinetic energy. At the end of an oscillation, when the spring is compressed or extended by the maximum amount, the p.e. of S is a maximum and the k.e. of M is zero. In the centre of the oscillation, when S is not extended, the p.e. of S is now zero and the k.e. of M is a maximum. At other points in the oscillation there is an 'exchange' between p.e. and k.e. The *total* energy, p.e.+k.e., is constant. This is also illustrated in Fig 2.15(*a*) and (*b*).

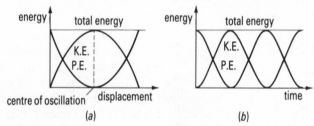

Fig 2.15 Energy in s.h.m.

We can find values for the energy from our previous results. At a displacement x, the velocity is $v = \omega\sqrt{r^2 - x^2}$. So

$$\text{kinetic energy} = \tfrac{1}{2}mv^2 = \tfrac{1}{2}m\omega^2(r^2 - x^2) \qquad (1)$$

The potential energy is the work done in stretching the spring a distance x = average force $\times x = \tfrac{1}{2}kx \times x = kx^2$. Now as we showed on p56, $\omega = \sqrt{k/m}$ or $k = m\omega^2$. So

$$\text{potential energy} = \tfrac{1}{2}kx^2 = \tfrac{1}{2}m\omega^2 x^2 \qquad (2)$$

From (1) and (2),

$$\text{k.e.} + \text{p.e.} = \tfrac{1}{2}m\omega^2 r^2 = \text{constant}$$

**MECHANICAL
RESONANCE. FORCED
AND DAMPED
OSCILLATIONS**

If the mass at the end of the spring in Fig 2.16 is displaced slightly and then released, it vibrates with *free* or *natural* oscillations. The period T is then given, as we have seen, by $T=2\pi\sqrt{m/k}$. The *natural frequency f* is $1/T$ and is the number of complete oscillations per second. So

$$f=\frac{1}{2\pi}\sqrt{\frac{k}{m}}$$

The *amplitude* of the oscillation is the maximum displacement. Owing to the air friction (viscosity), the amplitude diminishes slowly as the mechanical energy is used to overcome friction. Eventually, when all the mechanical energy is transferred to heat in the air, the mass comes to rest.

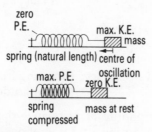

Fig 2.16 Transfer of
potential and kinetic energy

Figure 2.17(*a*) shows an *undamped* or free oscillation, when no friction is present. The amplitude is then constant. Figure 2.17(*b*) shows a highly damped oscillation, when the frictional force is high. The amplitude dies quickly to zero, as shown by the broken curve. With small damping or low friction, the amplitude would decrease more slowly.

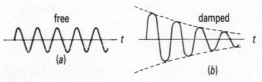

Fig 2.17 Free and damped
oscillations

FORCED OSCILLATIONS. RESONANCE AND SHARPNESS

Suppose the mass at the end of the spring is now forced to vibrate by an external (outside) periodic force F. If F has a frequency different from the natural frequency, the amplitude of oscillation will be small and the mass will now be forced to vibrate at a different frequency called the **forcing frequency** (Fig 2.18(a)). If, however, the forcing frequency becomes equal to the natural frequency then vibrations of large amplitude will occur and this is said to be **resonance** (Fig 2.18(b)).

Fig 2.18 Forced and resonant oscillations

At resonance: (i) forcing frequency=natural frequency,
              (ii) amplitude is a maximum

The amplitude at resonance depends on the amount of energy dissipated during each vibration. Systems which dissipate a large amount of energy are said to be heavily *damped*. If little energy is dissipated the damping is small. Figure 2.19 shows how the amplitude varies with forcing frequency for a system. Two graphs are shown, one, A, where the damping is small, the other, B, where it is larger.

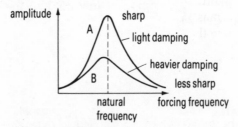

Fig 2.19 Resonance graphs

The peak (top) of curve A is sharper than that of curve B. This is because there is much less resistance, or light damping, in case A than in case B, where there is more resistance or heavier damping. Similar graphs are obtained for radio circuits where electrical oscillations occur (page 165) and in oscillations of string such as guitar strings (page 228).

WORKED EXAMPLES ON S.H.M.

2.6 PERIOD AND VELOCITY IN S.H.M.

A mass of 0.2kg oscillates in s.h.m. with an amplitude of 0.1m. At the extreme end of the oscillation, a force of 1N acts on the mass.

Calculate (a) the period and (b) the maximum velocity.

(**Overview** (i) For (a), we need to use $T=2\pi/\omega$.

(ii) From $F=ma$, we can use the force to find the acceleration at the end of the oscillation.

(iii) At the end of the oscillation, $a=a_{max}=\omega r^2$, where r is the amplitude.)

At extreme end, maximum displacement, force $F=1N=ma=0.2\,a$.

So
$$a=\frac{1}{0.2}=5\text{m s}^{-2}$$

Now maximum acceleration $a=-\omega^2 r=-\omega^2\times0.1$

so
$$5=\omega^2\times0.1 \text{ or } \omega=\sqrt{5/0.1}$$

Hence
$$T=\frac{2\pi}{\omega}=2\pi\sqrt{\frac{0.1}{5}}=0.9s \qquad (a)$$

Also   maximum velocity$=\omega r=\sqrt{5/0.1}\times0.1=0.7\text{m s}^{-1}$ (b)

2.7 OSCILLATING MASS ON SPRING

A spring suspended vertically has a force constant of $40N\ m^{-1}$ and a mass of 0.1kg is suspended from the spring. The mass is pulled down a distance of 5mm and then released.

Find (a) the period of oscillation, (b) the maximum acceleration of the mass and (c) the net force acting on the mass when it is 2mm below the centre of oscillation. (Assume $g=10N\ kg^{-1}$.)

(a)   From our previous result on page 56, the period T is given by

$$T=2\pi\sqrt{\frac{m}{k}}=2\pi\sqrt{\frac{0.1}{40}}=0.3s$$

(b)   Maximum acceleration $a=\omega^2 r=(k/m)\times r=(40/0.1)r=400r$

since, from page 56, $\omega^2=k/m$
Now $r=5mm=5\times10^{-3}m$. So

maximum acceleration $a=400\times5\times10^{-3}=2\text{m s}^{-2}$

(c)   The tension T' in the spring is greater than mg when the mass is 2mm below the centre of oscillation O. See Fig 2.13(b). So net force upward$=T'-mg$.

The weight $mg=0.1\times10=1N$. If 40N extends the spring by 1m, then 1N extends the spring by a length e which is

$$e=\frac{1}{40}\times1m=0.025m$$

At 2mm or 0.002m below e, the extension$=0.025+0.002=0.027m$. So

$$T'=40N\times0.027=1.08N$$

Hence   net force on mass$=T'-mg=1.08-1=0.08N$

2.8 MASS-SPRING ENERGY IN S.H.M.

A mass of 0.2kg is attached to the end of a suspended vertical spring of force constant 40N m^{-1}. The mass is pulled down 1cm and then released. Calculate (a) the maximum kinetic energy of the oscillating mass and (b) the potential energy of the spring when the mass is 0.5cm below the centre of oscillation (assume $g=10N$ kg^{-1}).

(**Overview** (i) Maximum k.e.$=\frac{1}{2}mv_{max}^2=\frac{1}{2}mr^2\omega^2$.
      (ii) p.e.$=\frac{1}{2}kx^2$ (x is the extension from the *original* length of spring).)

(a)   From page 000, $\omega^2=k/m=40/0.2=200$

So

$$\text{maximum k.e.}=\frac{1}{2}mv^2=\frac{1}{2}m\omega^2r^2=\frac{1}{2}\times0.2\times200\times(1\times10^{-2})^2$$
$$=2\times10^{-3}J \tag{1}$$

(b)   The weight mg (2N) pulls the spring a distance given by

$$\frac{2N}{40N}\times1m=0.05m$$

So 0.5cm or 0.005m below the centre of oscillation, the extension x of the spring from its original length is

$$x=0.05+0.005=0.055m$$

Hence potential energy of spring$=\frac{1}{2}kx^2=\frac{1}{2}\times40\times0.055^2$
$$=6.05\times10^{-2}J$$

ROTATIONAL DYNAMICS

So far in dynamics we have considered small objects which have their mass concentrated at a point and their linear motion. In this section we consider large objects, which have their mass distributed over a volume, and their *rotation about an axis*. Examples of such objects are a wheel in a machine rotating about its centre, a ballet dancer spinning on her toes about a vertical axis through her body and the Earth turning about its axis.

TORQUE AND ANGULAR ACCELERATION; MOMENT OF INERTIA

We have seen that if a force F acts on a mass m, the linear acceleration a produced is given by $F=ma$. In rotational dynamics, *angular acceleration* α is produced by a *torque T* discussed shortly and it can be shown that

$$T=I\alpha \qquad (1)$$

where I is the *moment of inertia* about the axis of rotation. So T is analogous to F, I to m and α to a in the formula $F=ma$.

The moment of inertia I of an object about its axis of rotation depends on its mass distribution and on the position of the axis. A uniform sphere of mass M and radius r spinning about a diameter through its centre has a value $I=2Mr^2/5$. A solid cylinder of mass M and radius r spinning about its axis has a value $I=Mr^2/2$.

The unit of I is kg m^2. From $T=I\alpha$, the moment of inertia I may be defined as

$$I=\frac{T}{\alpha}=torque/angular\ acceleration$$

TORQUE VALUES

The torque T of a force about an axis O is defined as its **moment** about O, that is, torque=force×perpendicular distance from O to the line of action of the force. Torque is therefore measured in newton metre, N m. In Fig 2.20(a), a disc W is rotated by a force of 2N acting tangentially to its rim. If the radius of W is 0.3m, the torque=2N×0.3m=0.6N m. Figure 2.20(b) shows an object turning about an axis O under the action of its weight of 5N acting at G. If OG=0.4m and the angle made by OG with the horizontal through O is 60°, the torque about O=5N×OP=5N×0.4 cos 60°m=1.0N m. Figure 2.20(c) shows a **couple** consisting of two equal forces of 2N separated by a perpendicular distance of 0.2m. The torque (moment) due to the two forces =one force×perpendicular distance between forces=2N × 0.2m =0.4N m.

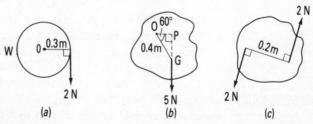

(a)   (b)   (c)

Fig 2.20 Torque and measurement

We can now apply equation (1), $T=I\alpha$. In Fig 2.20(a), suppose that the disc W has a moment of inertia I of 0.2kg m^2 about O and a constant torque of 0.6N m is applied to W. Then its angular acceleration α about O is

$$\alpha=\frac{T}{I}=\frac{0.6}{0.2}=3 \text{ rad s}^{-2}$$

ANGLE OF ROTATION

In linear motion, we have seen that the final velocity v of a uniformly accelerated object is given by $v=u+at$. In rotational motion a similar equation applies. If ω_0 is the initial angular velocity of a rotating object which then has a uniform angular acceleration of α for a time t, its final angular velocity ω is given by

$$\omega=\omega_0+\alpha t$$

Suppose a torque T of 0.6N m acts on the disc W in Fig 2.20(a) for 10 s. With a moment of inertia I of 0.2kg m^2 about O, then $\alpha=3$ rad s^{-2} as shown before. So if W was initially at rest, the angular velocity ω reached by W in 10 s is

$$\omega=\omega_0+\alpha t=0+(3\times10)=30 \text{ rad s}^{-1}$$

The *average* angular velocity during 10 s$=\frac{1}{2}(0+30)=15$ rad s^{-1}. So the angle of rotation θ after 10 s$=15\times10=150$ rad. One revolution is equivalent to an angle of 2π rad. So the number of revolutions made by W$=\theta/2\pi=150/2\pi=24$ revs.

KINETIC ENERGY OF ROTATION

The kinetic energy of a small mass moving with velocity v is $\frac{1}{2}mv^2$. Replacing m by I and v (linear velocity) by ω (angular velocity),

$$\text{kinetic energy of rotation}=\tfrac{1}{2}I\omega^2$$

The Earth has a moment of inertia about its axis of about 1.0×10^{18}kg m^2 and an angular velocity ω of about 7.3×10^{-5} rad s^{-1}. If we consider the Earth to be roughly a uniform sphere, then

$$\text{kinetic energy of rotation}=\tfrac{1}{2}I\omega^2$$
$$=\tfrac{1}{2}\times1.0\times10^{18}\times(7.3\times10^{-5})^2=2.7\times10^{29}\text{J}$$

The worked example on page 69 shows how the kinetic energy of rotation of a wheel is found after a torque is applied to the wheel.

WORK DONE BY TORQUE

Consider a wheel X of radius r which is rotated through an angle of θ about its centre O by a force in a rope round the wheel (Fig 2.21). If F is constant and tangential to X as it is turned, the torque is constant and equal to $F.r$. The work done, W, by F in moving a distance s from A to B$=F \times s = F \times r\theta$, since $s=r\theta$. Now the torque $T=F \times r$. So

$$\text{work done, } W=T \times \theta \tag{1}$$

Although this is a simple case, the relation in (1) is generally true. Suppose a constant torque of 2N m turns a wheel through 3 revolutions, that is, $\theta=3 \times 2\pi$ rad. Then

$$\text{work done}=T \times \theta=2 \times 3 \times 2\pi=37.7\text{J}$$

The worked example on page 69 shows how the kinetic energy of a rotating wheel can be found from the work done by a torque.

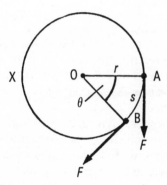

Fig. 2.20a Work done by torque

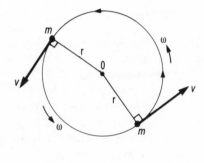

Fig. 2.21 Angular momentum in circular motion

ANGULAR MOMENTUM

In linear (translational) motion, an object has linear momentum mv where m is the mass and v is the velocity. In rotational motion, the **angular momentum** of an object about an axis is defined as the **moment of its momentum** about the axis.

In Fig 2.21 a particle of mass m is moving round a circle of centre O and radius r with a steady speed v. Its angular velocity ω is given by $v=r\omega$.

At any instant, the angular momentum about the centre of the mass is

$$\text{momentum} \times r = mv \times r = mr\omega \times r = mr^2\omega$$

mr^2 is actually the moment of inertia I of the mass about the centre, so the angular momentum is $I\omega$. Generally, for a large object rotating about an axis its angular momentum is given by

$$\text{angular momentum} = I\omega$$

TORQUE AND ANGULAR MOMENTUM

We have already seen that a torque T produces an angular acceleration α given by $T=I\alpha$. Suppose T is constant and produces an increase in angular velocity from ω_1 to ω_2 in a time t. Then $\alpha=(\omega_2-\omega_1)t$. So

$$T=I\alpha=\frac{I(\omega_2-\omega_1)}{t}$$

and $\quad T\times t=I\omega_2-I\omega_1=$change in angular momentum $\qquad$ (1)

$T\times t$ is analogous to $F\times t$ in linear motion, which produces a change in linear momentum.

It should be remembered that $T\times time=angular\ momentum$ change, whereas $T\times angle\ (\theta)=kinetic\ energy\ of\ rotation$ change, since this is the work done by a torque.

The worked example on page 69 shows how to apply the relation (1) to a rotating wheel when a torque acts on it.

CONSERVATION OF ANGULAR MOMENTUM

In the case of explosive or colliding objects in linear motion, we saw that the total momentum of the given remains constant if no external (outside) forces act on the system. A similar result holds for a rotating system: *The total angular momentum of a system about an axis is constant,* if no external torques act on the system.

So an ice skater, spinning on her toes, has an angular momentum about a vertical axis through the toes which remains constant since no external torque acts. With her arms folded, the moment of inertia about the axis changes from a value I when her arms were outstretched to a *smaller* value I_1 when the arms are folded, since the whole body is then nearer the vertical axis. If ω is the angular velocity with the arms outstretched and ω_1 with the arms folded, it follows from the conservation of angular momentum that

$$I_1\omega_1=I\omega$$

So $\qquad \omega_1=\frac{I}{I_1}\omega$

Since I_1 is less than I, the new angular velocity with arms folded, ω_1, is *greater* than with the arms outstretched. So the skater spins faster. Similarly, if a diver from the top board of a swimming bath curls his body in mid-air to reduce his moment of inertia about his centre, he will rotate faster and make more turns before entering the water.

The initial kinetic energy$=\frac{1}{2}I\omega^2$ and the final kinetic energy$=\frac{1}{2}I_1\omega_1^2$ in the case of the skater. So

$$\text{kinetic energy increase}=\frac{1}{2}I_1\omega_1^2-\frac{1}{2}I\omega^2=\frac{1}{2}\omega_1(I_1\omega_1)-\frac{1}{2}\omega(I\omega)$$

But $I_1\omega_1=I\omega$, and ω_1 is greater than ω. So we see that, in the above cases, some kinetic energy is gained. The gain in kinetic energy is equal to the work done by the skater in pulling in his or her arms.

CENTRAL FORCES AND ANGULAR MOMENTUM

As we showed on page 65, a small mass m rotating with constant speed v in a circle of radius r has an angular momentum about the centre $= mvr = mr^2\omega$, where ω is the angular velocity. Now the centripetal force F on m always acts *towards the centre* (Fig 2.22(a)). So F has no torque (moment) about the centre. Since there is no external force about the centre while the mass moves, it follows that

angular momentum about centre = constant

The worked example on page 68 shows how this conservation is applied to the motion of a mass connected to a point on a fixed table by a string whose length is altered.

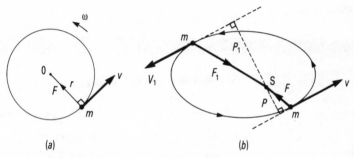

(a)   (b)

Fig 2.22 Central forces and angular momentum

Planets revolving round the Sun have a gravitational force F on them towards the Sun (Fig 2.22(b)). So there is no external torque about the Sun. Therefore $mvp = $ constant for a planet such as the Earth. This means that when p is less, v is greater. The Earth moves in an elliptical orbit round the Sun and when it is nearer the Sun (p less), the Earth's speed (v) increases.

ACCELERATION OF OBJECT ROLLING DOWN PLANE

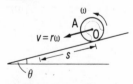

Fig 2.23 Acceleration of rolling ball

Consider a cylinder A rolling from rest down a plane inclined at an angle θ to the horizontal (Fig 2.23). The centre of mass O of the cylinder moves with a linear or translational velocity v down the plane and in addition the cylinder rotates round O. So

$$\text{total kinetic energy} = \text{linear} + \text{rotational} = \frac{1}{2}Mv^2 + \frac{1}{2}I\omega^2$$

where M is the mass of the cylinder, I is the moment of inertia about O and ω is the angular velocity about O. Now for rotation without slipping, $v = r\omega$, where r is the radius of the cylinder. So

$$\text{total kinetic energy} = \frac{1}{2}Mv^2 + \frac{1}{2}I \cdot \frac{v^2}{r^2} = \frac{1}{2}v^2\left(M + \frac{I}{r^2}\right)$$

Suppose the cylinder rolls down a distance s from rest. Then loss in potential energy $= Mgs \sin \theta =$ gain in kinetic energy $= \frac{1}{2}v^2(M + I/r^2)$, from above. So

$$v^2 = \frac{M}{(M + I/r^2)}.2gs \sin \theta$$

But the acceleration a down the plane is given by $v^2 = 2as$. Hence

$$a = \frac{Mg \sin \theta}{(M + I/r^2)}$$

For a hollow cylinder, $I = Mr^2$ or $I/r^2 = M$. For a solid cylinder of the same mass, $I = Mr^2/2$, or $I/r^2 = M/2$. So a is *less* for a hollow cylinder. Released from rest at the same place on an inclined plane, the solid cylinder will reach the bottom first.

WORKED EXAMPLES ON ROTATIONAL DYNAMICS

2.9 CIRCULAR MOTION AND ANGULAR MOMENTUM

A small mass 0.2kg is connected to a fixed point O on a table by a string of length 1.0m. With the string taut, the mass is given an initial velocity of 2m s^{-1} in a direction perpendicular to the string.

Calculate (a) the initial tension in the string, (b) the initial angular momentum of the mass about O.

While the mass is rotating, the length of the string attached to O is shortened to 0.5m. Calculate (c) the new velocity of the mass, (d) the new tension in the string.

(**Overview** (a) The tension is a centripetal force of value mv^2/r or $mr\omega^2$.
(b) Use angular momentum $= mv \times r$.
(c) Angular momentum about O $=$ constant.
(d) New tension $=$ new mv^2/r.)

(a) $T = mv^2/r = 0.2 \times 2^2/1 = 0.8\text{N}$

(b)   initial angular momentum $= mv \times r = 0.2 \times 2 \times 1.0 = 0.4\text{kg m}^2 \text{ s}^{-1}$

(c)   Since angular momentum about O is constant,

$$mv \times r = mv_1 \times r_1$$

or $2(v) \times 1.0(r) = v_1 \times 0.5 \ (r_1)$

So $v_1 = 4\text{m s}^{-1}$

(d)   new tension $T_1 = mv_1^2/r_1 = 0.2 \times 4^2/0.5 = 6.4\text{N}$

2.10 TORQUE ON WHEEL AND ANGULAR MOMENTUM, KINETIC ENERGY

A constant torque of 40N m is applied to a wheel for 10 s and the angular velocity of the wheel then increases from zero to 20 rad s^{-1}. Calculate (a) the moment of inertia of the wheel, (b) the kinetic energy at the end of the 10 s.

If the torque is now replaced by an opposing torque and the wheel angular velocity reduces to half its initial value in 40 s, calculate (c) the opposing torque and (d) the number of revolutions made by the wheel in 40 s.

(Overview (a) Angular momentum change=torque×time.
(b) Kinetic energy formula needed.
(c) Use as in (a).
(d) Work done (torque×angle)=kinetic energy change.)

(a)   From $T \times t$=angular momentum change, we have

$$40 \times 10 = I(20-0)$$

So
$$I = \frac{400}{20} = 20\text{kg m}^2$$

(b)
$$\text{Kinetic energy} = \frac{1}{2}I\omega^2 = \frac{1}{2} \times 20 \times 20^2 = 4000 \text{ J}$$

(c)   Opposing torque $T \times 40 = I(20-10)$

So
$$T = \frac{20 \times 10}{40} = 5\text{N m}$$

(d)   Work done $= T \times \theta = \frac{1}{2}I.20^2 - \frac{1}{2}I.10^2$

$$= \frac{1}{2} \times 20(20^2 - 10^2) = 3000 \text{ J}$$

So
$$\theta = \frac{3000}{5} = 600 \text{ rad}$$

Hence   number of revs $n = \frac{\theta}{2\pi} = \frac{600}{2\pi} = 95.5$ revs

1 CIRCULAR MOTION

Angular velocity, ω=angle (θ)/time (t) (Unit: rad s^{-1})

Velocity in circle, $v=r\omega$

Acceleration in circle, $a=v^2/r$ or $r\omega^2$

Centripetal force=force *towards centre*

$=mv^2/r$ or $mr\omega^2$

Conical pendulum or banked track problem, find component F of force *towards centre* of circle and use $F=mv^2/r$ or $mr\omega^2$.

For conical pendulum or banked track, $\tan\theta=v^2/rg$.

Owing to rotation of Earth, weight of an object changes from mg at the poles to mg' at the equator where $g'=g-r\omega^2$.

2 GRAVITATION

Newton's law: $F=Gm_1m_2/r^2$

At Earth's surface $g=GM/r_E^2$

Above Earth, $g' \propto 1/r^2$, where r is distance from centre of Earth

Satellite in Earth orbit:

(a) $F=mg'=mv^2/r$

(b) $g'/g=(1/r^2)\div(1/r_E^2)$

Parking orbit: $T=24\text{h}=2\pi r/v$.

In orbit,

potential energy$=-GMm/r$, kinetic energy$=+GMm/2r$

Continued on p.72

DIAGRAM SUMMARY

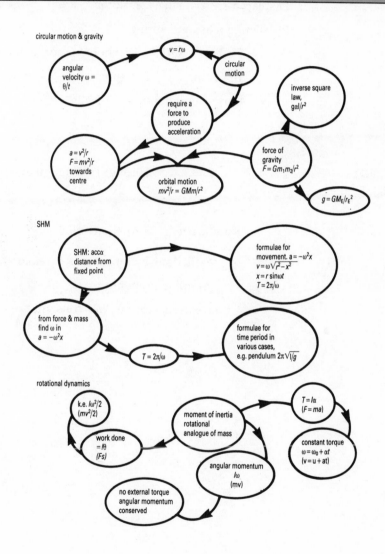

circular motion & gravity

angular velocity $\omega = \theta/t$

$v = r\omega$

circular motion

inverse square law, $g \propto 1/r^2$

require a force to produce acceleration

$a = v^2/r$
$F = mv^2/r$
towards centre

orbital motion $mv^2/r = GMm/r^2$

force of gravity $F = Gm_1m_2/r^2$

$g = GM_E/r_E^2$

SHM

SHM: acc $\propto$ distance from fixed point

formulae for movement. $a = -\omega^2 x$
$v = \omega\sqrt{r^2 - x^2}$
$x = r\sin\omega t$
$T = 2\pi/\omega$

from force & mass find ω in $a = -\omega^2 x$

$T = 2\pi/\omega$

formulae for time period in various cases, e.g. pendulum $2\pi\sqrt{l/g}$

rotational dynamics

k.e. $I\omega^2/2$
$(mv^2/2)$

work done $= I\theta$
(Fs)

moment of inertia rotational analogue of mass

$T = I\alpha$
$(F = ma)$

constant torque $\omega = \omega_0 + \alpha t$
$(v = u + at)$

angular momentum $I\omega$
(mv)

no external torque angular momentum conserved

3 SIMPLE HARMONIC MOTION

Definition:

Acceleration $\propto$ displacement and directed towards a fixed point

Equations:

acceleration $a=-\omega^2 x$; $\quad a_{max}=-\omega^2 r$, where r is amplitude

velocity $v=\omega\sqrt{r^2-x^2}$; $\quad v_{max}=\omega r$

Displacement $x=r \sin \omega t$. $T=2\pi/\omega$

Oscillating mass and spring:

$T=2\pi\sqrt{m/k}$ ($F=-kx$ where k is spring force constant)

In s.h.m., $\quad$ k.e. (mass)+p.e. (spring)=constant

Simple pendulum: $\quad T=2\pi\sqrt{l/g}$

4 ROTATIONAL DYNAMICS

Torque, $T=I\alpha$ (angular acceleration)

$$\omega=\omega_0+\alpha t$$

Kinetic energy of rotation$=\frac{1}{2}I\omega^2$

Work done by torque$=T\times\theta$ (radians)

Angular momentum$=I\omega$

If no external torque, $I\omega=$constant

2 QUESTIONS

1 The bob of a simple pendulum moves simple harmonically with amplitude 8.0cm and period 2.00 s. Its mass is 0.50kg. The motion of the bob is undamped. Calculate maximum values for (*a*) the speed of the bob, and (*b*) the kinetic energy of the bob.   (L)

2 (*a*)   A particle of mass *m* moves with a constant speed *v* in a circular path of radius *r*. Show that a force is necessary to maintain the circular motion and derive an expression for its magnitude. State clearly the direction of the force on the particle.

(*b*)   A small mass hangs by a string from a fixed point and moves in a circular path at constant speed in a horizontal plane. Draw a diagram showing the forces acting on the mass and derive an equation showing how the angle of inclination of the string to the vertical depends upon the speed of the mass and the radius of the circle in which it moves.

(*c*)   In the arrangement described in (*b*) the mass is 0.50kg, the radius of the circle is 1.00m and the mass makes 30 revolutions per minute. The string suddenly breaks and the mass falls freely to the ground through a vertical distance of 1.00m. Calculate
(i) the inclination of the string to the vertical immediately before the string breaks,
(ii) the horizontal distance travelled by the mass from its position when the strong breaks to the point of impact,
(iii) the change in kinetic energy of the mass during its free fall. (JMB)

3 A satellite of mass 66kg is in orbit round the Earth at a distance of 5.7*R* above its surface, where *R* is the mean radius of the Earth. If the gravitational field strength at the Earth's surface is 9.8N kg^{-1}, calculate the centripetal force acting on the satellite.
Assuming the Earth's mean radius to be 6400km, calculate the period of the satellite in orbit in hours.   (L)

4 (*a*)   A body moves at a constant speed along a circular path.
(i) Explain why a force is required to maintain the circular motion.
(ii) State the direction in which the force acts, and give an expression for its magnitude, identifying any symbols used.

(*b*)   The diagram (Fig 2.24) shows part of a fairground roundabout. AB is a horizontal arm which rotates about a fixed point A. The chair C, which is attached to B by a cable, moves in a horizontal circle of radius 5.0m completing each revolution in 6.0 s. The weight of chair and occupant is 800N.

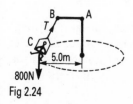

Fig 2.24

(i) Calculate the speed of the chair.

(ii) Determine the force required to maintain the circular motion of C and state how this force is provided.

(iii) Calculate the tension T in the cable (acceleration of free fall, g, is 10m s^{-2}). (*AEB* 1985)

5 (i) Show that the ratio (*radius of orbit*)^3/(*time for one complete orbit*)^2 is constant for all satellites moving in circular orbits around the Earth.

(ii) The Moon moves around the Earth in an orbit which is approximately circular. The radius of the orbit is 4.0×10^5km and the time for one complete orbit is 29.5 days. Find the height above the Earth's surface of a communications satellite which orbits the Earth once a day, given that the radius of the Earth is 6.4×10^3km.

The satellite will be most effective if it remains over the same place on the Earth as the Earth rotates. What is the other condition which must be satisfied if this is to be the case? (*AEB* 1984)

6 (*a*) What is meant by *simple harmonic motion*? The equation $x = a \sin 2\pi ft$ can represent the motion of a body executing simple harmonic motion where x represents the displacement of the body from a fixed point at time t. Sketch two cycles of the motion beginning at $t=0$, clearly labelling the axes of the graph. Use the graph to explain the physical meanings of a and f.

Explain how you could obtain from the graph the speed of the body at any instant.

(*b*) In order to check the timing of a camera shutter a student set up a simple pendulum of length 99.3cm so that the bob swung in front of a horizontal metre scale. The bob was observed to swing between the 40.0cm and 60.0cm marks at its extreme positions. The camera was mounted directly in front of the scale, set for an exposure time (time for which the shutter is open) of $\frac{1}{50}$ s and a photograph taken. The resulting photograph showed the bob to have moved from the 51.0cm mark to the 51.6cm mark while the shutter was open.

What is the percentage error in the exposure time indicated on the camera? (Period of oscillation of a simple pendulum of length l may be taken as $T = 2\pi \sqrt{l/g}$ where g is the acceleration of free fall.)

(*L*)

7 State the law describing the gravitational force between two point masses M and m a distance r apart. Two alternative units for gravitational field strength are N kg^{-1} and m s^{-2}. Use the method of dimensions to show they are equivalent.

State the general relationship between the field strength at a point in a field of force and the potential gradient at the point. Write down an expression for the gravitational potential at a point distant r from a mass M. Distinguish between *gravitational potential* and *gravitational potential energy*.

The curve in Fig 2.25 shows the way in which the gravitational energy of a body of mass m in the field of the Earth depends on r, the distance from the centre of the Earth, for values of r greater than the Earth's radius R_E. What does the gradient of the tangent to the curve at $r = R_E$ represent?

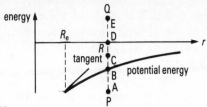

Fig 2.25

The body referred to above is a rocket which is projected vertically upwards from the Earth. At a certain distance R from the centre of the Earth, the *total* energy of the rocket (i.e. its gravitational potential energy plus its kinetic energy) may be represented by a point on the line PQ. Five points A,B,C,D,E have been marked on this line. Which point (or points) could represent the total energy of the rocket
(a)   if it were momentarily at rest at the top of its trajectory,
(b)   if it were falling towards the Earth,
(c)   if it were moving away from the Earth, with sufficient energy to reach an infinite distance?
In each case, explain briefly how you arrive at your answer.
(C)

8   State the relationship between the forces on a body and the distance of the body from a fixed position when the body is executing simple harmonic motion about that position.
Show that a body of mass m suspended by a light elastic string for which the ratio of tension to extension is λ will execute simple harmonic motion when given a small vertical displacement from its equilibrium position. Find the period of the motion for the case of $m=0.1$kg and $\lambda=20$N m^{-1}.
A second 0.1kg mass is attached to the first by a light inextensible wire and hangs below it. The system is allowed to come to rest, and at time $t=0$ the wire is cut. Calculate the position, velocity and acceleration of the first 0.1kg mass at time $t=1.05$s, assuming no resistance to the motion.
Give expressions for the kinetic and potential energy of the system at time t. Show that the total energy is independent of time. Outline qualitatively what would happen to the total energy of such a system set oscillating in a laboratory.   (O&C)

9   The moment of inertia of the Earth about its axis of rotation is 8.0×10^{37}kg m^2. Estimate (a) the Earth's angular momentum, and (b) the Earth's angular kinetic energy.
Due to the frictional effect of the tides on the ocean bed, the length of the Earth's day is very slowly increasing. What effect does this have on the kinetic energy and the angular momentum of the earth?
(L)

10 (a)   Giving the meanings of the symbols used,
            (i) write an equation expressing Newton's second law for linear motion,
            (ii) write the corresponding equation for rotational motion.
     (b)   A flywheel of radius 0.250m is mounted on a horizontal axis

of radius 0.015m. The moment of inertia of the system about its axis of rotation is 0.225kg m², and the frictional couple at the bearings is negligible. A constant force of 60N is applied tangentially to the axle for 4.0 s starting from rest.

Calculate
(i) the angular acceleration during the first 4.0 s,
(ii) the angular velocity and kinetic energy after 4.0 s,
(iii) the constant tangential braking force which must then be applied to the rim of the flywheel to bring it to rest in 10 revolutions.

(c)   Discuss briefly whether or not the system loses energy at a constant rate when slowing down under the action of a constant force as in (b)(iii).   (JMB)

STATIC ELECTRICITY

CONTENTS

CHARGE: FORCES, FIELD AND POTENTIAL

INTRODUCTION

Electric charge can be produced by rubbing insulating materials. In this process electrons (negative charges) are transferred between the insulating material A and the rubbing material B. If A loses electrons to B then A will now be positively charged. If electrons are transferred from B to A then A becomes negatively charged. The actual direction of electron transfer depends on the nature of the materials of A and B.

The simple 'O' level facts about electrostratics are dealt with in the authors' *Pan Study Aids – Physics*. It is assumed that the reader is familiar with them.

COULOMB'S LAW OF FORCE

The force F between two point charges Q_1, Q_2 separated by a distance r is found experimentally to be given by

$$F \propto \frac{Q_1 Q_2}{r^2}$$

The constant of proportionality depends on the medium in which the charges are situated. With the charges Q_1, Q_2 measured in coulombs and the distance r in metres, the constant of proportionality in a vacuum is written as $1/4\pi\varepsilon_0$, where $\varepsilon_0 = 8.85 \times 10^{-12} \text{F m}^{-1}$. ε_0 is called the **permittivity of free-space (*vacuum*)**. So

$$F = \frac{Q_1 Q_2}{4\pi\varepsilon_0 r^2}$$

F is attractive if Q_1 and Q_2 have opposite sign; it is repulsive if Q_1 and Q_2 have the same sign.

The formula also holds for charges distributed over small objects such as spheres, provided that the separation between their centres, r, is much greater than their linear dimensions. To a good approximation note that $\frac{1}{4\pi\varepsilon_0}$ is numerically equal to 9×10^9, which is useful in calculations.

ELECTRIC FIELD. INTENSITY DUE TO A POINT CHARGE

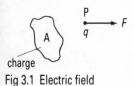

charge

Fig 3.1 Electric field

An **electric field** is a region in space where electric forces can be experienced. In Fig 3.1 the charge on an object A produces an electric field all around A. The field can then exert forces on small charges, such as q, at P.

The **intensity** E of the electric field at any point such as P, is defined as *force per coulomb* on a small test charge placed at that point. The direction of the intensity is in the direction of the force on a positive charge.

So if F is the force on a test charge q at a point where the intensity is E, then by definition $E=F/q$.

So $F=qE$

From this definition the unit of E is $N\,C^{-1}$. We shall see later that an equivalent unit $V\,m^{-1}$ is more often used.

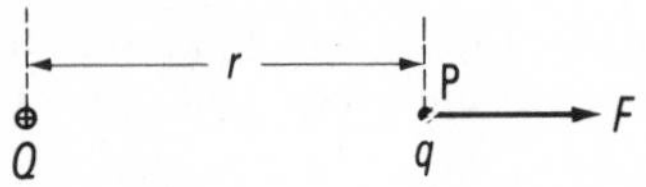

Fig 3.2 Field of a point charge

We can find the intensity E due to a charge Q at a point P a distance r from Q, Fig 3.2. With a test charge q at P, the force, F, on q is given by

$$F=\frac{Qq}{4\pi\varepsilon_0 r^2}$$

But $E=F/q$, so the intensity E at P is given by

$$E=\frac{Q}{4\pi\varepsilon_0 r^2}$$

Note that intensity E is a *vector*; its direction is the direction of the force on a positive charge, as we have said.

ELECTRIC POTENTIAL DIFFERENCE

As in the gravitational field, points in an electric field can have an electric **potential**. The *electric potential difference* between two points is defined as the *work done per coulomb* in taking a small +ve charge between the points. The unit of potential difference (p.d.) is the *volt* and from the above definition, 1 volt is 1 joule per coulomb.

The **electric potential**, V, at a point is the potential difference between infinity and that point. This means that infinity is taken to be the 'zero' of potential. Alternatively, we may say that the electric potential at a point is the work done per coulomb in bringing a small +ve charge from infinity to that point.

From these definitions, it can be seen that the p.d. between two points 1 and 2 where the potentials are V_1 and V_2 respectively is (V_2-V_1), and that the work done in taking Q coulombs from point 1 to point 2 is given by

$$W=Q(V_2-V_1)$$

or $W=QV$

where V is the p.d. between the points.

POTENTIAL DUE TO A POINT CHARGE

To calculate the potential at P due to a point charge $+Q$ (Fig 3.3), we must find the work done in bringing 1 coulomb of positive charge from infinity to P. The force F on the charge at a distance x from Q is given by

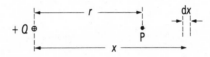

Fig 3.3 Potential of a point charge

$$F=\frac{Q\times1}{4\pi\varepsilon_0x^2}=\frac{Q}{4\pi\varepsilon_0x^2}$$

So $V=$ work done per coulomb $=\int_\infty^r -\frac{Q}{4\pi\varepsilon_0x^2}dx$

where the $-$ve sign indicates that F and dx are oppositely directed.

Hence $V=\left[+\frac{Q}{4\pi\varepsilon_0x}\right]_\infty^r=\frac{Q}{4\pi\varepsilon_0r}$

Note that (a) V is positive when Q is +ve, and negative when Q is $-$ve. (b) V is a scalar quantity, so we add potentials without worrying about direction. Electric intensity is added vectorially as we have seen.

VARIATION OF *E* AND *V* FOR CHARGED HOLLOW SPHERE

Outside a charged sphere, E and V vary with distance r from the centre of the sphere just as if all the charge Q were concentrated at the centre.

Inside the charged sphere, however, $E=0$ everywhere. This is true inside any charged hollow conductor because the charges are present on their outside surfaces and not inside. So sensitive electrical equipment is placed inside hollow conductors to shield it from an electric field.

If there is no field, there is no force on any charge inside the sphere. So no work is done in moving the charge. Thus the p.d. between *any* two points inside the sphere is zero. Thus the potential is the same everywhere inside and equal to the potential at the surface.

These results for E and V are shown graphically in Fig 3.4.

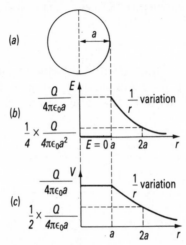

Fig 3.4 Field and potential
for a hollow sphere

FIELD INTENSITY AND POTENTIAL GRADIENT

Suppose q coulombs of charge moves a small distance δx between two points in the direction of an electric field. Then the work done per coulomb, by definition, is the p.d. δV between those points, that is,

$$\delta V = \frac{W}{q}$$

But the work done on the charge, $W = \text{force} \times \text{distance} = -F\delta x = -qE\delta x$. The negative sign shows that the work is done *on* the charge when E and δx are oppositely directed. Hence,

$$\delta V = W/q = -E\delta x$$

So
$$E = -\frac{\delta V}{\delta x} = -\frac{dV}{dx}$$

as δx becomes very small. E is therefore the negative *potential gradient* of V with distance x in the field.

In the graphs for a charged sphere shown in Fig 3.4, the electric intensity graph, E–r, at any point is minus the gradient of the potential graph, V–r. Inside the charged sphere V is constant and so has zero gradient. Outside the sphere the gradient of the $1/r$ graph for

V varies as $1/r^2$ as can be seen by differentiating $1/r$. This shows that E varies as $1/r^2$.

Because of the relation $E=-\dfrac{dV}{dx}$, E is numerically equal to the potential gradient at the point in the field and so E has units $V\,m^{-1}$. The negative sign shows E is directed in the direction of *decreasing* potential.

FIELD AND POTENTIAL FOR PARALLEL PLATES

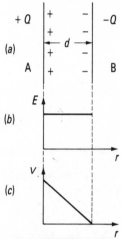

Fig 3.5 Field and potential for parallel plates

Figure 3.5(*a*) shows two parallel metal plates A and B separated by a distance d and carrying equal and opposite charges $+Q$ and $-Q$. Between the plates the field is *uniform*. Hence E is constant (Fig 3.5(*b*)), and so the potential has a constant gradient as shown in Fig 3.5(*c*). In this special case,

$$E=-\frac{dV}{dr}=\frac{V}{d}\ \text{numerically}$$

where V is the p.d. between the plate.

By treating the p charges on the plates as very many tiny separate charges and integrating their effect, it can be shown that

$$E=\frac{\sigma}{\varepsilon_0}$$

where σ is the *surface charge density*, that is, Q/A, where Q is the charge on a plate and A is the area. σ is therefore measured in $C\,m^{-2}$.

The formula $E=\sigma/\varepsilon_0$ can be used to find the field near any part of a charged surface where the charge density is σ.

ELECTRIC FIELD LINES AND EQUIPOTENTIALS

Electric fields can be mapped out by field lines called lines of **electric flux**. The tangent to a field line at any point gives the direction of the field or intensity at that point.

An **equipotential surface** is the surface connecting all points where the potential is the same. It can be labelled with the value of the potential of that surface.

Since E is equal to the potential *gradient*, there can be no component of electric field intensity along an equipotential surface because the potential does not change along the surface. Thus field lines and equipotential surfaces always cross *at right angles*. Figure 3.6 shows the equipotentials and electric field lines around (*a*) a spherical charge X, (*b*) parallel plates L, M, (*c*) two equal and opposite charges A, B.

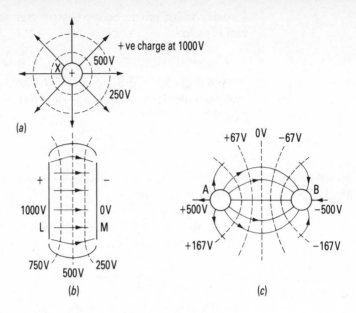

Fig 3.6 Field lines and
equipotentials

GRAVITATIONAL AND ELECTRIC FIELDS

There is a close similarity between the behaviour of gravitational and electric fields. This analogy is set out in the following table.

	Electric	Gravitational
Definition of field	$F=QE$	$F=mg$
Unit of field	$N\,C^{-1}$	$N\,kg^{-1}$
Force between spheres	$F=Q_1Q_2/4\pi\varepsilon_0 r^2$	$F=Gm_1m_2/r^2$
Direction of force	Attr: unlike Repulsive: alike	Always attractive
Field of single sphere	$E=Q/4\pi\varepsilon_0 r^2$	$g=Gm/r^2$
Potential of single sphere	$V=Q/4\pi\varepsilon_0 r$	Potential$=-Gm/r$
Unit of potential	$J\,C^{-1}=(\text{volt})$	$J\,kg^{-1}$

Note that the sign for the potential in the gravitational field is negative. This is because the zero of potential is defined at infinity (see page 80). Thus as a mass moves towards the Earth from a large distance its potential decreases and so becomes more and more negative.

In the case of electric potential the sign depends on the sign of Q, since the force on unit positive charge can be attractive (Q negative) or repulsive (Q positive).

Note that the only case of a uniform gravitational field is that near the Earth's surface where the distances involved are relatively too small to notice the effect of the inverse square law. In this case the motion of a mass under gravity is an exact parallel of the motion of a charge in a uniform electric field (see page 252).

CAPACITANCE

CAPACITORS

A capacitor consists of two parallel plates each made from a conducting material and which are separated by an insulator, called a **dielectric**. The circuit symbol for a capacitor is shown in Fig 3.7(a), and Fig 3.7(b) shows the symbol for an electrolytic capacitor. This type of capacitor *must* be connected with its positive terminal to the positive of the supply, since the layer of insulator builds up by electrolytic action when the capacitor is connected.

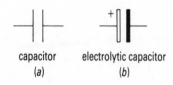

capacitor   electrolytic capacitor
(a)   (b)

Fig 3.7 Capacitor symbols

Capacitors are usually described by their dielectric, for example, mica, polycarbonate, paper or air.

CAPACITANCE

When a p.d. V from a battery, for example, is connected to the plates of a capacitor, electrons are pulled from one plate, leaving it positively charged, and pushed on to the other plate making that negatively charged. So there is a charge $+Q$ on one plate and $-Q$ on the other, and *a charge Q has flowed round the external circuit* to produce this situation.

The capacitance C of the capacitor is defined to be the ratio Q/V, or the charge separation produced by a p.d. of 1 volt. In general, then,

$$C = Q/V \text{ or } Q = CV$$

C is measured in units called farads (F). From the formula $1F = 1C\,V^{-1}$. Since 1 farad is the value for an extremely large capacitor it is more usual to use subunits, the microfarad, μF ($10^{-6}F$); the nanofarad, nF ($10^{-9}F$); and the picofarad, pF ($10^{-12}F$).

CHARGING AND DISCHARGING CIRCUITS

Consider the circuit of Fig 3.8(a) which consists of a capacitor C in series with a resistor R and a battery of p.d. V. At the moment when the switch is closed, there is no charge on the capacitor, so the p.d. across it is zero. Since $V_C+V_R=V$, then at $t=0$ all the supply p.d. V appears across the resistor R. A current $I=V/R$ thus flows. The *slope* of the charge against time graph is hence also V/R since $I=dQ/dt$ (page 111). As current flows the capacitor starts to charge and so V_C rises ($V_C=Q/C$), and V_R falls. The slope of the graph therefore continually decreases. Eventually (in theory after an infinite time) the current flow ceases and the capacitor is said to be 'fully charged'. The variation of Q and I with time are shown in the respective graphs of Fig 3.8(c) and (d), up to the vertical dotted line.

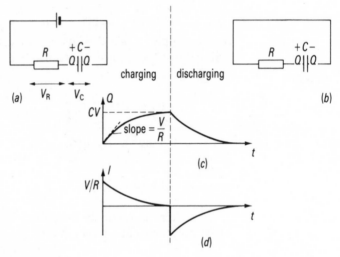

Fig 3.8 Charge and discharge of a capacitor

If the battery is now removed from the circuit, the charges are isolated and remain on the plates of the capacitor. They will stay there until we connect C and R together, as shown in Fig 3.8(b). Now the p.d. across the capacitor forces the current to flow through R in the opposite direction, as shown by the current in Fig 3.8(d) becoming negative beyond the vertical dotted lined. As the current flows the charge Q on the capacitor decreases and so the p.d. also falls. This means that the current falls and so the capacitor is discharging at an ever-decreasing rate, as the graphs show.

TIME CONSTANT

On discharge (Fig 3.8(b)), where there is no p.d. from the battery, we see that

$$V_C+V_R=0$$

or $\quad \dfrac{Q}{C}+IR=0$

so that $\quad I=\dfrac{dQ}{dt}=-\dfrac{Q}{CR}$

This equation can be integrated to give

$$Q=Q_0e^{-t/CR}$$

CR is called the **time-constant** of the circuit and can be shown to have units of seconds. When $t=CR$ we have

$$Q=Q_0e^{-1}=Q_0/e$$

So the time constant is the time taken for the charge to fall to 1/e of its original value, that is to about 1/3. (e=2.718...)
   On charge, Q rises to about $\frac{2}{3}$ of its final value in CR seconds. (More exactly, to (1−1/e) of its final value.)

PARALLEL-PLATE CAPACITOR

Consider a capacitor consisting of two plates of area A separated by a distance d. When a battery of p.d. V is connected to them a charge Q flows. We have already seen (page 83) that the electric field E between the plates is given by

$$E=\dfrac{V}{d} \text{ or by } E=\dfrac{\sigma}{\varepsilon_0}=\dfrac{Q}{\varepsilon_0 A}$$

Hence $\quad \dfrac{V}{d}=\dfrac{Q}{\varepsilon_0 A}$

So $\quad C=\dfrac{Q}{V}=\dfrac{\varepsilon_0 A}{d}$

Suppose $A=100\text{cm}^2=10^{-2}$ m^2 and $d=1\text{mm}=10^{-3}$m. Then, since $\varepsilon_0=8.85\times10^{-12}$F m^{-1},

$$C=\dfrac{10^{-2}\times8.85\times10^{-12}}{10^{-3}}=8.85\times10^{-11}\text{F}$$

This is a very small capacitance. In practice capacitors are made larger (a) by using thin foil for plates which can be of very large area A and subsequently rolled up, (b) by making d very small by using a thin layer of insulating material or dielectric. As we shall now see, using a dielectric between the plates increases the capacitance.

RELATIVE PERMITTIVITY

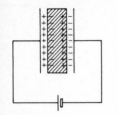

Fig 3.9 Effect of a dielectric

Figure 3.9 shows a parallel plate capacitor with a slab of dielectric between the plates. The electric field between the plates causes the dielectric to become 'polarized'. This means that within each molecule of the dielectric its −ve charge is displaced by the field slightly to the left, and its +ve charge slightly to the right. Thus there are *surface charges* on the dielectric as shown.

The p.d. between the plates is determined by the effective charge on the plates, that is, by the free charge on the metal plates together with the polarization charges on the surfaces of the dielectric. These polarization charges are opposite in sign to the free charges. So for a given p.d. we can have *more* free charge on the plates to give the same 'total' charge. Thus when a capacitor with a dielectric is charged, more charge flows round the external circuit than is the case when a vacuum is between the plates. So the capacitance has been increased.

If C is the capacitance with air (or, more strictly, a vacuum) between the plates, and C' is the capacitance with a dielectric, then the relative permittivity ε_r of the dielectric material is defined as $\varepsilon_r = \dfrac{C'}{C}$. Clearly ε_r has no dimensions; it is a number usually between 1 and 10. ε_0, on the other hand, is a dimensional constant and has a value $8.85 \times 10^{-12} \text{F m}^{-1}$.

When a dielectric is present, the formula for the capacitance of a parallel plate capacitor becomes

$$C = \frac{\varepsilon_r \varepsilon_0 A}{d}$$

MEASUREMENT OF CAPACITANCE BY VIBRATING REED SWITCH

In the circuit of Fig 3.10 a coil L produces an alternating magnetic field which causes the vibrating reed Z to move to and fro between the contacts X and Y with a frequency f equal to that of the a.c. supply, B. The diode D is present otherwise Z may vibrate at twice the frequency of the supply.

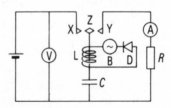

Fig 3.10 Capacitance by vibrating reed switch

When Z touches X, the battery is connected to the capacitor, C. So C charges up to a p.d. V, measured on the voltmeter. When Z touches

Y, the capacitor discharges through the ammeter and a charge Q flows. The resistor R is present to increase the time-constant of the discharge circuit, as the ammeter is more likely to read the average current correctly if the current bursts are not too short in duration. R must not be too high, however, as otherwise the capacitor will not have time to discharge fully.

Since Q coulombs discharge through the ammeter f times per second, the average current I is given by

$$I=Qf$$

or $$Q=\frac{I}{f}$$

So the capacitance of the capacitor $C=\dfrac{Q}{V}=\dfrac{I}{fV}$.

MEASUREMENT OF ε_0

The circuit in Fig 3.10 may be used to measure ε_0 if C is a large parallel plate capacitor with air between the plates. The plates are kept parallel by small, thin pieces of insulating spacers, and the area A of the plates and their separation D is measured. The current I is measured as in the last experiment. Then

$$C=\frac{I}{fV}=\frac{\varepsilon_0 A}{d}$$

So $$\varepsilon_0=\frac{Id}{fVA}$$

and ε_0 can be calculated from the values of I, V, f, d and A.

**CAPACITORS IN SERIES
AND IN PARALLEL**

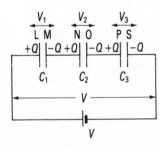

Fig 3.11 Capacitors in series

Figure 3.11 shows three capacitors C_1, C_2, C_3 in *series* with a battery of p.d. V. When the battery is connected, plate L gains a charge $+Q$ and hence plate S gains a charge $-Q$. This means that the charge on M must be $-Q$ and on P $+Q$. Hence the charges on the plates are as shown in Fig 3.11. All capacitors have the same charge Q. If C is the effective or total capacitance of the combination then

$$V=V_1+V_2+V_3$$

so $$\frac{Q}{C}=\frac{Q}{C_1}+\frac{Q}{C_2}+\frac{Q}{C_3}$$

and $$\frac{1}{C}=\frac{1}{C_1}+\frac{1}{C_2}+\frac{1}{C_3}$$

Figure 3.12 shows three capacitors C_1, C_2 and $C3$ in *parallel*. In this case the p.d. V across each capacitor is the same and there will be different charges on each. The total flowing through the external circuit will be $Q_1+Q_2+Q_3$, and so if C is the effective capacitance of the combination

$$Q=Q_1+Q_2+Q_3$$

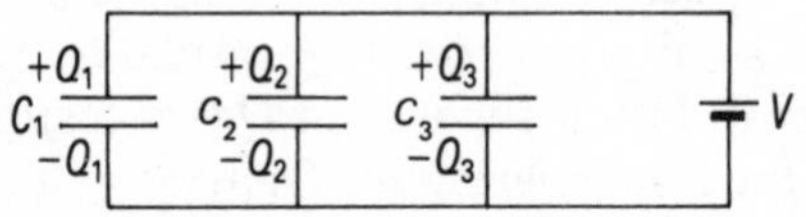

Fig 3.12 Capacitors in parallel

Hence $CV=C_1V+C_2V+C_3V$

or $C=C_1+C_2+C_3$

Note that the formulae for capacitors in series and parallel are the opposite way round compared with the formulae for resistors.

CHARGE SHARING BETWEEN CAPACITORS

Suppose a capacitor C, charged to a p.d. of V volts, is connected across an uncharged capacitor C'. To calculate what happens to the charge and p.d. we use the fact that no charge is lost, that is, the charge is conserved. The p.d. will fall to a new value V', and so after connection the charge on C is CV', and on C' the charge is $C'V'$. Before connection the charge on C was CV. Hence, from the conservation of charge

$$CV=CV'+C'V'$$

So $$V'=\frac{C}{C+C'}V$$

Note: If C' is much greater than C, then V' is very small and the p.d.

falls to a low value. In this case almost all the charge originally on C is now on C'.

If C' is much less than C, then $V \simeq V'$ and very little of the charge on C has flown on to C'.

MEASUREMENT OF CHARGE

It is possible to measure charge using a voltmeter of very high resistance. Such an instrument may be a digital voltmeter with an FET input stage.

In order to measure charge, a capacitor C' is connected across the input terminals of the voltmeter. If C' is charged and the resultant reading on the voltmeter is V, then the charge Q on C' is given by

$$Q = C'V$$

If, for example $C = 10^{-8}$F and $V = 0.4$ volt, then $Q = 4 \times 10^{-9}$C.

If C' is charged from another capacitor, not all the charge is necessarily transferred to C', as discussed below.

USE OF HIGH IMPEDANCE VOLTMETER TO MEASURE CAPACITANCE

The capacitor C under test is first connected to a battery and charged to a known p.d. V volts (Fig 3.13(a)). C is then isolated from the supply and the charge on it is measured by connecting it to the circuit as shown in Fig 3.13(b). Note that if most of the charge is to be transferred to C, then C' must be very much greater than C and hence V must be much greater than the f.s.d. reading on the voltmeter. (See note to 'charge sharing', page 90.) If this is not the case, charge will remain on C and so there will be error. Knowing Q, C is calculated from $C = Q/V$.

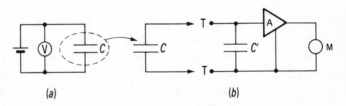

(a)   (b)

Fig 3.13 Measurement of capacitance

ENERGY STORED IN A CAPACITOR

When a capacitor is charged it stores energy which can be regained on discharge. As the capacitor is charged the p.d. across it changes, so we must integrate to find the total energy stored. Now energy = charge × p.d. (page 81). So the energy to charge a capacitor to a final p.d. of V_0 volts is given by

$$\text{Energy stored}=\int_0^{V_0}V\,dQ=\int_0^{V_0}CV\,dV$$

$$=\tfrac{1}{2}CV_0{}^2=\tfrac{1}{2}Q_0V_0=\tfrac{1}{2}\frac{Q_0{}^2}{C}$$

where Q_0 is the final charge.

CHANGES OF ENERGY WHEN PLATES MOVED

Consider a parallel plate capacitor charged to a p.d. V and then isolated from the supply (Fig 3.14(a)). The charge Q on the plates cannot change. Suppose the separation, d, of the plates now increases. Then C decreases and hence the p.d. V across the plates increases from $V=Q/C$. Thus the energy stored in the capacitor rises, since energy=$\tfrac{1}{2}QV$. We can also see this from the energy formula $\tfrac{1}{2}Q_0{}^2/C$. This can be used to find the new energy stored since Q_0 is constant and the new value of C can be found from the new separation. The increase in energy in the capacitor comes from the work done in separating the plates against the attractive forces between the opposite charges on them.

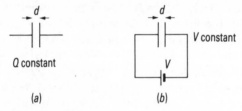

Fig 3.14 Energy changes when plates moved

In Fig 3.14(b), the p.d. V remains constant as the capacitor is connected permanently to a battery. Here when d is increased, C is decreased. So Q falls since the p.d. is constant and a current flows in the circuit. The energy stored in the capacitor *falls* since the energy=$\tfrac{1}{2}QV$. The simplest way of calculating the new energy is to use $\tfrac{1}{2}CV^2$, with the new value of C calculated from the plate separation.

WORKED EXAMPLES

3.1 FORCE BETWEEN CHARGES

A copper coin has a mass of 3.1g. Being electrically neutral it contains equal amounts of positive and negative electricity. The positive charge on the nucleus of a copper atom is 4.6×10^{-18} coulomb.
(a)   How many copper atoms are there in the coin?
(b)   What is the total positive charge in the coin?
(c)   If the total positive and total negative charges in a copper coin

are separated to a distance such that their force of attraction is 4.5N, how far apart must they be?

(d)   What would the force between the total positive and negative charge be if they were placed 1 metre apart?

(e)   What do your answers suggest about the possibility of removing all the electrons from the copper coin? (Take $1/4\pi\varepsilon_0=9\times10^9$ and the Avogadro constant$=6\times10^{23}\mathrm{mol}^{-1}$.)

(Overview This question illustrates the huge values of equal positive and negative charges that are present in matter. It also shows that only a minute fraction of this charge is affected when the object is electrically charged.)

(a)   63.5g of copper, 1 mole, contain 6×10^{23} atoms. So 3.1g contain $3.1\times6\times10^{23}/63.5=2.9\times10^{22}$ atoms.

(b)   The total positive charge$=$no of atoms$\times$charge on each nucleus

$$=2.9\times10^{22}\times4.6\times10^{-18}$$

$$=133\times10^3 \text{ coulomb}$$

(c)   The value of each charge would be 133×10^3 coulomb.

From $F=\dfrac{Q_1Q_2}{4\pi\varepsilon_0 r^2}$

we have $4.5=\dfrac{9\times10^9\times133\times133\times10^6}{r^2}$

so $r^2=2\times10^{15}\times133\times133$

and $r=5.9\times10^9\mathrm{m}$

(d)

$$F=\frac{Q_1Q_2}{4\pi\varepsilon_0 r^2}=\frac{9\times10^9\times133\times133\times10^6}{1^2}=1.6\times10^{20}\mathrm{N}$$

(e)   It is only possible to remove a very small fraction of the total number of electrons in a metal because of the huge forces involved to separate more than a small number of the electrons.

3.2 VECTOR NATURE OF ELECTRIC FIELD INTENSITY

Two point charges of $+10^{-6}$ and -10^{-6} coulomb are situated 10cm apart. Calculate the electric field (a) mid-way between them, (b) on their perpendicular bisector 5cm from the point mid-way between the charges.

(Overview To add electric field intensity at a point due to several charges, the intensity of each charge is calculated in magnitude and direction. The net intensity is then found by adding the intensities vectorially.)

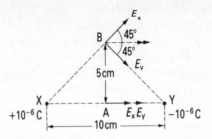

Fig 3.15 Worked example

(a)   At A, Fig 3.15 the field intensity E_x due to X is to the right as X is +ve. The field intensity E_y due to Y is to the right as Y is −ve. So total field$=E_x+E_y=2E_x$, as A is equidistant from X and Y,

$$=2.\frac{Q}{4\pi\varepsilon_0 r^2}$$

$$=\frac{2\times9\times10^9\times10^{-6}}{(0.05)^2}=7.2\times10^6\,\mathrm{N\,C^{-1}}$$

(b)   At B the field intensities E_x, E_y due to X and Y are in direction shown and these must be added as vectors to find the total field. E_x and E_y both act at 45° to the line XY, and since they are numerically equal their components perpendicular to XY cancel.

So total field $=E_x\cos 45°+E_y\cos 45°=2E_x\cos 45°$

$$=\frac{2\times9\times10^9\times10^{-6}}{XB^2}\cos 45°$$

But $XB^2=(0.05)^2+(0.05)^2=2\times(0.05)^2$

S $E=\dfrac{2\times9\times10^9\times10^{-6}}{2\times(0.05)^2}\cos 45°=2.5\times10^6\,\mathrm{N\,C^{-1}}$

The direction of the resultant field at B is parallel to XY and is directed towards the right.

3.3 FIELD AND POTENTIAL OF A SPHERE; CORONA DISCHARGE

Corona discharge through the air takes place when the electric field intensity exceeds 3×10^6 V m^{-1}. Calculate the greatest charge and greatest potential of an isolated metal sphere of radius (a) 25cm (b) 1m. (Take $1/4\pi\varepsilon_0=9\times10^9$ numerically.)

(**Overview** This calculation shows that the maximum potential which can be produced on a Van der Graaff generator is limited by the size of the spherical high voltage terminal. (It is possible, however, to increase the field intensity at which air conducts by *increasing* the air pressure around the sphere.))

(a)   The air conducts when E exceeds $3\times10^6 \text{V m}^{-1}$. But

$$E=Q/4\pi\varepsilon_0 r^2$$

So $3\times10^6 = 9\times10^9\times Q/(0.25)^2$

Simplifying, we find

$$Q=2\times10^{-5}\text{C}$$

The potential V is given by

$$V=\frac{Q}{4\pi\varepsilon_0 r}=\frac{9\times10^9\times2\times10^{-5}}{0.25}=750\,000\text{V}=750\text{kV}$$

(b)   When $r=1\text{m}$, a similar calculation shows that the maximum charge is $3.3\times10^{-4}\text{C}$, and the maximum potential is $3\times10^6\text{V}$.

3.4 COMPARISON OF VECTOR NATURE OF INTENSITY WITH SCALAR NATURE OF POTENTIAL

Two charges of 10^{-6}C are placed 10cm apart. Calculate (a) the electric field intensity and (b) the potential at a point P which completes an equilateral triangle with the charges in the case when (i) both charges are +ve, (ii) one charge is +ve the other −ve.

(**Overview** This question contrasts the vector nature of electric field intensity and the scalar nature of electric potential. In each case the contribution of each charge to the value at a particular point is calculated. For intensity these contributions are added vectorially. For potential they are added algebraically.)

In case (i), the field

$$E_x=\frac{Q}{4\pi\varepsilon_0 r^2}=\frac{9\times10^9\times10^{-6}}{0.1^2}=9\times10^5\text{N C}^{-1} \text{ (Fig 3.16(a))}$$

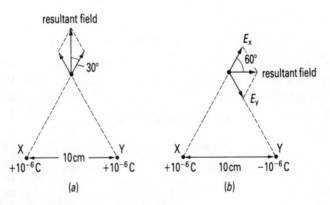

Fig 3.16 Worked example

Similarly $E_y=9\times10^5$. The total or resultant field is upwards as the sideways components cancel. The total field is given by

$$E=2E_x\cos 30°$$

$$=2\times9\times10^5\times\cos 30°=15.6\times10^5\text{N C}^{-1}$$

The potential at P=potential due to X+potential due to Y

$$=\frac{Q_x}{4\pi\varepsilon_0 r}+\frac{Q_y}{4\pi\varepsilon_0 r}$$

$$=\frac{9\times10^9\times10^{-6}}{0.1}+\frac{9\times10^9\times10^{-6}}{0.1}$$

$$=18\times10^4\text{V}$$

In case (ii) E_x and E_y have the same value but are directed as shown in Fig 3.16(b). Now there is only a sideways component to the total electric field intensity and this is given by

$$E=2E_x\cos 60°$$

$$=2\times9\times10^5\times\tfrac{1}{2}$$

$$=9\times10^5\text{N C}^{-1}$$

The potential at P is zero since the two opposite charges make equal and opposite contributions to the potential at P ($+9\times10^4$V and -9×10^4V).

3.5 INTENSITY AND FORCE IN A UNIFORM FIELD

An oil drop carrying a charge of 3.2×10^{-18}C is held stationary between horizontal parallel plates 5mm apart when a p.d. of 500V is applied between the plates. Calculate the mass of the drop.

(**Overview** A charged oil drop is suspended between parallel plates so that the weight of the drop is balanced by the electric force. This idea is used in Millikan's experiment (see page 255).)

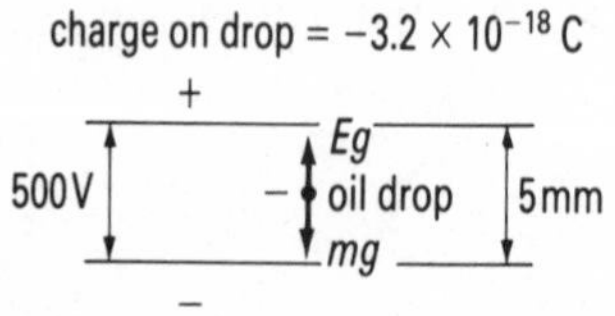

charge on drop = -3.2×10^{-18} C

Fig 3.17 Worked example

The arrangement is shown in Fig 3.17. the weight of the drop, mg, is balanced by the upward electric force Eq. But $E=V/d$ and hence

$$\frac{V}{d}\cdot q=mg$$

So $\qquad m=\dfrac{Vq}{gd}=\dfrac{500\times3.2\times10^{-18}}{9.8\times5\times10^{-3}}=3.3\times10^{-14}\text{kg}$

3.6 USE OF INTENSITY AND POTENTIAL TO CALCULATE FORCE AND ENERGY

An α-particle of charge 3.2×10^{-19}C and kinetic energy 8×10^{-13} J approaches a nucleus head on. The nucleus contains 80 protons each of positive charge equal numerically to e, 1.6×10^{-19}C. Calculate (a) the closest distance of approach of the α-particle (b) the force on it at this distance.

(Overview When two like charges approach, the k.e. in the system is reduced and the p.e. increases. This is used to find how near an α-particle can get to a nucleus before all its k.e. is transferred to p.e. (It would then rebound.) The force on the α-particle at this distance is then obtained from the inverse square law.)

(a)   The charge, Q_1, on the α-particle is 3.2×10^{-19}C, and on the nucleus, Q_2, is $80\times1.6\times10^{-19}$C. The α-particle will come to rest when all its kinetic energy has been transferred to electric potential energy in the field round the nucleus. Using work done$=Q\times V$, this occurs when

$$8\times10^{-13}=Q_1V=Q_1\times(Q_2/4\pi\varepsilon_0r)$$

where V is the potential at a distance r from the nucleus. Hence r is given by,

$$8\times10^{-13}=3.2\times10^{-19}\times\dfrac{9\times10^9\times80\times1.6\times10^{-19}}{r}$$

Simplifying $r=4.6\times10^{-14}$m

(Note that the diameter of an atom is about 10^{-10}m.)

(b)   At this distance the force is given by

$$F=\dfrac{Q_1Q_2}{4\pi\varepsilon_0r^2}$$

$$=\dfrac{9\times10^9\times2\times1.6\times10^{-19}\times80\times1.6\times10^{-19}}{(4.6\times10^{-14})^2}$$

$$=17.4\text{N}$$

This shows that there is an enormous force on the tiny α-particle when it is very close to the nucleus.

3.7 USE OF V/d AND σ/ε_0 TO FIND INTENSITY AND CHARGE FOR PARALLEL PLATES

A pair of parallel plates of area 100cm^2 and separated by 1cm are connected to a p.d. of 1000V. Find (a) the electric field intensity between the plates (b) the charges on the plates (c) the force of attraction between the plates. ($\varepsilon_0=8.85\times10^{-12}$F m^{-1}.)

(**Overview** Electric field between parallel plates can be found from V/d. The electric field can then be used to find the charge producing the field. Finally force can be found from $F=QE$.)

(a)   The field intensity

$$E=V/d=1000/10^{-2}=10^5\text{V m}^{-1}$$

(b)   Also, the field intensity E is given by σ/ε_0

Hence $\sigma=\varepsilon_0 E=8.85\times10^{-12}\times10^5=8.85\times10^{-7}$C m^{-2}

Thus the charge on each plate ($+$ve on one, $-$ve on the other) is given by

$$Q=\sigma A$$
$$=8.85\times10^{-7}\times10^{-2}\,(100\text{cm}^2=10^{-2}\text{m}^2)$$
$$=8.85\times10^{-9}\text{C}$$

(c)   Since the field intensity is contributed equally by each plate the intensity due to one plate is $\frac{1}{2}\times10^5$V m^{-1}. This intensity will produce the force of attraction on the other plate. (Note that we cannot use the total field intensity since a plate cannot exert a force on itself.)

Hence   force on each plate$=$charge$\times$field intensity (qE)

$$=8.85\times10^{-9}\times\tfrac{1}{2}\times10^5$$
$$=4.43\times10^{-4}\text{N}$$

3.8 A HIGH IMPEDANCE VOLTMETER USED TO MEASURE CAPACITANCE

A capacitor C is charged to 1000 volts and is then connected to a very high impedance voltmeter with a capacitor of 10^{-8}F across its input terminals. The voltmeter reads 0.61 volt. What is the value of C?

(**Overview** In this case charge, and hence capacitance from $C=Q/V$, is measured by finding the p.d. it sets up across a capacitor of known size. In this question it is reasonable to assume that all the charge is transferred to the known capacitor.)

The charge on the 10^{-8}F capacitor is $10^{-8}\times0.61=6.1\times10^{-9}$ coulombs.
     Since the p.d. now is very much less than 1000V it is correct to assume that the input capacitor of 10^{-8}F is much greater than C, and

that very nearly all the charge was transferred from C. Thus the charge initially on $C=6.1\times10^{-9}$ coulombs.

So $\quad C=Q/V=6.1\times10^{-12}F$

3.9 ERRORS CAUSED IN MEASURING CAPACITANCE BY VOLTMETER METHOD

A capacitor C is charged to 5V and is then connected to a high impedance voltmeter with an input capacitor of $10^{-8}F$. The voltmeter reads 0.82 volt. Estimate the value of C. Why is the result only an estimate?

(**Overview** In this instance the p.d. on the unknown capacitor is not very much greater than the final p.d. across the known capacitor. (Compare example 3.8 above.) Here it is not reasonable to assume that all the charge is transferred.)

The charge of the $10^{-8}F$ capacitor is now $10^{-8}\times0.82=8.2\times10^{-9}$ coulombs.
Hence IF all the charge originally on C were transferred then the charge on C was 8.2×10^{-9} coulombs before connection. So

$$C=Q/V=8.2\times10^{-9}/5=1.64\times10^{-10}F$$

Since in this case C is NOT much less than $10^{-8}F$, not all the charge on C would be transferred. In fact the $10^{-8}F$ capacitor is about 6 times larger than C, so that 6/7 of the charge would be transferred and 1/7 would remain on C. This would give an error of about 1 part in 7, or 14%.

3.10 PARALLEL PLATE CAPACITOR AND CHARGE SHARING

A parallel plate capacitor with air as dielectric has plates of area $4\times10^{-2}m^2$ which are 2mm apart. It is charged by connecting it to a 100V battery. It is then disconnected from the battery and connected in parallel with a similar uncharged capacitor with plates of half the area which are twice the distance apart. Calculate the final charge on each capacitor. Edge effects may be neglected. (Permittivity of air- $=8.8\times10^{-12}F\,m^{-1}$.)

(**Overview** This example uses the standard formula for the capacitance of a parallel plate capacitor. Charges are then shared when the capacitors are connected together. The p.d. across each is the same, and the charges are shared in the ratio of their capacitance.)

The capacitance of the first capacitor,

$$\frac{\varepsilon_0 A}{d}=\frac{4\times10^{-2}\times8.8\times10^{-12}}{2\times10^{-3}}$$

$$=1.76\times10^{-10}F$$

The capacitance of the second capacitor

$$=\frac{2\times10^{-2}\times8.8\times10^{-12}}{4\times10^{-3}}$$

$$=0.44\times10^{-10}\text{F}$$

The initial charge Q on the first capacitor$=CV=$ $1.76\times10^{-10}\times100=1.76\times10^{-8}$ coulomb. This charge is shared after connecting the second capacitor. Since the capacitors are in the ratio 4:1, and the p.d. across each is the same, then the charges must also be shared in this ratio. So

$$\text{final charge on larger capacitor}=\frac{4}{5}\times1.76\times10^{-8}$$

$$=1.41\times10^{-8}\text{ coulomb}$$

$$\text{The final charge on the smaller capacitor}=\frac{1}{5}\times1.76\times10^{-8}$$

$$=0.35\times10^{-8}\text{ coulomb}$$

3.11 ENERGY STORED IN A CAPACITOR

A capacitor of capacitance C is fully charged by a 200V battery. It is then discharged through a small coil of resistance wire embedded in a thermally insulated block of specific heat capacity $2.5\times10^2\text{J kg}^{-1}$ K^{-1} and of mass 01.kg. If the temperature of the block rises by 0.4K, what is the value of C?

(**Overview** This question is solved by calculating the heat produced since mass, specific heat capacity and rise in temperature are known. This can then be equated to the energy stored in the capacitor.)

The heat energy gained by the block

$$=mc\theta$$

$$=0.1\times2.5\times10^2\times0.4=10\text{J}$$

This energy was stored in the capacitor when charged to 200V.

Hence $\frac{1}{2}CV^2=\frac{1}{2}C\times(200)^2=10\text{J}$

So $C=5\times10^{-4}\text{F}=500\mu\text{F}$

3.12 RELATIVE PERMITTIVITY, CHARGE SHARING AND ENERGY STORED

Two identical capacitors C of 10^{-8}F are each charged to 100V, separated from their supply and then connected as shown in Fig 3.18. A slab of dielectric of relative permittivity 3 is now placed between the plates of *one* capacitor.

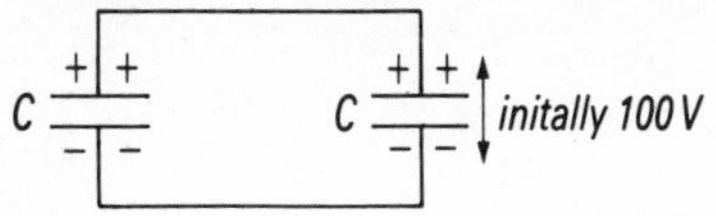

Fig 3.18 Worked example

(a)   Calculate the new p.d. across the plates. (Use charge sharing method.)

(b)   Calculate the total energy before and after putting the dielectric in place.

(**Overview** This is related to example 3.10. The p.d. across each capacitor is the same and charge is conserved. There is an energy loss in the process when the charge redistributes itself, as explained at the end of the example.)

(a)   The capacitances become 10^{-8}F and 3×10^{-8}F when the dielectric is in place. The charges therefore share in the ratio $\frac{1}{4}:\frac{3}{4}$. Initially the total charge is $2\times$charge on each capacitor$=2\times CV=2\times 10^{-8}\times100=2\times10^{-6}$ coulomb. Hence the charge on the 10^{-8}F capacitor is now $\frac{1}{4}\times2\times10^{-6}=0.5\times10^{-6}$ coulomb. The charge on the 3×10^{-8}F capacitor is 1.5×10^{-6} coulomb.

The p.d. V across 3×10^{-8}F capacitor $=\dfrac{Q}{C}=\dfrac{1.5\times10^{-6}}{3\times10^{-8}}=50$V

(Similarly the p.d. V across the 10^{-8}F capacitor$=Q/C=\dfrac{0.5\times10^{-6}}{10^{-8}}$

$=50$V, as it must be since the capacitors are in parallel.)

(b)   The energy on each capacitor before introducing the dielectric$=\frac{1}{2}CV^2=\frac{1}{2}\times10^{-8}\times(100)^2=0.5\times10^{-4}$J. Hence the total energy initially is 10^{-4}J.

After connecting the two capacitors the total energy

$$=\tfrac{1}{2}\times10^{-8}\times50^2+\tfrac{1}{2}\times(3\times10^{-8})\times50^2$$

$$=0.5\times10^{-4}J$$

Hence half the initial electrical energy is lost. This is due to the fact that heat has been produced in the connecting wires by the flow of current when the charge is redistributed. Work is also done when the dieletric is introduced, as it can be shown that an attractive force pulls the dielectric into the space between the capacitor plates.

1 LAW OF FORCE

$$F=Q_1Q_2/4\pi\varepsilon_0 r^2$$

2 ELECTRIC FIELD INTENSITY

Intensity=Force per coulomb on +ve charge. $F=qE$.

For a point charge $E=Q/4\pi\varepsilon_0 r^2$, between parallel plates $E=\sigma/\varepsilon_0$.

Electric field intensity E is a vector.

3 ELECTRIC POTENTIAL DIFFERENCE

p.d.=work done in moving 1 coulomb of +ve charge from one point to the other, $W=QV$.

Electric potential at a point=work done in moving 1 coulomb of +ve charge from infinity to the point.

Potential due to a point charge, $V=Q/4\pi\varepsilon_0 r$.

Inside a hollow sphere, $E=0$, V=constant.

Electric field=potential gradient=$-dV/dr=-V/d$ for uniform field.

Equipotential surfaces and field lines cross at right angles.

Potential is a scalar quantity.

4 CAPACITANCE

C defined as ratio Q/V.

Capacitance of parallel capacitor=$\varepsilon_r\varepsilon_0 A/d$.

Relative permittivity=(C with dielectric)/(C with vacuum)=ε_r.

Vibrating reed switch can be used to measure (a) capacitance, $C=I/fV$, (b) ε_0 from $\varepsilon_0=Id/fVA$.

Capacitors in series $1/C=1/C_1+1/C_2$, in parallel $C=C_1+C_2$.

Charge sharing: No charge lost and p.d. falls to $CV'/(C+C')$.

Very high impedance voltmeter (or electrometer) can measure charge by measuring p.d. across a capacitor of known value.

Energy stored in a capacitor $= CV^2/2 = QV/2 = Q^2/2C$.

DIAGRAM SUMMARY

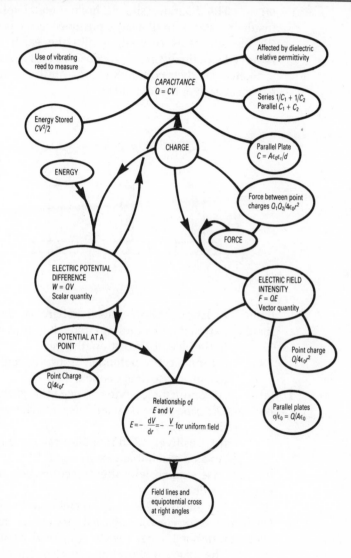

3 QUESTIONS

1 (*a*)   The diagrams show a hollow metal sphere supported on an insulating stand. In (i) a large positive charge is near to the sphere; in (ii) the sphere is earthed; in (iii) the earth connection has been removed and finally in (iv) the positive charge has been removed.

Sketch the distribution of charge on the sphere which you would expect at each of the four stakes.

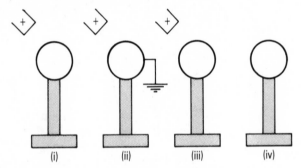

Fig 3.19

(*b*)   A large, hollow, metal sphere is charged positively and insulated from its surroundings. Sketch graphs of
(i) the electric field strength, and
(ii) the electric potential, from the centre of the sphere to a distance of several diameters.   (*AEB* 1985)

2 (*a*)   A conductor carrying a negative charge has an insulating handle. Describe how you would use it to charge
(i) negatively,
(ii) positively a thin spherical conducting shell which is isolated and initially uncharged.
In each case explain why the procedure you describe produces the desired result.
(*b*)   For **one** of these cases, sketch graphs showing how the electric field and the electric potential vary along a line outwards from the centre of the shell, when the charging device has been removed.
(*c*)   The diagram shows an arrangement of two point charges in air, Q being $0.30\mu C$.
(i) Find the electric field strength and the electric potential at P.

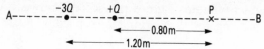

Fig 3.20

(ii) Find the point on AB between the two charges at which the electric potential is zero.

(iii) Explain why the potential on AB to the left of the $-3Q$ charge is always negative.

Take $\varepsilon_0 = 8.8 \times 10^{-12}$ m^{-1} or $= \dfrac{1}{36\pi} \times 10^{-9}$ F m^{-1} (*JMB*)

3   (*a*)   Define the terms *potential* and *field strength* at a point in an electric field. The diagram shows two horizontal parallel conducting plates in a vacuum.

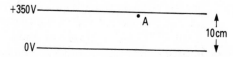

Fig 3.21

A small particle of mass 4×10^{-12} kg, carrying a positive charge of 3.0×10^{-14} C is released at A close to the upper plate. What *total* force acts on this particle?

Calculate the kinetic energy of the particle when it reaches the lower plate.

(*b*)   The diagram shows a positively charged metal sphere and a nearby uncharged metal rod.

Fig 3.22

Explain why a redistribution of charge occurs on the rod when the charged metal sphere is brought close to the rod.

Copy this diagram and show on it the charge distribution on the rod. Sketch a few electric field lines in the region between the sphere and the rod.

Sketch graphs which show how
    (i) the potential relative to earth, and
    (ii) the field strength vary along the axis of the rod from the centre of the charged sphere to a point beyond the end of the rod furthest from the sphere.

How is graph (i) related to graph (ii)?

How will the potential distribution along this axis be changed if the rod is now earthed?   (*L*)

4   (*a*)   A capacitor is made from two parallel metal sheets each of area A. The sheets are connected into the circuit at C. S is a switch

which vibrates f times per second between contacts X and Y, fully charging and discharging the capacitor.

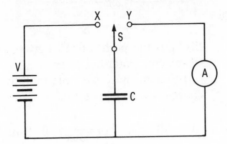

Fig 3.23

(i) The distance d between the sheets is varied and the discharge current is recorded by the microammeter A. Explain how the capacitance C of the plates is related to the ammeter reading. Draw a graph to show the expected relationship between the current and the plate separation.

(ii) Suggest modifications to this experiment to investigate how the capacitance of the capacitor depends on quantities other than the distance between the plates.

(iii) The ammeter is replaced by an identical capacitor to the one already in the circuit. The capacitor is initially uncharged. Find the p.d. across its plates, in terms of V, after the switch S has passed from X to Y four times and has returned to X. How many times does the switch have to vibrate before the charge on this second capacitor is within 0.01 of its final charge?

(b)   One of the parallel plate capacitors is removed from the circuit, charged from the battery V, and then isolated. A sheet of plastic of area A, relative permittivity ε, and thickness d, equal to the separation of the plates, is inserted between them.

Explain what changes the insertion of the plastic causes to
(i) the charge on the plates,
(ii) the p.d. between the plates,
(iii) the capacitance, and
(iv) the energy stored in the system.   (O&C)

5   In an experiment to investigate the discharge of a capacitor through a resistor, the circuit shown in Fig 3.24 was set up. The battery had an e.m.f. of 10V and negligible internal resistance. The switch was first closed and the capacitor allowed to charge fully. The switch was then opened (at time $t=0$), and Fig 3.25 shows how the milliammeter reading subsequently changed with time.

(a)   Use the graph to estimate the initial charge on the capacitor. Explain how you arrived at your answer.
(b)   Use your answer to (a) to estimate the capacitance of C.
(c)   Calculate the resistance of R.

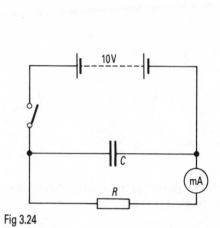

Fig 3.24

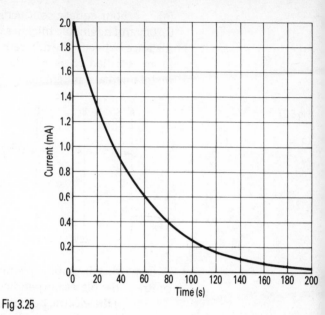

Fig 3.25

6 A charged capacitor of capacitance $100\mu F$ is connected across the terminals of a voltmeter of resistance $100k\Omega$. When time $t=0$, the reading on the voltmeter is 10.0V. Calculate

(a)   the charge on the capacitor at $t=0$.
(b)   the reading on the voltmeter at $t=20.0s$,
(c)   the time which must elapse, from $t=0$, before 75% of the energy stored in the capacitor at $t=0$ has been dissipated.   (JMB)

7 Examine the circuit illustrated and calculate

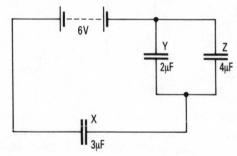

Fig 3.26

(a)   the potential difference across capacitor X,
(b)   the charge on the plates of capacitor Y,
(c)   the energy associated with the charge stored in capacitor Z.

(L)

8 (a)   Explain what is meant by electric field intensity E and electric potential V. State a relation between these two quantities for an electric field of uniform intensity.

(b)   Four square conducting plates, A, B, C and D, each of side 0.70m and negligible thickness, are arranged as shown in Fig 3.27. The distance between the adjacent plates is 15mm. The outer plates A and D are earthed, B is maintained at a potential of +30V, and C is maintained at a potential of −50V.

Fig 3.27.

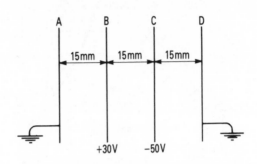

Draw sketch-graphs to show the variation along a line through the centre of the plates perpendicular to their plane of:
          (i) the electric potential;
          (ii) the electric field intensity.
Indicate clearly along the axes the magnitudes and the units of the quantities involved.
(c)   Plate C is now disconnected from its supply and joined to plate B, which is still maintained at +30V (see Fig 3.28).
          (Take the permittivity ε_0 of free space to be 8.9×10^{-12}F m^{-1}.)

Fig 3.28.

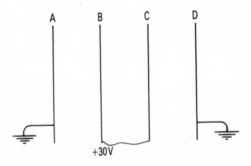

Calculate:
          (i) the capacitance of the system;
          (ii) the total charge on plates B and C;
          (iii) the energy stored in the system.
(d)   The inner plates, at 30V, are now disconnected from the supply. Plates A and D are drawn apart until the separation between A and B, and between C and D is 50mm.
          Calculate:
          (i) the new potential of plates B and C;
          (ii) the new energy of the system.
          Explain why there has been a change of electrical energy.
                                                                      (O)

CURRENT ELECTRICITY

CONTENTS

CHARGE, CURRENT AND POTENTIAL DIFFERENCE

We begin with some basic principles about electrical quantities and their measurement.

CURRENT AND CHARGE

The basic electrical quantity nowadays is **electric current**. It is measured in units called **amperes** (A), and the definition of an ampere is given on page 153. Electric current is the rate of flow of electric *charge*. An electric charge of 1 **coulomb**, (C), is the charge which flows when a current of 1 ampere flows for 1 second. In general if a *steady* current I (A) flows for a time t (s), the charge Q (C) flowing is given by

$$Q = It$$

or $$I = \frac{Q}{t}$$

If the current is *not* steady, then

$$I = \text{rate of flow of charge} = \frac{dQ}{dt}$$

POTENTIAL DIFFERENCE, p.d. (VOLTAGE)

The **potential difference** (p.d.) in *volts* between two points in a circuit is the energy transfer which occurs when one coulomb of charge flows between those points. In the circuit of Fig 4.1, for each coulomb of charge that flows 3J of energy are transferred to heat in the resistor R, since the p.d. across it is 3V. 7J of energy are transferred to (mainly) mechanical energy in the motor M, since the p.d. across it is 7V. Since the energy is supplied from the battery it must provide 10J for each coulomb flowing. Hence the p.d. across the battery must be 10V as shown. Note that **voltmeters** are connected in *parallel* between the points where the p.d. is to be measured. **Ammeters** are connected in *series* with the part of the circuit where the current is to be measured, as shown.

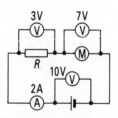

Fig 4.1 Potential difference in series circuit

**CHARGE TRANSPORT
EQUATION**

Suppose a current I is carried by charges of e coulombs moving with an average drift speed v. In a metal, for example, the charges would be the free electrons each of charge -1.6×10^{-19}C. If the wire has cross sectional area A and there are n charges per cubic metre, then nA will be the number of charges per metre length of the metal. In one second the charges move v metres and hence the number of charges passing a point per second is $v\times nA$. Hence $I=$charge flow per second passing a point$=e\times$number of charges per second$=evnA$. Hence

$$I=Aevn \qquad\qquad (1)$$

If $I=1$A and $A=1\text{mm}^2=10^{-6}$ m^2, $e=1.6\times10^{-19}$C and n (for copper) is about 8×10^{28}m^{-3} (see page 141), then (1) gives v about 1/10mm s^{-1}, which is a very low speed.

In a circuit when a switch is closed, the charges in the circuit all start moving together. One electron causes adjacent ones to move and this 'knock-on' effect can be shown to travel with the speed of light. The actual electron speed is, however, very small as we have just seen. Thus when a light is switched on in a distant lamp, the light comes on at the same moment as the switch is closed, although the electron speed in the wires is small.

ELECTRICAL POWER AND ENERGY

From the definition of p.d., V joules of energy are transferred for each coulomb of charge flowing between two points where the p.d. is V volts. The current I gives the number of coulombs flowing each second. Thus VI gives the number of joules of energy transferred each second. This is the power, P, and is measured in joules/second or watts (W).

So   power, $P=VI$

If V and I are constant for t seconds, the total electrical energy transferred is given by

energy$=VIt$ (J)

Larger units of power and energy are: 1kW$=1000$W, and 1MW$=1000000$W; 1kJ$=1000$J and 1MJ$=1000000$J.

In the circuit of Fig 4.1 the resistor R is transferring $VI=3\text{V}\times2\text{A}=6$ joules of electrical energy per second to heat, and the motor is transferring $7\times2=14$ joules of electrical energy per second to mechanical energy (and to heat, as the motor will not be 100% efficient).

RESISTANCE

Devices which transfer electrical energy *entirely* to heat are called **resistors**. The **resistance**, R, in ohms (Ω) of a resistor is defined by

$$R=\frac{V}{I}$$

For a given p.d. a high value of resistance permits a smaller current than a low resistor. In the circuit of Fig 4.1, the value of the resistor R is $V/I= 3/2=1.5\Omega$. We *cannot* say that the resistance of the motor is $7/2=3.5\Omega$, since the motor is not transferring all the electrical energy to heat. In motors, **electromagnetic induction** is the main factor governing the value of the current which flows, as we shall see later.

OHM'S LAW

It can be shown experimentally that for many materials a change in the p.d. V across a wire of that material causes a proportional change in current I, provided that other physical conditions, such as temperature, do not alter. If $I \propto V$ for a material, under constant physical conditions, then that material is said to obey *Ohm's law* and is called an 'ohmic conductor'. Resistors made of such materials will have a *constant* resistance, R, when their temperature is constant, since, if $V \propto I$, then V/I remains the same as V and I vary, even when the p.d. V is reversed.

In the case of a resistor obeying Ohm's law, the power VI can be written $IR \times I=I^2R$. Hence for a resistor,

$$\text{power}=I^2R,$$

or

$$\text{power}=VI=V.\frac{V}{R}=\frac{V^2}{R}$$

These are two very useful formulae for power.

SERIES AND PARALLEL CIRCUITS

SERIES CIRCUITS

(*a*)   For *any* components in series the current through each component at any instant *is the same*. This is true as charge cannot be lost or created.

(*b*)   The total p.d. across several components at any instant is the sum of the p.d.s across each component at that instant. This is true since the total energy per coulomb is the sum of the separate energies per coulomb in each component.

These ideas are used to show that, for two resistors R_1, R_2 in series, the total or effective resistance R is given by

$$R=R_1+R_2$$

The student should make sure he/she can prove this result, using total p.d. $V=V_1+V_2$.

PARALLEL CIRCUITS

(a)   For each component in parallel the p.d. across each component at any instant is the same. This is because each component is connected to the same two points.

(b)   The total current flowing into a parallel combination at any instant is the sum of the currents in the separate components at that instant. Figure 4.2 shows such a parallel circuit.

Since the p.d. V is the same across each resistor you should be able to show that for two resistors R_1, R_2 in parallel, the total effective resistance, R, is given by

$$\frac{1}{R}=\frac{1}{R_1}+\frac{1}{R_2} \text{ or } R=\frac{R_1R_2}{R_1+R_2}$$

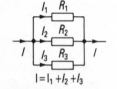

Fig 4.2 Resistors in parallel

TEMPERATURE COEFFICIENT; RESISTIVITY

For a material which obeys Ohm's law, the resistance normally depends on temperature. For many materials, over a fairly small range of temperatures the resistance R at a temperature t can be related to the resistance at 0°C, R_0, by

$$R=R_0(1+\alpha t)$$

where α is a constant called the **temperature coefficient of resistance**. For metals α is a positive quantity and the resistance increases with temperature. This is due to increased thermal vibrations impeding the electron flow when the temperature increases. The number of free electrons in a metal does not change with temperature.

For cabon, which is often used in the manufacture of resistors for electronic circuits, α is negative because the reisistance decreases with temperature. For pure semiconductors the resistance decreases (non-linearly) with temperature rise. This is caused by increased thermal vibration due to the temperature rise which produce a greatly increased number of charge carriers, as we shall see later.

RESISTIVITY

If a wire of length l and cross-sectional area A has a resistance R, then the *resistivity*, ρ, of the material is given by

$$R=\rho\frac{l}{A}$$

Note that R is a property of the whole wire, whilst ρ is a property of the *material* of which the wire is made. ρ has units Ωm.

Often it is more appropriate to use the conductivity, σ, of a material. σ is defined $1/\rho$. σ is higher for a good conductor like copper, than a poorer conductor like carbon.

E.M.F. AND INTERNAL RESISTANCE

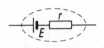

Fig 4.3 Symbol for internal resistance

The **electromotive force**, e.m.f., of a cell can be defined as the total energy transformed by a cell when one coulomb of charge flows through it. This means that if I amperes flow through a cell of e.m.f. E, the energy is transformed at a rate of EI in watts (W).

Some of the energy may be transformed to heat within the cell if it has **internal resistance**. The chemicals which make the cell have some resistance and heat will be produced in them when a current flows through the cell. The internal resistance, r, of a cell can be represented as shown in Fig 4.3 and separated from the e.m.f. E, but it should be remembered that r is actually *between* the terminals.

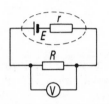

Fig 4.4 E.m.f. and terminal p.d.

Consider the circuit of Fig 4.4. From the conservation of energy, energy per second supplied by the cell=energy per second transformed internally+energy per second transformed externally

Hence $EI=I^2r+I^2R$

so $\qquad E=Ir+IR$

or $\qquad E=I(r+R)$ $\hfill (1)$

Further, the p.d., V, across the external resistor R, *which is also the p.d. across the terminals of the cell*, is given by

$\qquad V=IR$ $\hfill (2)$

Substituting from (2) in (1)

$\qquad\qquad E=Ir+V$

and hence $\qquad V=E-Ir$ $\hfill (3)$

These equations are used in worked example 1.

POWER AND TERMINAL p.d. VARIATION WITH R

By repeating these calculations for various values of R for a fixed e.m.f. E and internal resistance r, the following graphs of power against R, and terminal p.d. V against R, are obtained (Fig 4.5). From these graphs we can draw two important conclusions.

1 The maximum power is delivered to the external resistor when $R=r$ (the internal resistance). In this case equal power is dissipated internally and externally, so the efficiency of the cell is 50%.

2 The greatest p.d. occurs across the cell when $R=\infty$, that is, when the cell is an 'open circuit'. This greatest p.d. is equal to the e.m.f. of the cell.

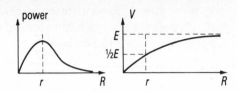

Fig 4.5 Power and terminal
p.d.

So it is only possible to measure e.m.f. when the cell delivers no current. In practice very high resistance voltmeters can be connected across the terminals of the cell. In this case very little current flows so that their reading can be taken to be the e.m.f. of the cell. Alternatively, as we shall see later, a potentiometer can be used to measure the e.m.f., because this instrument draws no current from the terminals of the cell under test.

NETWORKS; KIRCHHOFF'S LAWS

It is not always possible to solve all circuit problems using the simple methods discussed previously. To solve problems with more complex arrangements of components needs Kirchhoff's laws.

KIRCHHOFF'S FIRST LAW

This is a statement of the conservation of charge. In a network there can be no build up of charge at any point. Thus the total current flowing towards a point such as A, Fig 4.6, is balanced by the total current leaving A. Thus

$$I_3 + I_4 + I_5 = I_1 + I_2$$

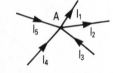

Fig 4.6 Kirchhoff's first law

Alternatively the law may be stated more formally by assigning a sign to the current. Currents flowing towards a point may be counted as positive, and away from the point as negative. In this case we can say that the *algebraic sum of the currents at any point is zero.* Or

$$-I_1 - I_2 + I_3 + I_4 + I_5 = 0$$

KIRCHHOFF'S SECOND LAW

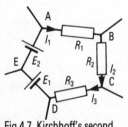

Fig 4.7 Kirchhoff's second law

In the part of a network shown in Fig 4.7, the e.m.f. of the two cells are both acting in the same direction in the loop ABCDE. Thus the total e.m.f. in the loop is $E_1 + E_2$. This must be equal to the fall in p.d. across the resistors, so

$$E_1 + E_2 = I_1R_1 + I_2R_2 + I_3R_3$$

This is an example of Kirchhoff's Second Law which states that the algebraic sum of the e.m.f.s and the products of current and resistance in a closed loop in a circuit is zero. That is,

$$E_1+E_2+I_1R_1-I_2R_2-I_3R_3=0$$

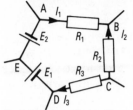

Fig 4.8 Signs in Kirchhoff's second law

If the circuit loop were as shown in Fig 4.8, then E_2 is acting opposite to E_1. Also I_2 is flowing in a counter-clockwise direction. In this case Kirchoff's Second Law would give

$$E_1-E_2=I_1 R_1-I_2 R_1+I_3 R_3$$

In practice it does not matter if the direction of the current is assumed to act in the wrong direction. When the equations resulting from Kirchhoff's Laws are solved, the current would come out to have a negative value.

To use Kirchhoff's Laws in a simple case:

(*a*)   Draw a circuit diagram and put in all the known and unknown quantities.

(*b*)   Look at each point where currents meet and write down the equations resulting from Kirchhoff's First Law.

(*c*)   Take any closed loops and write down equations resulting from the second law.

In choosing which loops to take, you need to make sure that each known quantity comes into an equation. Make sure you have enough equations to solve for the unknowns. This method is illustrated in worked example 4.

WHEATSTONE BRIDGE

The circuit of Fig 4.9 shows a Wheatstone Bridge. It is used to measure *resistance* and operates when no current flows through G. In this case the p.d. across BD=0 and thus the p.d. across AB=p.d. across AD, so

$$I_1R_1=I_2R_3 \dots\dots\dots\dots\dots(1)$$

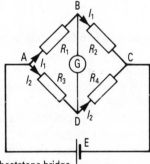

Fig 4.9 Wheatstone bridge

I_1 flows through R_2, as no current flows through G. Similarly I_2 flows through R_4. As before, the p.d. across BC=p.d. across DC so that

$$I_1R_2=I_2R_4 \dots\dots\dots\dots\dots(2)$$

Dividing (1) by (2)

$$R_1/R_2 = R_3/R_4$$

This is the equation used in the balanced Wheatstone Bridge.

Suppose, for example, R_2 was a standard 1000Ω resistor, R_1 was a standard 10Ω resistor, R_3 was unknown and R_4 a standard resistance box whose value could be changed in steps of 1Ω. If no current was obtained on G when R was set at 1934Ω, then

$$R_1/R_2 = R_3/R_4 \text{ gives}$$

$$10/1000 = R_3/1934$$

So $R_3 = 19.34\Omega$

THE POTENTIOMETER

THEORY OF THE POTENTIOMETER

A potentiometer is an instrument for measuring p.d. accurately. Basically it consists of a driver cell such as an accumulator which can maintain a constant current I in a uniform wire and a galvanometer G, as shown in Fig 4.10.

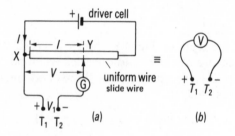

Fig 4.10 Theory of the potentiometer

The uniform wire can be a slide wire, usually 1m long, along which a slider or jockey can be moved. Another form is a helical potentiometer where a slider moves along a helix, generally ten turns long. The slider is connected to a vernier scale which can measure the position of the slider to one part in 1000.

As we shall see, the potentiometer acts like a voltmeter. In Fig 4.10(a) it is connected into a circuit to measure the p.d. between two terminals T_1 and T_2 exactly as the voltmeter V in Fig 4.10(b).

The p.d., V, across the length l of resistance wire of resistivity ρ is given by

$$V = IR = \frac{I\rho l}{A}$$

Thus provided that the current from the driver cell remains unaltered and the wire has a uniform cross-sectional area A, then $V=I\rho l/A$ and so $V\propto l$. So in these conditions $V=kl$ where k is a constant.

If an unknown p.d. of size V_1 is connected with its positive pole joined to T_1 and its negative pole to T_2, then in general a current will flow through G. If Y is near to X, V_1 will be *greater* than the p.d. across XY and so a current will flow one way through G. If Y is near the other end of the potentiometer wire and V_1 is *less* than the p.d. of the driver cell, then the p.d. across XY will be greater than V_1 and the current will flow the other way through G. If, however, Y is positioned so that the unknown p.d. V_1 is *equal* to V, the p.d. across XY, then the applied p.d. is balanced by the fall in p.d. across the length l of wire, and *no current* flows through G. In this case

$$\text{applied p.d. } V_1=V=kl$$

Note that it will be impossible to obtain a balance if:

1 the driver cell p.d. is less than the p.d. to be measured, as explained in the last paragraph;
2 the +ve pole of the unknown cell is connected to T_2 instead of T_1. In this case the p.d. across G from V_1 would add to the p.d. across XY and so it would be impossible to balance the two p.d.s.

CALIBRATION OF THE POTENTIOMETER

A potentiometer is calibrated using a *standard cell* such as a Weston Cell, which has an accurately known e.m.f. of 1.0186 volts at a particular temperature. The standard cell, e.m.f. E_0, is connected with its +ve pole to T_1 and its −ve pole to T_2. The corresponding balance length l_0, is measured. Then

$$E_0=kl_0$$

$$\text{and so }\quad k=\frac{E_0}{l_0}$$

The unknown e.m.f. E is now used to replace E_0 and the new balance length l is measured. Now

$$E=kl$$

$$\text{So}\qquad E=\frac{E_0}{l_0}l$$

from which E can be found.

PRECAUTIONS IN USING A POTENTIOMETER

1 Move the slider or jockey on to each end of the slide wire to check that the galvanometer deflects in opposite directions. Two reasons why it might not do this have already been discussed.
2 The driver cell must provide a *constant* current. Checking the balance

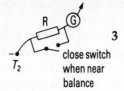

close switch
when near
balance

Fig 4.11 Protective resistor

3 length for a given e.m.f. at various times will check if there is any 'drift' in the driver cell current.

A protective resistor, R, is used in series with the galvanometer to protect it from damage, as shown in Fig 4.11. When the balance point has been found approximately, R can then be shorted out and the precise (exact) balance point can be found.

USES OF A POTENTIOMETER

Essentially a potentiometer is a device for measuring p.d. But

(a)   using $I=V/R$ it can be used to measure current when R is known and,

(b)   using $R=V/I$ it can be used to compare resistances provided I is the same through each resistor.

We discuss first how the potentiometer is used to measure a small p.d.

1 MEASURING A SMALL p.d. (e.g. FROM A THERMOCOUPLE)

Suppose we need to measure an e.m.f. from a thermocouple whose maximum value is 10mV. Then the p.d. across the potentiometer wire must be reduced to this value by placing a *high resistance R_1 in series* with the wire, as shown in Fig 4.12. if the wire has a resistance of 5Ω then the current through it must be $I=V/R=0.010/5=0.002A=2mA$.

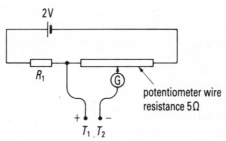

potentiometer wire
resistance 5Ω

Fig 4.12 Measurement of small e.m.f. by potentiometer

Assuming that the driver cell has an e.m.f. of 2V and negligible internal resistance, the resistor R_1 then has a p.d. of $2-0.01=1.99V$ across it. The current required through R_1 is 2mA and so

$$R_1=\frac{V}{I}=\frac{1.99}{0.002}=995\Omega$$

So to measure a small p.d. a *large* resistance is included in series with the potentiometer wire. As we have just shown, its value can be chosen to make the maximum p.d. across the wire any desired small value and the e.m.f. of a thermocouple can then be found from the balance length of the wire.

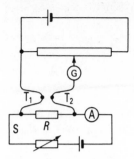

Fig 4.13 Measurement of current by potentiometer

2 DETERMINING CURRENT (OR CALIBRATING AN AMMETER)

The lower circuit in Fig 4.13 contains the ammeter A to be calibrated and includes a *standard resistance*, R (that is one whose value is accurately known such as 1.000Ω). The potentiometer, previously calibrated by using a standard cell, has its terminals connected across R and is used to determine the p.d., V, across R. The current flowing is then accurately given by

$$I = V/R$$

3 COMPARING RESISTANCES

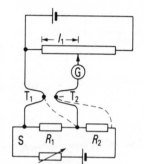

Fig 4.14 Comparison of resistances by potentiometer

The lower circuit S in Fig 4.14 contains the two resistors, R_1 and R_2, whose values are to be compared. The potentiometer is first used to measure the p.d., V_1 across R_1. It is then transferred to measure the p.d. V_2 across R_2 (as shown by the dotted lines). The current through each resistor is the same so

$$V_1 = IR_1 = kl_1$$

and $\quad V_2 = IR_2 = kl_2$

Hence $\dfrac{R_1}{R_2} = \dfrac{l_1}{l_2}$

4 MEASURING INTERNAL RESISTANCE OF A CELL

The e.m.f., E, of the cell is first measured by connecting it directly to the potentiometer and gives a balance length l_0 (Fig 4.15(a)). The cell is then connected to a resistance box R, when the p.d. across its terminals falls to V, given by

$$V = IR = \frac{E}{(R+r)} R$$

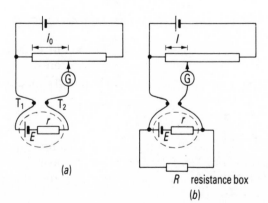

(a)

R   resistance box
(b)

Fig 4.15 Internal resistance by potentiometer

This gives a smaller balance length l (Fig 4.15(b)). Since

$$R+r=\frac{ER}{V}$$

$$r=\frac{ER}{V}-R=\frac{(E-V)R}{V}=\frac{(l_0-l)R}{l}$$

The experiment is now repeated for various values of R, each time measuring the corresponding value of l. A graph is plotted of (l_0-l) against l/R. Since

$$l_0-l=r(l/R)$$

the slope of the graph gives the internal resistance of the cell.

HEATING EFFECT OF A CURRENT

The heating effect of a current can be calculated from the power formula, already discussed, applied to a resistor. For a resistor, R,

$$\text{power}=IV=I^2R=V^2/R$$

WORKED EXAMPLES

4.1 E.M.F. AND INTERNAL RESISTANCE

A cell of e.m.f. 10V and internal resistance 2Ω is connected to a 3Ω resistor. What is the p.d. across the 3Ω resistor and the power dissipated in it?

(**Overview** We can use the equations derived on page 115.)

From $\quad E=I(R+r)$

$\qquad 10=I(2+3)$

giving $\quad I=2A$

From $\quad V=IR$

$\qquad V=2\times3=6V$

So power dissipated in $R=IV=2\times6=12W$.

4.2 CIRCUIT CALCULATIONS

In Fig 4.16, what is the reading on the high resistance voltmeter V?

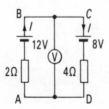

Fig 4.16 Worked example

(**Overview** Use net e.m.f.$=I\times$total resistance to find I. The p.d. across V can then be found.)

The net e.m.f. E in the circuit is $12-8=4$V. The total resistance in the circuit is $2+4=6\Omega$. (The circuit is a series one, as no significant current flows through the high resistance voltmeter.)

Hence current $I=\dfrac{E}{R}=\dfrac{4}{6}=\dfrac{2}{3}$A

The p.d. across AB$=12$V$-$p.d. across 2Ω resistor

$$=12-IR$$

$$=12-\tfrac{2}{3}\times2=10\tfrac{2}{3}\text{V}$$

Through CD the current is flowing opposite to the direction in which one would normally expect current to flow from the 8V cell. This is because the 12V cell has the greater p.d. Hence if we use $V=E-Ir$ (page 115) for the 8V cell, we must take I as negative.

So   p.d. across CD$=8+IR$

$$=8+\tfrac{2}{3}\times4=10\tfrac{2}{3}\text{V}$$

This agrees with our previous result, as the p.d. across AB must equal that across DC. Hence the voltmeter reads $10\tfrac{2}{3}$V.

4.3 RESISTORS IN SERIES A cable, each wire of which has a resistance of 2Ω per 100m, joins two points 400m apart. A fault develops and the wires are joined by a resistance R, as shown in Fig 4.17. The resistance measured from A is 10Ω, and from B is 8Ω. Calculate the distance x in metres of the fault from A, and the resistance R which joins the cables.

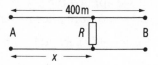

Fig 4.17 Worked example

(**Overview** There are two separate pieces of information giving the resistance from A and from B. These can be used to generate two equations for the unknowns R and x.)

The resistance of the cable from A in ohms is $\left(\dfrac{2x}{100}\times2\right)+R.$

Hence $\dfrac{x}{25}+R=10$ \hfill (1)

The resistance from B is $\left(\dfrac{2(400-x)}{100}\times2\right)+R$

Simplifying $16+R-\dfrac{x}{25}=8$ $\qquad$ (2)

Adding (1) and (2) to eliminate x,

$$16+2R=18$$

So $\qquad R=1\Omega$

Substituting in (1), $\qquad x=25\times9=225\text{m}$

Hence the cables are joined by a 1Ω resistor 225m from end A.

4.4 USE OF KIRCHHOFF'S LAWS

Calculate the currents flowing in the circuit of Fig 4.18.

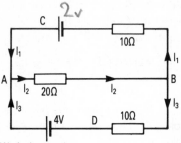

Fig 4.18 Worked example

(**Overview** There are three unknown currents. The aim must be to get three independent equations from Kirchhoff's Laws and then solve them simultaneously.)

At A from Kirchhoff's First Law

$$I_1+I_3=I_2 \qquad (1)$$

(Note that if we applied the law at B we would get precisely the same equation and so no extra information.)

Applying Kirchhoff's Second Law to the top loop ABC

$$2=20I_2+10I_1 \qquad (2)$$

And to the lower loop ADB,
$$4=20I_2+10I_1 \qquad (3)$$

We now have three equations for three unknowns and there is no need to obtain more equations. (Note that we could write down Kirchhoff's Second Law for the outside loop CADB. In this case we get

$$4-2=-10I_1+10I_3.$$

This gives no additional information, however.)

Solving equations (1), (2) and (3) gives $I_1=-1/25\text{A}$, $I_2=3/25\text{A}$ and $I_3=4/25\text{A}$. Notice that I has come out to the negative, and so must flow in the opposite direction to that shown in the diagram. This is because of the effect of the larger 4V cell acting in this direction.

4.5 POTENTIOMETER TO MEASURE SMALL E.M.F.

In the circuit of Fig 4.19, a balance is obtained on the potentiometer (i) when C touches A, and the balance length is 20cm, and (ii) when C touches B, and the balance length is 60cm. In each case the resistance box is set at 1018Ω. Calculate (a) the e.m.f. of the thermocouple, (b) the value of resistor R_2, given that the driver cell has negligible internal resistance.

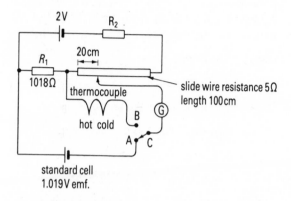

Fig 4.19 Worked example

(**Overview** The thermocouple balances the p.d. across 20cm of wire. The resistance of this wire can be found easily, so to find the p.d. it is first necessary to calculate the current. R_2 can be found by using e.m.f.=current×total resistance for the driver circuit.)

(a)   When C touches A the p.d. across the standard cell, 1.019V, balances the p.d. across R_1 and the resistance of 20cm of wire. Since the total wire resistance is 5Ω, 20cm has a resistance of 1Ω. Thus the p.d. 1.019V balances the p.d. across 1018+1=1019Ω.

Hence the current flowing$=V/R=\frac{1.019}{1019}=0.001A=1mA$

When C touches B the p.d. across the thermocouple balances the p.d. across 60cm of wire. This section has resistance 3Ω. Hence p.d.$=IR=0.001×3=0.003V=3mV$. Thus the e.m.f. of the thermocouple=3mV.

(b)   The total resistance of the driver circuit is $1018+5+R_2$. But using

e.m.f. of driver cell$=I×$(total resistance)

we have $2=0.001(1018+5+R_2)$

giving $\qquad$ $2000=1023+R_2$

So $\qquad$ $R_2=977\Omega$

4.6 POTENTIOMETER TO CALIBRATE AMMETER

A potentiometer gives a balance length of 34.2cm when connected to a standard cell of 1.018V. A ammeter reading 0.500A measures the current through a standard resistor of resistance 2.000Ω. The potentiometer when connected across the resistor gives a balance length of 32.8cm. Calculate the percentage error in the meter.

(**Overview** Use the standard cell information to calculate k from the equation $E=kl$. Then find the p.d. across the standard resistor and use this to calculate the current through it.)

Using $\qquad$ $E=kl_0$

$\qquad$ $1.018=k\times34.2$

So $\qquad$ $k=2.98\times10^{-2}$ V/cm

Hence with a balance length of 32.8cm, the p.d. is

$\qquad$ $V=kl$

$\qquad$ $=2.98\times10^{-2}\times32.8=0.976V$

Thus the true current, I, through the ammeter is given by

$\qquad$ $I=V/R$

$\qquad$ $=0.976/2.000=0.488A$

Hence the ammeter reads 0.012A too high, and thus the error is

$\qquad$ $\dfrac{0.012}{0.488}\times100=2.5\%$

4.7 HEATING EFFECT OF A CURRENT

Two heating coils, which, when correctly connected in parallel to the 250V mains dissipate 3kW of power, are wrongly wired in series. They then dissipate 500W of power. Find the resistance of the two coils (a) when used in parallel (b) when wired in series.

(**Overviews** Use V/R since the p.d. and power are known and we want to find the resistance.)

In each case the p.d. is the same, 250V. Hence the most appropriate formula to use is power$=V^2/R$.

So, in parallel $\quad$ $3000=\dfrac{250^2}{R}$

giving $\qquad$ $R=250/12=20.8\Omega$

And in series $\qquad 500 = \dfrac{250^2}{R}$

giving $\qquad\qquad R = 125\Omega$

It is left as an exercise for the student to now calculate the resistance R_1, R_2 of each coil by using $R_1 + R_2 = 125$ and $\dfrac{1}{R_1} + \dfrac{1}{R_2} = \dfrac{1}{20.8}$.

VERBAL SUMMARY	

1 CURRENT

I=rate of flow of charge=dQ/dt=Q/t (for constant current).

Charge transport equation: $I=Aevn$.

2 POWER AND ENERGY

Power=VI, energy=VIt.

3 RESISTANCE

R is defined by V/I. If R is a constant (temp. etc., constant), then the conductor obeys Ohm's Law (ohmic conductor).

Power formulae: power=$IV=I^2R=V^2/R$.

Resistors in series: $R=R_1+R_2$, in parallel $1/R=1/R_1+1/R_2$.

Resistivity of materials defined by $R=\rho l/a$.

Temperature coefficient of resistance α: $R_t=R_0(1+\alpha t)$.

4 E.M.F. AND INTERNAL RESISTANCE

$E=I(r+R)$, terminal p.d. $V=IR$, $E=V+Ir$.

$V=E$ when $R=\infty$, i.e. cell on open circuit.

Maximum power delivered to external circuit when $R=r$.

5 KIRCHHOFF'S LAWS

First Law. Algebraic sum of currents meeting at a point is zero.

Second Law. In a closed loop the sum of the e.m.f.s and products of current and resistance is zero.

6 WHEATSTONE BRIDGE

In the case of a balanced bridge $R_1/R_2=R_3/R_4$. Used for measurement of resistance.

Continued on p.130

DIAGRAM SUMMARY

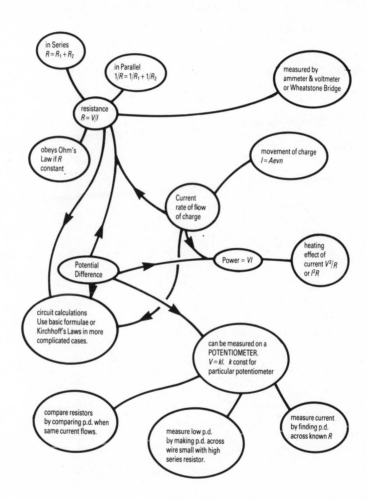

in Series
$R = R_1 + R_2$

in Parallel
$1/R = 1/R_1 + 1/R_2$

measured by
ammeter & voltmeter
or Wheatstone Bridge

resistance
$R = V/I$

obeys Ohm's
Law if R
constant

movement of charge
$I = Aevn$

Current
rate of flow
of charge

heating
effect of
current V^2/R
or I^2R

Power = VI

Potential
Difference

circuit calculations
Use basic formulae or
Kirchhoff's Laws in more
complicated cases.

can be measured on a
POTENTIOMETER.
$V = kl$. k const for
particular potentiometer

compare resistors
by comparing p.d. when
same current flows.

measure low p.d.
by making p.d. across
wire small with high
series resistor.

measure current
by finding p.d.
across known R

7 POTENTIOMETER

Measures p.d. $V=kl$

Calibrated using a standard cell, $k=E_0/l_0$.

Used to measure
(a)   small e.m.f. Include high resistance in series with potentiometer wire,
(b)   current. Measure p.d. across a known resistor,
(c)   resistance. Compare resistances by comparing p.d. across resistors in series,
(d)   internal resistance. Measure p.d. across cell for various values of external resistor.

8 HEATING EFFECT OF A CURRENT

Calculate using power formulae. Heat per second in a resistor$= I^2/R=V^2/R$.

4 QUESTIONS

1   State Ohm's law.

Give **two** examples in which Ohm's law is *not* obeyed, and sketch typical current-voltage characteristics for each.

2   (*a*)   Describe an experiment in which an unknown resistor may be compared with a standard resistor, where both resistances are less than one ohm. Include a suitable circuit diagram.

(*b*)   A galvanometer and several resistors are arranged to act as a multirange ammeter as shown in Fig 4.20.

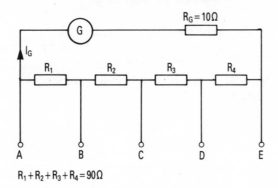

$R_1 + R_2 + R_3 + R_4 = 90\Omega$

The galvanometer has a resistance R_G of 10Ω and is connected across four resistors in series of total resistance 90Ω. The meter is to be used as an ammeter with full scale deflection currents of 1A, 100mA, 10mA or 1mA by using connections AB, AC, AD or AE respectively.

(i) Find the full scale deflection current I_G of the galvanometer.

(ii) Calculate the values of the four resistors.

(*O&C* part Q)

3 In the circuit (Fig 4.21), what must be the values of R_1 and R_2 for the two lamps A and B to be operated at the ratings indicated? If lamp A burns out, what would be the effect on lamp B? Give reasons.

(W)

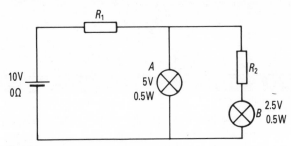

Fig 4.21.

4 A laboratory power supply is known to have an e.m.f. of 1000V. However, when a voltmeter of resistance 10kΩ is connected to the output terminals of the supply, a reading of **only 2V** is obtained.
   (a)   Explain this observation.
   (b)   Calculate
           (i) the current flowing in the meter;
           (ii) the internal resistance of the power supply.

(AEB 1984)

5 The 4.00-V cell in the circuits shown (Fig 4.22) has zero internal resistance.

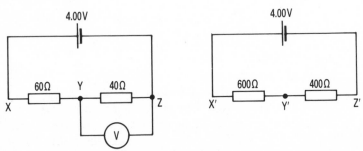

Fig 4.22

An accurately calibrated voltmeter connected across YZ records 1.50V. Calculate
   (a)   the resistance of the voltmeter,
   (b)   The voltmeter reading when it is connected across Y'Z'.
What do your results suggest concerning the use of voltmeters?

(L)

6 The slide connection J in the circuit (Fig 4.23) is moved along the 100cm potentiometer wire AB to find the point C at which the centre zero galvanometer registers zero current.
   (a)   Both cells have negligible internal resistance.
   Calculate the length AC

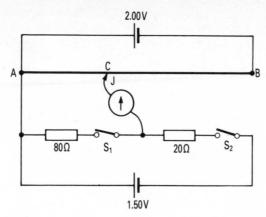

Fig 4.23

(i) with switches S_1 and S_2 both closed,
(ii) with switch S_1 open and switch S_2 closed.
(b)   The 1.50-V cell develops an internal resistance of a few ohms. Identify and explain, without calculations, any effect on the balance lengths determined in (a). $\qquad$ (L)
7   (a)   In the Wheatstone bridge arrangement shown (Fig 4.24), X is the resistance of a length of constantan wire and Y is the resistance of a standard resistor. Derive from first principles the equation relating X, Y, L_1 and L_2 when no current flows through the galvanometer, stating any assumptions you make.

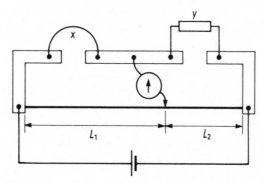

Fig 4.24

(b)   A student plans to use the apparatus described in (a) to determine the resistances of several different lengths of constantan wire of diameter 0.50mm. If a 2Ω standard resistor is available, suggest the range of lengths he might use given that the resistivity of constantan wire is approximately $5\times10^{-7}\Omega$m. Give reasons for your answers.
(c)   In the circuit (Fig 4.25) P, Q, R and S form part of a Wheatstone bridge network in which an operational amplifier circuit is used to detect the difference between the voltages V_1 and V_2.

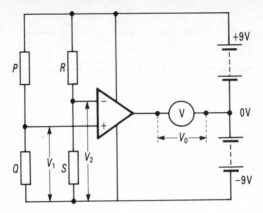

Fig 4.25

(i) The output voltage in this circuit, V_0, is given by $V_0 = A(V_1 - V_2)$ where A is the open-loop gain of the amplifier and V_0 must lie between 9V and -9V. Show that, if $A = 90\,000$, then V_0 should be either 9V or -9V if the difference between V_1 and V_2 exceeds $100\mu V$.

(ii) If the resistances P and Q are each $10k\Omega$ and R is the resistance of a variable standard resistor, outline how you would use the arrangement to determine the resistance S, which is of the order of a few $k\Omega$.

(iii) A typical centre zero galvanometer has a resistance of 40Ω and is graduated in divisions of 0.1mA. If you were to choose between such a meter and the above operational amplifier as a null detector, which would you choose, and why?

(JMB)

8   (a)   The flow of electric current in a metal wire is due to the movement of conduction electrons.

(i) What is meant by the term *conduction electrons*?

(ii) Under what circumstances will the movement of the electrons produce current flow?

(b)   The diagram (fig 4.26) shows a length L of conductor of cross-sectional area A. The conductor contains charged particles free to move from left to right as shown.

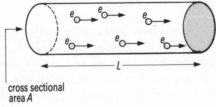

cross sectional
area A

Fig 4.26

(i) If there are n such particles per unit volume each with charge e, derive an expression for the total charge of these particles in the length of conductor.

(ii) If the particles are each moving with a drift velocity v in the direction shown, write down an expression for the time taken for all the particles to pass through the shaded area.

(iii) Use your answers to (i) and (ii) to show that the current I flowing in the conductor is given by the equation:

$$I = nAve$$

(c) A current of 1.0A flows in a 1.0m length of copper wire of cross-sectional area 1.0mm^2. Given that each cubic metre of copper contains 9.0×10^{28} conduction electrons each of charge -1.6×10^{-19}C, calculate the average drift velocity of the electrons.

(AEB part Q 1984)

ELECTROMAGNETISM

CONTENTS

FORCE ON WIRES AND MOVING CHARGES IN MAGNETIC FIELDS

MAGNETIC FLUX DENSITY

The force, F, on a wire of length l carrying a current I placed at an angle θ to a magnetic field is shown by experiment to be given by

$$F \propto Il\sin\theta$$

The constant of proportionality depends on the strength of the magnetic field, called the **magnetic flux density**, B. We can write

$$F = BIl\sin\theta$$

If I is measured in amperes, l in metres and F in newtons, B is measured in teslas (T). From the last equation it can be seen that we can define a field of flux density 1 tesla as one which will produce a force of 1N on a wire 1m long carrying a current of 1A when the wire is placed perpendicular to the field.

If I and B are parallel, then $\theta=0°$, and hence $\sin\theta=0$. Thus $F=0$. So the direction of a magnetic field is the direction in which a current must flow so as to experience *no* magnetic force in the field.

In general the direction of the force F is perpendicular to the current I and perpendicular to the field B, and is given by Fleming's left hand rule, illustrated in Fig 5.1.

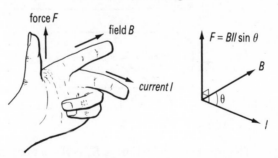

Fig 5.1 Fleming's left hand rule

FORCE ON MOVING CHARGE IN MAGNETIC FIELD

Suppose there are N charges in a wire of length l and that each charge has an average drift velocity v along the wire (Fig 5.2). The time taken for the charges to move through the wire is l/v, and the total charge flowing in this time is Nq. Hence the current, I, is $Q/t=Nq/(l/v)=Nqv/l$

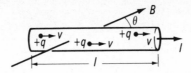

Fig 5.2 Force on a moving charge

The force on the wire $=F=BIl\sin\theta$

$$=\frac{BNqv}{l}l\sin\theta=NBqv\sin\theta$$

The force on the wire is the total force on all the electrons. Thus if f is the force on each electron, then $F=Nf$. Comparing this with the last equation we see that

$$f=Bqv\sin\theta$$

When the charge q is moving *perpendicular* to the field B, then $\sin\theta=\sin 90°=1$. In this case the force on the charge $=Bqv$.

HALL EFFECT

When a magnetic field is applied perpendicular to a conductor which is carrying a current, a p.d. is set up *perpendicular* to the current flow. This is called the **Hall effect**. The value, V_H, of the p.d. can be calculated as follows for a rectangular cross-section conductor (Fig 5.3), where we assume the current is carried by the movement of $+$ve charge carriers (for example, $+$ve holes in a semiconductor).

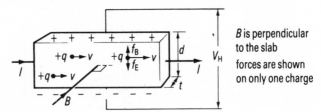

B is perpendicular to the slab

forces are shown on only one charge

Fig 5.3 Hall effect

The force on each charge, f_B, is Bqv since $\sin\theta=1$. By Fleming's left hand rule, this force acts in an upwards direction. Thus positive charge tends to accumulate on the top face, leaving a negative charge on the bottom face. This charge separation sets up a downward electric field, E, and this gives a downward force, f_E, on each charge. The charge migration will cease when $f_E=f_B$. Now from Chapter 3, $f_E=qE=qV_H/d$ where V_H is the **Hall voltage**. Hence when $f_E=f_B$,

$$\frac{qV_H}{d}=Bqv$$

So $V_H=Bvd$

Using the result of Chapter 4, page 112, $I=Anqv$, where A is the cross-sectional area and n is the number of charges per metre^3. So $v=I/Anq$ and

$$V_H=\frac{BId}{Anq}$$

But $A=td$, so $V_H=\frac{BI}{ntq}$

If the charges were −ve, they would be moving through the conductor in the opposite direction to that shown in Fig 5.3. In $f=Bqv$, both q and v would have changed sign and so f would be in the same direction. Thus −ve charge would accumulate on the top face and the Hall voltage would be *reversed* in sign. The sign of the Hall voltage thus enables the sign of the predominant charge carriers in a conductor to be determined.

Consider copper, for example, where there is one free electron per atom. Here $n=$number of atoms per metre^3. Now in 63.5g of copper there are 6×10^{23} atoms. Hence in 1kg of copper there are $6\times10^{26}/63.5$ atoms. But the density of copper is 9000kg/m^3, so in 1m^3 there are $9000\times6\times10^{26}/63.5$ atoms, that is, 8.5×10^{28} atoms/m^3. If $B=0.1$T, $I=10$A, $t=1$mm$=10^{-3}$m, and $q=1.6\times10^{-19}$ coulomb, then

$$V_H=\frac{0.1\times10}{8.5\times10^{28}\times10^{-3}\times1.6\times10^{-19}}=7.3\times10^{-8}\text{ volts}$$

This is an extremely small p.d. and not easily measurable. In general the Hall voltage is very small in *metals*.

SEMICONDUCTORS AND HALL VOLTAGE

The Hall voltage is *much* larger in semiconductors because the number of charges per m^3 is very much smaller. In this case the speed of the charge carriers is much greater for any given current, and hence so is the magnetic force which produces the charge migration.

In semiconductors the Hall effect may be used:

1   to measure B, since $B\propto V_H$ if other factors are kept constant (e.g. the current through specimen);

2   to study the properties of semiconductor materials. If, for example, B, V_H, I, t and q are known, then n can be calculated. In semiconductors n varies considerably with temperature and this variation can yield useful theoretical information about the semiconductor.

COUPLE ON A COIL

In ammeters and in motors, a current in a coil situated in a magnetic field causes the coil to turn. The rotation is due to a **couple** or **torque** acting on the coil, which we now calculate.

Suppose PQRS is a coil of N turns carrying a current I and a magnetic field B acts at an angle α to its normal (Fig 5.4(a)). By Fleming's left hand rule, the forces on PS and QR are as shown in the plan (Fig 5.4(b)). The force F on PS or QR, length a, is given by,

$$F = BIl \sin \theta. \ N \text{ (as there are } N \text{ turns)}$$

$$= BIaN, \text{ as } B \text{ is perpendicular to PS and QR}$$

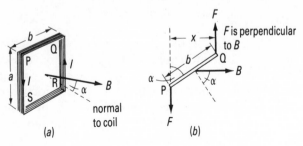

Fig 5.4   Couple on a coil

The torque provided by these forces is given, if PQ has a length b, by

$$\text{torque} = F \times x = Fb \sin \alpha = BIaNb \sin \alpha$$

$$= BANI \sin \alpha,$$

since $ab = A$, the area of the coil.

The maximum couple occurs when $\alpha = 90°$, that is, when the normal to the coil is at right angles to B (and so the plane of the coil is *parallel* to B). The coil will then turn and eventually come to equilibrium when its plane is perpendicular to B.

MOTOR AND MOVING COIL AMMETER

A *d.c. motor* makes use of this turning effect. A continuous rotation of the coil is obtained by use of a commutator, as described in 16+ texts.

In a *moving coil ammeter*, shown in Fig 5.5, a *radial field*, B, is used. In this case the field remains parallel to the plane of the coil as the coil rotates. Hence α is 90° at every position of the coil and so, since $\sin \alpha = \sin 90° = 1$,

$$\text{torque on coil} = BIAN$$

Here the opposing torque due to the springs is proportional to θ, that is, opposing torque $= c\theta$ where c is the opposing torque per radian. The coil and pointer will come to rest when these turning effects balance and so

$$BIAN = c\theta$$

and thus $I \propto \theta$

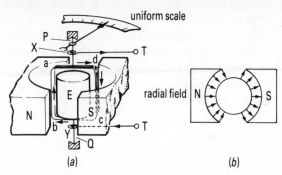

Fig 5.5 Moving coil ammeter

So in this case the angle of deflection will be proportional to the current. The meter scale will therefore be *linear*, which is a great advantage.

Damping of the coil is provided by winding the coil on a metal (aluminium) former. This brings the coil to rest quickly after it is deflected.

Greater sensitivity, that is, a greater deflection for a given current, can be obtained by suspending the coil by a very fine wire. Here the torsion in the wire provides the restoring torque on the coil, but the torque needed to produce a certain deflection can be made much smaller than by the use of springs. Such a meter is shown in Fig 5.6.

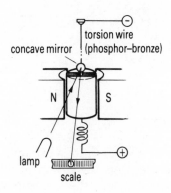

Fig 5.6 Suspended coil meter

ELECTROMAGNETIC INDUCTION

FARADAY'S LAW AND LENZ'S LAW

The *magnetic flux* Φ through an area A is defined as $BA \cos\theta$ where θ is the angle between the magnetic flux density, B, and the normal to the area (Fig 5.7). $B \cos\theta$ is the component of B perpendicular to the area, so that Φ can be thought of as total magnetic influence which passes through the area. Φ has units called 'webers' (Wb). If the circuit

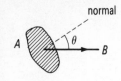

Fig 5.7 Magnetic flux

contains N turns then the total flux through the area of the circuit is $NAB \cos \theta$.

Experiment shows that an e.m.f. is induced in this circuit when the flux Φ through the circuit changes. **Faraday's law** states that the induced e.m.f. is proportional to the *rate of change of magnetic flux*. In the units which we use it can be shown that the constant of proportionality is unity, so that the induced e.m.f., E, is given by

$$E = -\frac{d\Phi}{dt}$$

The negative sign shows that the induced current flows in such a direction as to *oppose* the change producing it. This is called **Lenz's law**. In fact Lenz's law is a direct consequence of the conservation of energy as we can now show.

In Fig 5.8, the flux through the coil circuit changes as the magnet NS approaches and so there is an induced current in the coil. According to Lenz's law this current flows so that the right hand end of the coil behaves as a N-pole of the magnet and so opposing its motion. If Lenz's law were *not* true and the opposite happened, the magnet would be attracted, causing it to approach the coil faster. This would produce a bigger induced current and an even greater attraction. Thus mechanical and electrical energy would be continuously produced without any expenditure of energy, which is impossible.

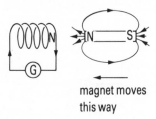

magnet moves
this way

Fig 5.8 Lenz's law

It can be seen from this example that the energy of the induced current comes from the energy pushing the magnet towards the coil. Work is done against the repulsive force set up between the coil and magnet when the induced current flows, and so electrical energy is obtained.

INDUCED E.M.F. IN STRAIGHT WIRE

The formula for the induced e.m.f. in a straight wire is useful in calculating the voltage produced by a dynamo. Two ways of deriving the formula are set out below side by side. One relies on Faraday's law, the other uses $F = BIl \sin \theta$. Consider a three-sided rectangular frame on which a wire PQ can move (Fig 5.9(a)). A force F pulls the wire along with a constant speed v. This induces a current I, which produces an opposing force BIl on the wire.

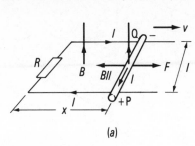
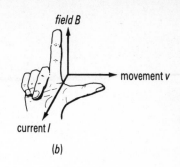

Fig 5.9 Induced current and
e.m.f. in a straight wire

$$E = -\frac{d\emptyset}{dt}$$

$F = BIl$

$\emptyset = BA$ (B is perpendicular to area)
  $= Bxl$
So induced voltage E is given by

$$e = \frac{d\emptyset}{dt} \text{ (numerically)}$$

$$= \frac{d}{dt}(Bxl)$$

$$= BI\,dx/dt = Blv$$

If the induced e.m.f.$=E$ the power produced$=EI$. But
this power is provided by the external force F. From
page 26, mechanical power$=Fv$ so

$$EI = Fv,$$

or $EI = BIlv$

as $F = BIl$ for a steady speed

So $E = Blv$

It should be noted that, in using this formula more generally, v must
be the component of velocity *perpendicular* to B.

By Lenz's law, BIl acts to the left in Fig 5.9(*a*) so as to oppose F.
Fleming's left hand rule shows that I must flow from Q to P for this to
be so. Hence if the wire is considered as a source of electrical energy,
similar to a battery, P acts as the *positive* pole from which current
flows to an external circuit R, and Q as the negative pole, as shown.

The direction of the induced current in a straight wire can also be
found more directly from Fleming's *right* hand rule, which is illustrated in Fig 5.9(*b*).

A.C. DYNAMO

Figure 5.10 shows the basic construction of an a.c. dynamo. We
assume that the reader is already familiar with the method of operation. The sides ab, cd of the coil produce the e.m.f. as the flux
between NS is cut. Figure 5.11(*a*) shows an end-on view after the coil
has rotated with a constant angular velocity ω for t seconds, and so
has turned through an angle ωt radians. The speed of ab (or cd) is ωx
where $2x$ is the length ad, and the component of this velocity perpen-
dicular to B is $\omega x \cos \omega t$. Hence the induced e.m.f. in ab is $Blv = B.y.\omega x$
$\cos \omega t$, where y is the length of ab.

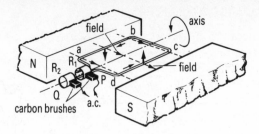

Fig 5.10 a.c. dynamo

The e.m.f. in ab *and* cd is just double this. Also, there are N turns on the coil. So the total e.m.f. is

$$E=2NBy\omega x \cos \omega t$$

$$=NB\omega 2yx \cos \omega t$$

But $2yx$ is the area A of the coil, so

$$E=NB\omega A \cos \omega t=BAN\omega \cos \omega t$$

From this we note:

1　The peak value of the e.m.f. is $BAN\omega$ as the greatest value of $\cos \omega t$ is 1.

2　This peak value occurs when $\cos \omega t=1$, that is, when $\omega t=0$, π, etc. At this moment the sides ab, cd of the coil are slicing across field lines at the greatest rate, and the plane of the coil is *parallel* to the field. In fig 5.11(*b*), $t=0$ corresponds to the peak value.

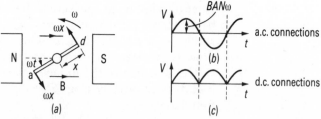

Fig 5.11 Dynamo calculation
and connections

The formula for the induced e.m.f. can also be found from Faraday's law by using calculus. At the instant shown in Fig 5.11(*a*),

we have　　　　$\Phi=NBA \cos (90°-\omega t)=NBA \sin \omega t$

so　　　　　　$E=\dfrac{d}{dt}(NBA \sin \omega t)$　　　　　(numerically)

　　　　　　　$=BAN\omega \cos \omega t$

A.C. AND D.C. CONNECTIONS

The a.c. connections to a dynamo are *via* two *slip rings* R_1, R_2, which are attached to the wires leading from the coil and which turn with the coil. See Fig 5.10. These slip rings press against two fixed *brushes* which connect the coil to the outside circuit. The a.c. voltage variation is shown in Fig 5.11(*b*). To obtain d.c. a *split-ring commutator* is used, which then gives an output as shown in Fig 5.11(*c*), since the commutator reverses the connections from the coil each half revolution.

INDUCED E.M.F. IN A D.C. MOTOR

When a motor turns, the coil rotates in a magnetic field. So an induced e.m.f. is produced in the coil. Let E be the average value of this induced or 'back' e.m.f. The power supplied to the motor is VI where V is the applied p.d. The power wasted as heat is I^2R, and the mechanical power produced is EI. Thus, by the conservation of energy,

$$VI = EI + I^2R$$

or

$$V = E + IR$$

$$\text{The efficiency of the motor} = \frac{\text{mechanical power out}}{\text{power in}}$$

$$= \frac{EI}{VI} \times 100\% = \frac{E}{V} \times 100\%$$

Thus a motor is most efficient when E is as near equal to V as possible. This means that IR must be small and so it is important that the resistance of the windings should be low.

When a motor starts up, the low resistance R can, however, cause problems. Since E is propotional to the motor speed, on start up $E=0$. Hence at the start, $V=IR$ and I can therefore be very large. In large motors it is necessary to have some means (for example, a variable resistance in series) to prevent the current damaging the motor on start. In small motors it is necessary to avoid running them too long at low speeds otherwise they will burn out.

THE TRANSFORMER

A transformer is a device which uses electromagnetic induction to change the magnitude of an alternating (a.c.) voltage. A simplified transformer is shown in Fig 5.12. It is assumed that the reader is familiar with the basic operation. At any instant let Φ be the flux in the soft iron core. In the primary coil of n_p turns the flux is $n_p\Phi$ and so the induced e.m.f. Ep is as shown in Fig 5.12. As in the case of the motor we can write, for the primary circuit,

V_p=applied e.m.f.

$$=E_p+I_pR_p \tag{1}$$

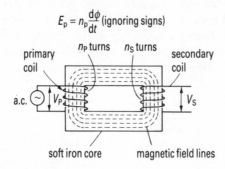

Fig 5.12 The transformer

where I_p and R_p are the current and resistance for the primary coil. Now in an ideal transformer $R_p=0$, so that $V_p=E_p$. Now the secondary voltage induced will be

$$V_s=n_s\frac{d\varPhi}{dt}$$

Hence $\dfrac{V_s}{V_p}=\dfrac{n_sd\varPhi/dt}{n_pd\varPhi/dt}=\dfrac{n_s}{n_p}$

This proof assumes:
1 the primary coil has zero resistance,
2 the flux linking the primary and secondary coils is the same.

When the secondary circuit delivers current there will be a reduction in the p.d. across its terminals unless the secondary coil also has very low resistance.

The power input to the transformer is V_pI_p and the power output is V_sI_s. Hence the efficiency is

$$\frac{V_sI_s}{V_pI_p}\times100\%$$

For 100% efficiency, $V_sI_s=V_pI_p$. So

$$\frac{I_s}{I_p}=\frac{V_p}{V_s}=\frac{n_p}{n_s}$$

Hence in a step-up transformer, when V_s is greater than V_p, the secondary current is less than the primary current.

LOSSES IN A TRANSFORMER

Factors which reduce efficiency are:
1 The resistance of the primary and secondary coils, which dissipate energy as heat.

2   Losses due to induced currents within the iron core, called **eddy currents**, which again produce heat. These currents are reduced by laminating the iron core and separating the laminations or sheets of iron by insulating paint so that the induced (eddy) currents can not flow across the sheets.

3   Hysteresis loss. When a magnetic material is alternately magnetized and demagnetized (as in the case in a transformer core) it absorbs energy in the process and heat is produced. This is called hysteresis loss. The amount of energy loss depends on the type of material used, and can be minimized by careful choice of the correct iron alloy.

In a transformer the current flowing in the primary depends on the load placed across the secondary. If the resistance across the secondary is lowered, the current it supplies increases. This increases the flux produced by the secondary coil which, by Lenz's law, opposes the flux in the iron. The net flux in the iron thus falls, and hence the induced e.m.f. in the primary also falls. From equation (1), page 148, the primary current will rise until a new state of equilibrium is reached.

FLUX CHANGE AND CHARGE

Consider a coil connected in a circuit of *total* resistance R. Suppose initially there is a flux Φ_1 through the coil and that this changes to Φ_2. During the time the flux changes, there will be an induced e.m.f. E and we can derive a simple formula for the charge flow which this induced e.m.f. produces.

From page 111, $I=\dfrac{dQ}{dt}$

So $Q=\int I dt$

Here $Q=\int\dfrac{E}{R}dt=\dfrac{1}{R}\int_{\Phi_1}^{\Phi_2}-\dfrac{d\Phi}{dt}dt=\dfrac{1}{R}\Big[-\Phi\Big]_{\Phi_1}^{\Phi_2}$

So $Q=\dfrac{\Phi_1-\Phi_2}{R}=\dfrac{\text{flux change}}{R}$

By measuring the charge flow Q it is thus possible to measure flux change, and hence magnetic flux density B. This is illustrated in the worked example at the end of this section.

A type of galvanometer for measuring charge is called a *ballistic galvanometer*. The main differences between a ballistic galvanometer and a current measuring instrument are listed overleaf.

Current Galvo	Ballistic Galvo
1 Measures a steady current by measuring a steady deflection.	1 Measures a charge flowing in a short time by measuring the maximum deflection (or 'throw'). The coil then continues to oscillate.
2 The coil is critically damped to bring the pointer to rest in the shortest possible time. This is achieved by using an aluminium coil former. When the coil moves, the induced currents in the aluminium coil former oppose the motion and so produce damping.	2 The coil is damped as little as possible. The coil is wound on an insulating former so that no induced currents can be produced.
3 The inertia of the coil does not matter much provided it is not too large.	3 The inertia of the coil needs to be large so that the coil does not move far in the time the charge takes to flow.

FIELD PRODUCED BY CURRENTS

The value of the magnetic field near current carrying conductors can be investigated experimentally by either a search coil, or a Hall probe. The two experiments below give examples of some properties which may be investigated with these methods.

1 FIELD NEAR A STRAIGHT WIRE USING A SEARCH COIL

The apparatus of Fig 5.13 can be used to investigate the variation of B with distance r from a long straight wire. The search coil must be positioned so that the circular magnetic field lines pass through the coil at right angles to its plane. Also the reading on the a.c. ammeter in the circuit must be kept constant.

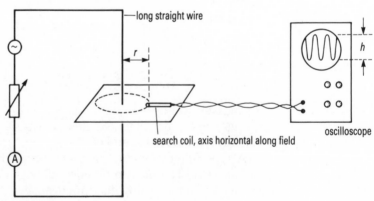

Fig 5.13 Field of a long
straight wire

    The alternating current in the straight wire sets up an alternating magnetic field. This field creates a changing flux through the search coil and so an induced e.m.f. is produced. The peak value of this e.m.f. is proportional to the peak value of the magnetic field at the point where the coil is placed. Thus h measured on the oscilloscope is proportional to the peak value of B.

Measurements of h are made for various values of r, and a graph of $1/h$ against r is plotted. This is a straight line from which it follows that $1/B$ is proportional to r, or B is proportional to $1/r$. (Note: It is more sensible to plot $1/h$ against r than to plot $1/r$ against h. It is left as an exercise to the reader to consider why this is so.)

The relationship between B and r is confirmed by full theory which shows that for an infinitely long straight wire

$$B=\frac{\mu_0 I}{2\pi r}$$

where I is the current flowing. μ_0 is a constant called the permeability of free space and has the numerical value of 4×10^{-7}. We shall see later that the units are henry metre^{-1} ($H\,m^{-1}$).

2 FIELD INSIDE A SOLENOID USING A HALL PROBE

In Fig 5.14 the Hall probe generates a voltage proportional to the value of B acting perpendicular to its plane. By moving the Hall probe keeping its plane perpendicular to the axis of the solenoid it can be shown that the field is uniform inside the solenoid.

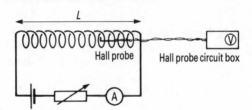

Fig 5.14 Field of a solenoid

The apparatus can also be used to investigate the variation of B with length L of the solenoid. The value of the Hall voltage V is measured for various values of L. Care is needed to see that the current in the solenoid remains constant and that the coils of the slinky are evenly spread and do not touch one another. By plotting a graph of V against $1/L$ a straight line is produced showing that B is proportional to $1/L$. This is consistent with the full theory which shows that

$$B=\mu_0\frac{NI}{L}$$

where N is the number of turns on the coil of length L carrying a current I. This can also be written

$$B=\mu_0 n I$$

where n is the number of turns per metre. This formula also applies to a torus or 'endless' solenoid.

Fig 5.15 Field of a circular coil

3 CIRCULAR COIL

B at centre of a circular coil of N turns, radius r, carrying a current I (Fig 5.15) is given by

$$B=\frac{\mu_0 NI}{2r}$$

FORCE BETWEEN STRAIGHT WIRES; THE AMPERE

Consider two infinitely long straight wires carrying currents I_1 and I_2 as shown in Fig 5.16. The field B_2 at wire 1 due to I_2 is given by

$$B_2=\frac{\mu_0 I_2}{2\pi r}$$

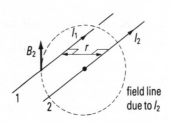

Fig 5.16 Force between straight wires

The force, F, on a length l of wire 1 is given by

$$F=B_2 I_1 l$$

$$=\frac{\mu_0 I_2 I_1 l}{2\pi r} \tag{1}$$

Fleming's left hand rule shows that currents which flow in the same direction attract, while currents flowing in *opposite* directions *repel*. Figure 5.17 shows these situations and the patterns of field lines in each case.

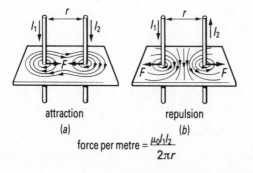

Fig 5.17 Fields between straight wires

attraction
(a)

repulsion
(b)

force per metre $=\dfrac{\mu_0 I_1 I_2}{2\pi r}$

The force between straight wires is used to define the ampere as follows.

DEFINITION OF AMPERE

1 ampere is that current which, flowing in each of two infinitely long parallel straight wires placed 1 metre apart in a vacuum, produces a force of 2×10^{-7} newton per metre length of the wires.

Thus in equation (1) page 152, when $I_1 = I_2 = 1A$, $l = r = 1m$, then $F = 2 \times 10^{-7}$N. Hence

$$2 \times 10^{-7} = \frac{\mu_0 \times 1 \times 1 \times 1}{2\pi \times 1}$$

giving $\mu_0 = 4\pi \times 10^{-7}$ numerically.

Thus the value μ_0 is fixed by the definition of the ampere. μ_0 is called the *permeability of free space*, and from the last equation its units are $N\,A^{-2}$, which can be shown to be equivalent to $H\,m^{-1}$ (henry/metre). So we write $\mu_0 = 4\pi \times 10^{-7} H\,m^{-1}$.

CURRENT BALANCE

A current balance is an instrument for measuring current based on the basic definition of the ampere just given. A simple form is shown in Fig 5.18. A wire frame is balanced on two knife edges at P and S. The current to be measured flows through the frame PQRS then via leads to the straight wire UV situated over the section QR of the frame. When the current is switched on there will be a repulsive force between the wires QR and UV and extra masses m will need to be added to the scale pan to restore balance and regain the initial separation r. The value of the repulsive force is mg if the scale pan and RQ are equidistant from the pivot. The current can then be calculated from

$$F = mg = \frac{\mu_0 I_1 I_2 l}{2\pi r} = \frac{\mu_0 I^2 l}{2\pi r}$$

So $$I = \sqrt{\frac{2\pi r m g}{\mu_0 l}}$$

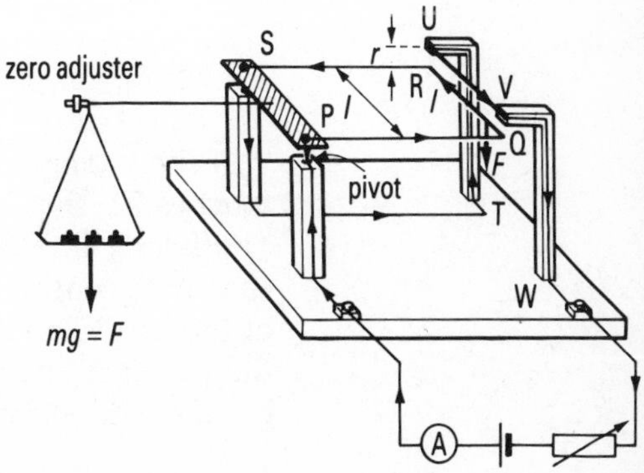

Fig 5.18 The current balance

SELF INDUCTANCE

When a current I flows through a coil it will produce a flux Φ through itself. The *self inductance* L of the coil is the flux (in Wb) produced by a current of 1A.

Hence $L = \dfrac{\Phi}{I}$

The unit of inductance is called the *henry* (H). From the definition just given 1H=1Wb/A.

If the current through an inductor changes then the flux Φ will change, and so there will be an induced e.m.f., E, in the coil. From previously,

$$E_b = \frac{d\Phi}{dt} = \frac{d(LI)}{dt} = L\frac{dI}{dt} \text{ numerically}$$

actually $-L\dfrac{dI}{dt}$ as back E.M.F. is set-up.

Consider a circuit with a battery e.m.f. V and negligible internal resistance connected to an inductance L and resistance R as in Fig 5.19(a). At the moment of switching on, when X touches Y, $I=0$. Hence $V_R(=IR)$ is also zero and thus $V_L=V$ at this moment. But $V_L=LdI/dt$ and hence $dI/dt=V/L$. The initial slope of the graph of current against time is therefore V/L. See Fig 5.19(b). As I builds up, V_R increases and therefore V_L must fall. Eventually V_L, and hence dI/dt, become zero. In this case $V_R=V$ so $I=V/R$, the value expected if only the resistor were present. The growth of current in the circuit is shown in Fig 5.19(b).

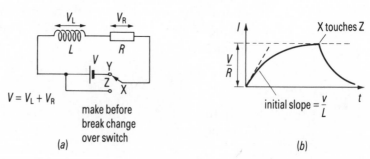

$V = V_L + V_R$

make before
break change
over switch

(a)

initial slope $= \dfrac{V}{L}$

(b)

Fig 5.19 *L-R* d.c. circuit

If the battery is removed but the circuit remains complete (i.e. X touches Z), the current in the circuit does not fall instantly to zero. The falling flux in the inductor produces an e.m.f. which causes a current to continue to flow for some time, as shown by the falling part of the graph in Fig 5.19(b). (Note the current can continue to flow as the circuit is complete. What happens when the circuit is suddenly broken is different, and is dealt with shortly.)

ENERGY STORED IN AN INDUCTOR

During the build up of current in an inductor, energy becomes stored in the magnetic field. The energy $=\int(\text{power}).dt$ with limits 0 and I_0 where I_0 is the final current.

$$\text{Hence energy}=\int_0^{I_0} EIdt=\int_0^{I_0} L\frac{dI}{dt}.Idt$$

$$=\int_0^{I_0} LIdI=\frac{1}{2}LI_0^2 \text{ joules}$$

If a circuit containing inductance is suddenly broken, there is a rapid decrease in current and a very high induced e.m.f. is produced. This can be high enough to cause a spark; the energy for the spark comes from the energy stored in the inductor. This can be a nuisance as switch contacts then soon wear out. To help minimize the wear, a capacitor is often included across switch contacts. The induced voltage when the circuit is broken causes the capacitor to charge, and so the energy from the inductor can be transferred to the capacitor without causing a spark.

MUTUAL INDUCTANCE

The mutual inductance M between two coils 1 and 2 is the flux produced in 2 by a current of one ampere in coil 1. Mathematically

$$\Phi_2=MI_1$$

The induced e.m.f. in coil 2, E_2, is given by

$$E_2=\frac{d\Phi_2}{dt}=M\frac{dI_1}{dt}. \qquad (1)$$

If E_1 is the induced e.m.f. in coil 1 produced by a changing current I_2 in coil 2, then it can be shown

$$E_1=M\frac{dI_2}{dt}$$

with the *same* value of M as in equation (1). A transformer is the most common example of a pair of coils having mutual inductance.

REACTANCE OF CAPACITOR AND INDUCTOR

1 CAPACITOR

Consider an alternating voltage $V=V_0 \sin \omega t$ applied to a capacitor of capacitance C. The charge on the capacitor is $Q=CV$ and the current flowing is

$$I=\frac{dQ}{dt}=C\frac{dV}{dt}=CV_0\omega \cos \omega t$$

From this we observe that the peak current I_0 (when cos $\omega t=1$) is given by $CV_0\omega$. The ratio of peak voltage/peak current (or r.m.s. voltage/r.m.s. current) is called the *reactance* of the capacitor, symbol X_C. Here then

$$X_C=\frac{V_0}{I_0}=\frac{1}{\omega C}=\frac{1}{2\pi fC}$$

Since the voltage varies as sin ωt, and the current as cos ωt we also observe that the current reaches its peak $\frac{1}{4}$ of a cycle ahead of the voltage. We say that I leads V in *phase* by $\frac{1}{4}$ of a cycle (or $\pi/2$ radians or 90°).

The variation of p.d. V across the capacitor with time t is shown in Fig 5.20(a). Since $Q=CV$ for a capacitor, the variation of charge Q with time follows a similar graph. The current is dQ/dt and so I is a maximum where the *gradient* of the charge-time graph is a maximum. Similarly I is zero when Q is not changing, that is, at the maxima and minima of charge-time graph. The variation of current I with time t is thus as in Fig 5.20(b). It can also be seen that these graphs confirm the fact that I leads V by $\frac{1}{4}$ of a cycle.

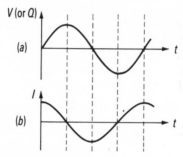

Fig 5.20 P.d. and current for a capacitor

2 INDUCTOR

Now consider an alternating voltage $V=V_0$ sin ωt applied to an inductor of inductance L.

Here $\quad V=L\dfrac{dI}{dt},$

so $\quad I=\dfrac{1}{L}\displaystyle\int V dt$

$$=\frac{1}{L}\int V_0 \sin \omega t.dt=-\frac{V_0}{\omega L}\cos \omega t$$

The peak current I_0 is now numerically $V_0/\omega L$, so that the *reactance* of the inductor, symbol X_L, is $V_0/I_0=\omega L=2\pi fL$. I now varies as $-\cos \omega t$ and so lags in phase by $\frac{1}{4}$ of a cycle, or $\pi/2$ radians or 90°.

The variation of p.d. V across the inductor is shown in Fig 5.21(a). Since $E=LdI/dt$, the current must vary with time so that the gradient of the current-time graph is proportional to the p.d. In other words the p.d. is greatest when the current is changing at the greatest rate, that is, when the current is zero. Figure 5.21(b) illustrates the variation of current I with time t and shows that I lags in phase behind V by $\frac{1}{4}$ of a cycle.

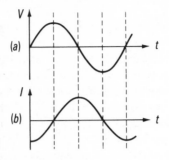

Fig 5.21 P.d. and current for an inductor

For a resistor, V/I is a constant equal to the resistance R, and so V and I are in phase with one another. These results are shown in the following table, which should be memorized. Note that X_C is in ohms when f is in Hz and C in farads (F) and that X_L is in ohms when f is in Hz and L in henrys (H).

Device	Resistance or reactance	Phase between V and I
Resistor	R	in phase
Capacitor	$X_C=\dfrac{1}{\omega C}$ or $\dfrac{1}{2\pi f C}$	I leads V by $\frac{1}{4}$ cycle (90°)
Inductor	$X_L=\omega L$ or $2\pi f L$	I lags V by $\frac{1}{4}$ cycle (90°)

Note that the reactance of the capacitor and inductor both vary with frequency – at high frequencies an inductor will have a high reactance and little current will flow when an alternating p.d. is connected to it. At high frequencies, however, there will be a large current in the circuit connecting an alternating p.d. to a capacitor, since the reactance of a capacitor falls as the frequency rises.

R.M.S. AND PEAK VALUES   The root mean square (r.m.s.) value of an alternating current is that value of steady direct current which gives the same heating effect. For a sine curve (sinusoidal) variation of a.c., it can be shown that

$$\text{r.m.s. value}=\frac{1}{\sqrt{2}}\times\text{peak value}$$

Generally, measuring instruments are calibrated to read r.m.s. values and in the rest of this section it will be assumed that currents and voltages are r.m.s. values unless otherwise stated. Since peak values are proportional to r.m.s. values, results which involve the ratio of voltage to current apply to peak values equally. The formula for power dissipated in a resistor, $P=IV$, however, is true only for r.m.s. values.

OSCILLATIONS OF L-C CIRCUIT

Figure 5.22 shows an L-C circuit with a capacitor C of $1000\mu F$ and an inductor L of 5H connected to the Y-plates of an oscilloscope with the time base switched off. The capacitor is charged by closing switch K, when the spot on the oscilloscope screen is displaced vertically. When K is now opened the spot on the screen oscillates at about 1.6Hz showing an *oscillating p.d.* across C. The vibrations decay quite quickly, as the inductor will not be pure but contain some resistance which dissipates energy as heat.

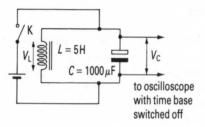

Fig 5.22 *L-R oscillatory* circuit

The formula for the frequency of oscillations can be found as follows. Once K has been opened there is no net p.d. in the circuit, so $V_L+V_C=0$. Hence

$$L\frac{dI}{dt}+\frac{Q}{C}=0, \text{ or } \frac{dI}{dt}=\frac{d^2Q}{dt^2}=-\frac{Q}{LC}$$

This is the equation for s.h.m., with a time period $T=2\pi\sqrt{LC}$. Thus the frequency of vibration, f, is given by

$$f=\frac{1}{2\pi\sqrt{LC}}$$

Substituting $C=1000\mu F=10^{-3}F$ and $L=10H$ gives $f=1.6Hz$ as stated above.

During the electrical oscillations of a circuit containing a pure capacitor and a pure inductor, no energy is dissipated and the electrical energy alternates between the capacitor and the inductor. The situation at various stages during a cycle of the oscillation is illustrated in Fig 5.23.

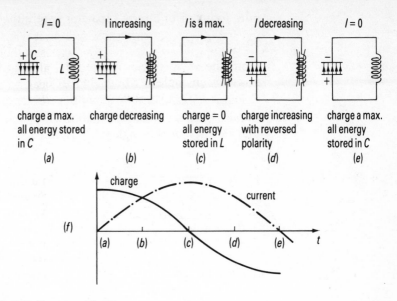

charge a max.   charge decreasing   charge = 0   charge increasing   charge a max.
all energy stored   all energy   with reversed   all energy
in C   stored in L   polarity   stored in C

(a)   (b)   (c)   (d)   (e)

Fig 5.23 Energy and field in
L-C circuit

L-R AND C-R CIRCUITS

For an inductor L and resistor R in series (Fig 5.24(a)), it is not possible to add just arithmetically the r.m.s. voltages across L and R as the voltages are not in phase. (Only the *instantaneous* voltages can be added arithmetically.) To take account of the phase difference we use the vector or phasor diagram method. In Fig 5.24(b) vector I represents the current. V_R is the p.d. across R and is in phase with I. V_R is equal to IR in length. V_L is the p.d. across L and leads the current by 90°. $V_L = IX_L = I\omega L$ since ωL is the reactance.

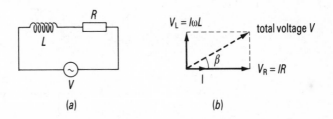

Fig 5.24 L-R a.c. circxuit

The total p.d. $V = vector\ sum\ of\ V_R\ and\ V_L$

$$= \sqrt{(IR)^2 + (I\omega L)^2}$$

$$= I\sqrt{R^2 + \omega^2 L^2}$$

The *impedance* Z of the circuit is defined as (r.m.s. p.d.)/(r.m.s. current). (It should be noted that the term 'reactance' only applies to a pure inductor or a pure capacitor where the phase difference is 90°.)

So $Z=V/I=\sqrt{R_2+\omega^2L^2}$

From the vector diagram the total voltage V leads the current I in phase by an angle β, where $\tan \beta=\omega L/R$.

Figures 5.25(a) and (b) show the corresponding diagrams for a C-R circuit. Here a similar calculation shows that $Z=\sqrt{R^2+(1/\omega^2C^2)}$ and that I leads V by the angle $\tan^{-1}(1/\omega CR)$.

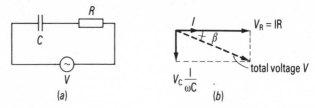

Fig 5.25 C-R a.c. circuit

L-C-R SERIES CIRCUIT

In this case we have three vector voltages to add (Fig 5.26(a)). V_C and V_L are opposite in direction so their total value is $(V_C-V_L)=(I/\omega C-I\omega L)$, assuming V_C is greater than V_L. The total p.d., V, is then, from Pythagoras, see Fig 5.26(b),

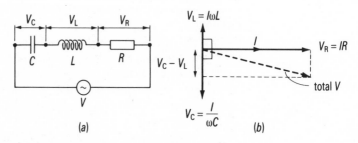

Fig 5.26 L-C-R series circuit

$$V=\sqrt{\left(\frac{I}{\omega C}-I\omega L\right)^2+I^2R^2}$$

since (V_C-V_L) is 90° to V_R. Hence the impedance Z is given by,

$$Z=\frac{V}{I}=\sqrt{\left(\frac{1}{\omega C}-\omega L\right)^2+R^2}$$

From this formula we see that Z is a *minimum* when $1/\omega C=\omega L$. In this case the impedance Z is simply R. When Z is a minimum the current I is a maximum and we have *resonance*. The variation of current I with frequency f is shown in Fig 5.27.

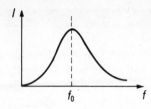

Fig 5.27 *L-C-R* resonance
graph

The resonant frequency, f_0, occurs when $1/\omega C = \omega L$. In this case we have,

$$\omega^2 = \frac{1}{LC}, \text{ so } \omega = \frac{1}{\sqrt{LC}}$$

Since $\omega = 2\pi f_0$ then $f_0 = 1/2\pi \sqrt{LC}$. It should be noted that this is the same value as the natural frequency of oscillations of an *L-C* circuit found on page 158. Just as in mechanical oscillations, resonance occurs when the forcing frequency and the natural frequency are the same.

From the vector diagram it is clear that at the resonant frequency $(V_C = V_L) = 0$, and so the voltages across L and C cancel. At resonance the circuit behaves simply as a pure resistor of magnitude R, so the maximum current $= V/R$ where V is the supply voltage.

POWER IN A.C. CIRCUITS   In an alternating current circuit, power is not absorbed in a pure capacitor or a pure inductor. For example, a capacitor stores electrostatic energy as it charges, but this energy is returned when it discharges. An inductor stores magnetic energy as the current and field build up, but this is returned as the current decreases. Power is absorbed only in *resistors*.

In the *L-R* series circuit of Fig 5.24(a), power, P, is absorbed only in the resistor. Here $P = IV_R$ where I is the current and V_R is the p.d. across R. But from the vector diagram in Fig 5.24(b) we see that $V_R = V \cos \beta$, where β is the phase angle between the total (applied) p.d. V and the current I.

Thus   power $= IV \cos \beta$

$\cos \beta$ is called the *power factor* of the circuit. V and I must be r.m.s. values in this formula and also when power is calculated for a resistor from $P = IV$ or $I^2 R$ or V^2/R.

WORKED EXAMPLES

5.1 FORCES ON CURRENT IN MAGNETIC FIELD

A wire frame ABCD is balanced as shown in Fig 5.28 and a current of 5A flows through AB whose length is 10cm. A uniform magnetic field of flux-density 0.01T is now switched on. It acts downwards over the left-hand area of the frame and is at 70° to the horizontal. Calculate (a) the force which acts on AB and (b) the mass which must be altered in the scale pan, stating whether mass is to be added or removed.

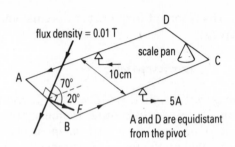

Fig 5.28 Worked example

(**Overview** (a) $F=BIl$ is used to calculate the force.

(b) The weight in the scale pan must be sufficient to balance the vertical component of F.)

(a)   The magnetic field is at right angles to AB so the force is given by

$$F=BIl$$

$$=0.01\times5\times0.1=0.005\text{N}$$

By Fleming's left hand rule, F acts downwards perpendicular to B and to the wire as shown in the diagram.

(a)   Only the vertical component of F will make the frame turn about the pivots and affect the scale pan mass. The downward component of F is

$$F\cos 70°=0.005\cos 70°$$

To balance this an *extra* mass m must be placed in the scale pan where

$$mg=0.005\cos 70°$$

So $m=\dfrac{0.005\cos 70°}{9.8}=0.17\times10^{-3}\text{kg}=0.17\text{g}$

5.2 THE HALL EFFECT

A specimen of semiconductor of cross-section 5mm×1mm carries a current mainly due to a flow of electrons. A magnetic field B of 0.1T is applied perpendicular to the slab and a Hall voltage of 0.4mV is set up (Fig 5.29). Calculate (a) the drift velocity of the electrons and (b) the current flowing in the specimen if the density of the electrons is 10^{25}m^{-3}.

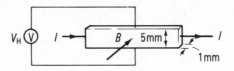

Fig 5.29 Worked example

(**Overview** (a) $V_H = Bvd$ can be used to find v.

(b) The current can then be obtained using the formula relating current and drift velocity (page 112).)

(a)   Using $V_H = Bvd$ (page 141),

$$0.4 \times 10^{-3} = 0.1 \times v \times 5 \times 10^{-3}$$

So $v = 0.8$m s^{-1}

(b)   Using $I = Anqv$, and assuming that the electrons each carry 1.6×10^{-19} coulomb,

$$I = (5 \times 10^{-3} \times 1 \times 10^{-3}) \times 10^{25} \times 1.6 \times 10^{-19} \times 0.8$$

$$= 6.4\text{A}$$

5.3 CHARGE AND FLUX CHANGE

A flat coil of 500 turns, resistance 100Ω and average area 5cm^2 is held with its plane perpendicular to the magnetic field between the poles of a magnet. The coil is connected to a ballistic galvanometer of sensitivity 5μC/cm and resistance 50Ω. When the coil is rapidly turned through 90° so that its plane is now parallel to the field, the galvanometer gives a 'throw' of 3cm.
(a)   Calculate the strength of the magnetic field and
(b)   find the throw of the galvanometer if the coil had been turned a *further* 90° so that its plane was again perpendicular to the field.

(**Overview** First find the flux change from the charge. The magnetic field can then be found using the relationship between flux and charge.)

(a)   Total $R = 100 + 50 = 150$Ω.
The charge flowing is $3 \times 5 = 15$μC $= 15 \times 10^{-6}$C
Now the flux change is given by

$$Q = \frac{\Phi_1 - \Phi_2}{R}$$

So $\Phi_1 - \Phi_2 = QR = 15 \times 10^{-6} \times 150 = 225 \times 10^{-5}$Wb

However $\Phi_2=0$ as the magnetic field is then parallel to the plane of the coil. Also $\Phi_1=BAN$. So

$$B(5\times10^{-4}\times500)=225\times10^{-5}$$

or $\qquad B(25\times10^{-2})=225\times10^{-5}$

So $\qquad\qquad\qquad B=9\times10^{-3}\mathrm{T}$

(b)   If the search coil had been turned through 180°, the flux change would be double. So the deflection would be 2×3=6cm.

5.4 INDUCED E.M.F.=RATE OF CHANGE OF FLUX

An alternating magnetic field of peak value 0.01T and frequency 50Hz is applied perpendicular to a coil of mean area 5cm² and having 100 turns.
(a)   Calculate the peak e.m.f. induced in the coil.
(b)   State and explain the effect of changing the frequency of the alternating field to 100Hz.

(**Overview** To find the peak e.m.f. it is necessary to find the formula for the e.m.f. as the time varies. This in turn is obtained using dΦ/dt. Hence the first task is to get Φ as a function of time.)

(a)   If $B=B_0\sin 2\pi ft$, then

$$B=0.01\sin 100\pi t$$

The flux, Φ through the coil is $NBA=100\times0.01\times5\times10^{-4}\sin 100\pi t$.

The induced e.m.f., $E=\dfrac{d\Phi}{d}$ numerically

$$=\frac{d}{dt}(100\times0.01\times5\times10^{-4}\sin 100\pi t)$$

$$=100\times0.01\times5\times10^{-4}\times100\pi\cos 100\pi t$$

$$=5\pi\times10^{-2}\cos 100\pi t$$

Hence the peak e.m.f.$=5\pi\times10^{-2}=0.16$ volt.

(b)   Doubling f would double the peak induced e.m.f., since the flux is changing twice as fast.

Note
Measuring the peak e.m.f. induced in a coil produced by an alternating magnetic field is often used as a method of measuring the peak value of the flux-density of the field. The coil used is called *search coil*.

5.5 INDUCED E.M.F. IN A STRAIGHT CONDUCTOR=Blv

An aeroplane flies at 200m s^{-1} in an easterly direction. Find the e.m.f. between the wing tips which are 20m apart. The Earth's magnetic field is 5×10^{-5}T and the angle of dip is 70°.

(**Overview** In using Blv it must be remembered that B, l, and v are in perpendicular directions. (The angle of dip is the angle the Earth's magnetic field makes with the horizontal.))

In the formula $E=Blv$, B, l and v are all perpendicular. Hence B must be the component of the Earth's magnetic field which is *perpendicular* to l and v, that is, the vertical component, B_v.

Now $B_v=B\sin 70°=5\times10^{-5}\sin 70°$

So $E=(5\times10^{-5}\times\sin 70°)\times20\times200$

         $=0.188$ volts

5.6 PRINCIPLE OF THE CURRENT BALANCE

A current I flows through a balanced wire frame and a solenoid in series (Fig 5.30). The solenoid is of length 20cm and has 100 turns. When the current is switched on, 2mg must be placed in the scale pan to restore balance. Find the current flowing.

Fig 5.30 Worked example

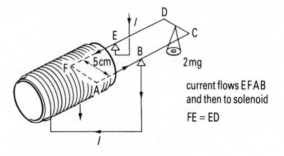

current flows EFAB
and then to solenoid
FE = ED

(**Overview** To find the force on the wire we can use BIl, but first we need to find B. This can be done using the formula for the field produced by a solenoid.)

The field produced by the solenoid is

$$B=\frac{\mu_0 NI}{l}=\frac{4\pi\times10^{-7}\times100}{0.2}I$$

The force on the section of wire $AF=BIl$

$$=\left(\frac{4\pi\times10^{-7}\times100}{0.2}\right)I\times I\times0.05$$

But this force is balanced by the weight of a 2mg (2×10^{-6}kg) mass, and so must be $(2\times10^{-6})\times9.8$N

Hence $2\times10^{-6}\times9.8=\frac{4\pi\times10^{-7}\times100\times0.05}{0.2}I^2$

so $I^2=6.2\times10^{-2}$

giving $I=0.25$A.

5.7 REACTANCE OF CAPACITOR AND INDUCTOR

Find the reactance of (i) a capacitor of $30\mu F$ and (ii) an inductor of 0.3H at frequencies (a) 10Hz, (b) 50Hz, (c) 1000Hz.

(**Overview** Use the formulae for reactance appropriate to each components.)

(a)   (i) Reactance of capacitor, $X_C = \dfrac{1}{2\pi fC} = \dfrac{1}{2\pi \times 10 \times 30 \times 10^{-6}}$
$$= 531\Omega.$$

(ii) Reactance of inductor, $X_L = 2\pi fL = 2\pi \times 10 \times 0.3 = 19\Omega.$

(b)   Similar methods show that at 50Hz the results are $X_C = 106\Omega$ and $X_L = 94\Omega$.

(c)   Again, at 1000Hz we obtain 5.3Ω for X_C and 1884Ω for X_L.
From results such as these it is possible to plot a graph of the reactance against frequency, f. The variations are illustrated in Fig 5.31. Note that $X_C \propto 1/f$ but $X_L \propto f$.

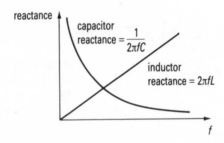

Fig 5.31 Variation of reactance with frequency

5.8 ENERGY STORED IN INDUCTOR AND CAPACITOR

(a)   An inductor of 5H carries a current of 2A. Calculate the energy stored.
(b)   A capacitor is to be included across switch contacts so that the p.d. across the terminals does not exceed 100V when the current in the inductor is switched off. Calculate the value of the capacitor required.

(**Overview** This is a direct use of $LI^2/2$ and $CV^2/2$. When the current is switched off the current will fall and an e.m.f. is produced which must not exceed 100V. The capacitor charges up and prevents a very rapid rate of decrease of current.)

(a)   The energy stored in the inductor
$$= \tfrac{1}{2}LI^2 = \tfrac{1}{2} \times 5 \times 2^2 = 10J$$

(b)   All of this energy must be transferred to the capacitor when the p.d. across the capacitor is 100V.

Hence $\frac{1}{2}CV^2=\frac{1}{2}C\times100^2=10J$

By calculation, $C=2\times10^{-3}F=2000\mu F$

5.9 L-R A.C. CIRCUIT

An inductor of 0.5H is connected in series with a resistor of 1000Ω, across an alternating supply of constant 10V r.m.s. but of variable frequency. Find the frequency at which the r.m.s. p.d. across the inductor is (a) 2V, (b) 5V. Find the p.d. across the resistor in these cases. Comment on the results.

(**Overview** A vector diagram is used in part (a) to find the p.d. across R and hence the current flowing. From $V_L=2\pi fLI$, f can be found.)

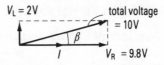

Fig 5.32 Worked example

The vector diagram for the circuit is shown in Fig 5.32.

(a)   When the p.d. across the inductor L is 2V, then $\omega LI=2V$ and hence the phase angle β is given by sin $\beta=2/10=0.2$, giving $\beta=11.5°$. So

$$V_R=10\cos\beta=9.8V$$

The current flowing is given by $V_R=IR$

so $9.8=I\times1000$

or $I=0.0098A=9.8mA$

Now $V_L=\omega LI=2\pi fLI$

so $2=2\pi f\times0.5\times0.0098$

Simplifying, $f=65Hz$

(b)   When the p.d. across the inductor is 5V, $\omega LI=5V$ and hence sin $\beta=5/10=0.5$, giving $\beta=30°$. So

$$V_R=10\cos\beta=8.7V$$

Similar methods to the above show that the current flowing is 8.7mA and that the frequency is 184Hz.

In cases (a) and (b) the p.d. across the resistor has already been calculated, being 9.8V and 8.7V respectively. The r.m.s. p.d. across L and R do *not* add to 10V in (a) or (b) since the voltages are out of phase. The instantaneous p.d.s do add arithmetically, but the r.m.s. voltages add vectorially, i.e. $\sqrt{2^2+9.8^2}=10$ for (a) and $\sqrt{5^2+8.7^2}=10$ for (b).

5.10 L-C-R RESONANCE

A coil of 1mH is connected in series with a capacitor of 10μF and a resistor of 100Ω. A signal generator of variable frequency giving a r.m.s. voltage of 10V is connected across the components. Calculate (a) the resonant frequency, (b) the p.d. across each component at this frequency, (c) the power dissipated in the circuit at resonance.

(**Overview** (a) The resonant frequency can be found from the standard formula.
(b) At resonance V_L and V_C have opposite phases and so all the applied p.d. appears across the resistor. This enables I to be found. P.d. can then be obtained from reactance×current.
(c) All the power is dissipated in the resistor, C and L can be ignored for the purposes of calculating power.)

(a)   The resonant frequency is given by

$$f_0=\frac{1}{2\pi\sqrt{LC}}$$

$$=\frac{1}{2\pi\sqrt{(10^{-3}\times10^{-5})}}=1590\text{Hz},$$

since $L=10^{-3}$H and $C=10^{-5}$F.

(b)   At this frequency the p.d.s V_L and V_C across the inductor and capacitor respectively are equal and opposite in phase. Thus the p.d. V_R across the resistor is the supply voltage 10V. So the current I in the circuit$=V/R=10/100=0.1$A. The reactance of the inductor X_L (=reactance of capacitor X_C)

$$=2\pi fL=2\pi\times1590\times10^{-3}=10\Omega$$

Hence the p.d. across the inductor=reactance×I

$$=10\times0.1=1\text{V}$$

The p.d. across the capacitor is the same but, at any instant, is opposite in sign. Hence the p.d.s are: capacitor and inductor each 1V, resistor 10V.

(c)   In a circuit with inductance, capacitance and resistance the power is dissipated only in the resistor. Hence, for the resistor R,

$$\text{Power}=IV=0.1\times10=1\text{W}$$

1 MAGNETIC FLUX DENSITY

$F=BIl \sin \theta$. Direction of force obtained from Fleming's left hand rule.

Force on a moving charge: $f=Bqv \sin \theta$.

Hall effect – due to force on charge carriers moving in a magnetic field.

Voltage set up across specimen, $V_H=BI/nqt$. V_H is small in metals, larger for semiconductors (n less). Hall probe is used to measure B.

Torque or couple on a coil, $T=BANI \sin \alpha$. Torque is a maximum when plane of coil is parallel to field. Effect used in ammeters and motors. In an ammeter there is a radial field, so $\alpha=90°$ for any position of the coil.

2 ELECTROMAGNETIC INDUCTION

Faraday's law: $E=-d\Phi/dt$.

Lenz's law: Induced current flows to *oppose* change producing it. It is a consequence of conservation of energy.

E.m.f. induced in a straight wire$=Blv$. Direction of current obtained from right hand rule.

A.c. dynamo: $E=BAN\omega \cos \omega t$.

Induced e.m.f. in d.c. motor$=$mechanical energy output per coulomb flowing. Efficiency of motor$=E/V \times 100\%$.

Transformer, $V_s/V_p=n_s/n_p$. Step up if $n_s>n_p$. If 100% efficient, $I_s/I_p =n_p/n_s$. Losses of energy due to (a) resistance of coils (b) eddy currents (c) hysteresis loss.

Flux change produces a flow of charge. $Q=(\Phi_1-\Phi_2)/R$. Q can be measured on a ballistic galvanometer.

3 FIELD PRODUCED BY CURRENTS

Straight wire: $B=\mu_0 I/2\pi r$.

Solenoid or torus: $B=\mu_0 NI/l=\mu_0 nI$.

Circular coil: $B=\mu_0 NI/2r$.

Force between straight wires $F=\mu_0 I_1 I_2 l/2\pi r$. Used to define the ampere by choosing $\mu_0=4\pi\times10^{-7}$H m^{-1}. Current balance used to measure current in terms of mechanical quantities.

4 INDUCTANCE

Self-inductance $L=\Phi/I=$flux produced by current of 1A. (Unit: henry (H).)

Induced e.m.f. in an inductor, $E=-LdI/dt$.

Energy stored in an inductor$=LI^2/2$.

Mutual inductance, $M=$flux produced in coil 2 current of 1A in coil 1.

5 A.C. CIRCUITS

Reactance of capacitor and inductor, X_C or $X_L=V_0/I_0=V/I$ (r.m.s.).

For capacitor $X_C=1/\omega C$, for inductor $X_L=\omega L$. Current leads by 90° in C, and lags by 90° in L.

R.m.s. value for sinusoidal a.c.$=$peak value/$\sqrt{2}$.

L-C circuits have natural frequency of oscillation $f=1/2\pi\sqrt{LC}$.

L-R circuit. Impedance$=\sqrt{R^2+\omega^2L^2}$, phase angle$=\tan^{-1}(\omega L/R)$.

C-R circuit. Impedance$=\sqrt{R^2+(1/\omega^2C^2)}$, phase angle$=\tan^{-1}(1/\omega CR)$.

L-C-R series circuit. Resonance at $f=1/2\pi\sqrt{LC}$, circuit behaves as pure resistor R at resonance, when p.d. across L and C are equal and opposite.

Power in a.c. circuits$=I^2R$, no power absorbed in L or C. Also power$=IV\cos\beta$ where I, V are r.m.s. values and β is phase angle.

DIAGRAM SUMMARY

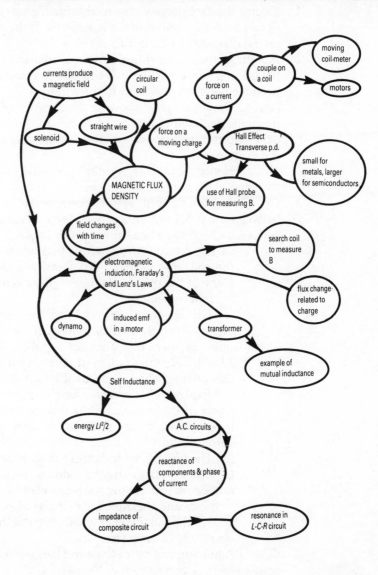

5 QUESTIONS

1   Taking the flux density due to the Earth's magnetic field as $100\mu T$, obtain a rough estimate of the magnitude of the magnetic force acting on the filament of a car headlamp bulb. (C)

2   If a straight wire carrying a current is subject to a uniform magnetic field acting perpendicularly to the wire, a potential difference will be produced across the *diameter* of the wire.
(*a*)   Draw a diagram showing the direction of the force (due to the magnetic field) which acts on a free electron which is moving with the drift velocity. (Indicate clearly the directions of the drift velocity and the magnetic field.)
(*b*)   Hence explain how the potential difference across the diameter of the wire arises. (L)

3   (*a*)   A moving coil meter possesses a square coil mounted between the poles of a strong permanent magnet. The torque on the coil is 4.2×10^{-9}N m when the current is $100\mu A$.
(i) The meter is designed so that whatever the deflection of the coil, the magnetic flux density is always parallel to the plane of the coil. Explain, with the aid of a labelled diagram, how this is achieved.
(ii) The restoring springs bring the coil to rest after it has turned through a certain angle. If the restoring couple per unit angular displacement applied by the springs is 3.0×10^{-9}N m per radian, through what angle, in radian, will the coil turn when a current of $100\mu A$ flows?
(iii) Explain what is meant by the *current sensitivity* of such a meter. If the pointer on the instrument is 7.0cm long, what length of arc on the scale would correspond to a change in current of $2\mu A$?
(iv) The instrument indicates full scale deflection for a current of $100\mu A$. What current produces full scale deflection if the number of turns in the coil is doubled?
    Increasing the number of turns also increases the resistance of the coil. Explain whether or not this change affects the sensitivity of the meter.
(*b*)   A moving coil meter has a resistance of 1000Ω and gives a full scale deflection for a current of $100\mu A$.
(i) What value resistor would be required to convert it to an ammeter reading up to 1.00A? Draw a circuit diagram

showing where the resistor would be connected. What form might this resistor have?

(ii) Draw a diagram showing the additional circuitry needed for the moving coil meter to be adapted to measure alternating currents. Mark clearly on the diagram the connecting points for the meter and for the a.c. supply.

What is the relationship between the steady current registered by the meter and the current from the a.c. supply?   (L)

4   A closed square coil consisting of a single turn of area A rotates at a constant angular speed, ω, about a horizontal axis through the midpoints of two opposite sides. The coil rotates in a uniform horizontal magnetic flux density, B, which is directed perpendicularly to the axis of rotation.

(a)   Give an expression for the flux linking the coil when the normal to the plane of the coil is at an angle α to the direction of B.

(b)   If at time $t=0$ the normal to the plane of the coil is in the same direction as that of B, show that the e.m.f., E, induced in the coil is given by

$$E = BA\,\omega \sin \omega t.$$

(c)   With the aid of a diagram, describe the positions of the coil relative to B when E is (i) a maximum (ii) zero. Explain your answer.

(JMB)

5   In the diagram (Fig 5.33) a uniform magnetic field of 2.0T exists, normal to the paper, in the shaded region. Outside this region the magnetic field is zero. X, Y and Z are three loops of wire.

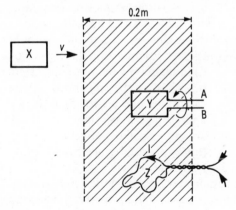

Fig 5.33

(a)   Loop X is a rigid rectangle in the plane of the paper with dimensions 50×100mm and resistance 0.5Ω. It is pulled through the field at a constant velocity of 20mm s^{-1} as shown. Its leading edge enters the field at time $t=0$. Sketch graphs, giving explanations, of how the following vary with time from $t=0$ to $t=20$s:

(i) the magnetic flux linking loop X,

(ii) the induced e.m.f. around the loop,

(iii) the current in the loop.

Neglect any effect of self-inductance.

(b)   Loop Y of the same dimensions as X is mounted on a shaft within the field region. It is rotated at a constant frequency f to provide a source of alternating e.m.f. At $t=0$ the loop is in the plane of the paper.

(i) Give an expression for the e.m.f. between the terminals A and B at any instant.

(ii) Calculate the frequency of rotation required to give an output of 3V r.m.s.

(c)   Loop Z, made of a fixed length 0.3m of flexible wire, rests on a smooth surface in the plane of the paper within the field region. A direct current I in the wire causes it to take up a circular shape.

(i) Explain this observation.

(ii) What would you expect to happen if the current I was reversed in direction?   (O&C)

6 (a)   State the laws of Faraday and Lenz that relate to electromagnetic induction. Describe experiments that you would perform to demonstrate the factors determining (i) the magnitude and (ii) the direction of induced e.m.f.s.

(b)   (i) With the aid of a labelled diagram, describe the construction and principles of operation of a d.c. generator utilizing electromagnetic induction.

(ii) Explain carefully why an increased torque has to be applied to the driving shaft of the generator when an increased current is drawn from its output terminals.

(c)   A large earth-mover of total mass 250000kg is driven by a diesel engine coupled to a d.c. generator which supplies current to d.c. electric motors attached to the wheels. To assist the brakes when the vehicle is descending an incline the motors can be disconnected from the generator and connected to a bank of resistors.

(i) Explain how the braking system operates.

(ii) What is the maximum speed at which the vehicle can safely descend a 1 in 10 gradient if the resistors can dissipate energy at a maximum rate of 2000kW? Neglect friction and any other braking effects.

(iii) Explain why such a retarding system cannot be used to bring the vehicle to a halt.   (O)

7 Explain in non-mathematical terms why the ratio of the potential differences across the primary and secondary coils of the transformer equals the ratio of the numbers of turns in the primary and secondary coils respectively.

Identify *two* assumptions relating to the transformer on which your explanation relies.   (L)

8 (i)   A flat coil of N turns and area A so that its plane is perpendicular to a magnetic field of flux density B. The coil is suddenly removed from the field. If the coil is in a closed circuit of resistance R

prove, from first principles, that the charge q which is circulated is given by

$$q = \frac{NBA}{R}$$

(ii)   A ballistic galvanometer has charge sensitivity 100mm μC^{-1} and a resistance of 100Ω. A square search coil of negligible resistance, of 25 turns, having sides of length 10mm is in series with the galvanometer. When the coil is removed from a magnetic field the deflection on the galvanometer is 250mm. Calculate the magnetic flux density.

(iii)   If the search coil is removed from the field in 0.5s, how much heat is generated by the circulating charge?

(iv)   How would a current balance rather than the search coil be used to measure the flux density?   (W)

9   A constant voltage source in the circuit illustrated (Fig 5.34) supplies a sinusoidal alternating e.m.f. The voltages marked against the other components are the peak values developed across each for a particular value of the capacitance of C.

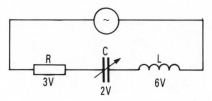

Fig 5.34

(a)   Determine
         (i) the peak value of the applied e.m.f.,
         (ii) the phase angle between the applied e.m.f. and the current.

(b)   If the variable capacitor C is now adjusted until the voltage across R is a maximum, the circuit is said to be resonant. Explain why the current in the circuit has its maximum value when this is so.   (L)

10 (a)   The alternating voltage across a mains supply has an r.m.s. value of 240V and a frequency of 50Hz.
         What is meant by *an r.m.s. value of 240V* and *a frequency of 50Hz*?
    (b)   In alternating current circuits what is meant by
         (i) the *reactance* of a capacitor;
         (ii) the *impedance* of a circuit?
    (c)   Figure 5.35 shows a resistor, a capacitor and a pure inductor connected in series to a sinusoidal alternating current source.
         A voltmeter is connected across each of the components in turn and the r.m.s. voltages obtained are shown on the diagram.

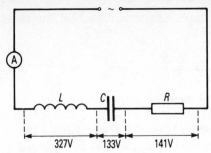

327V   133V   141V

Fig 5.35

(i) By using a phasor diagram or otherwise, calculate the r.m.s. voltage across the supply terminals.

(ii) If the resistor has a resistance of 68Ω, calculate the reading you would expect on the ammeter.

(iii) Calculate the values of the inductor and capacitor assuming that the frequency of the supply is 50Hz.

(*AEB* part Q 1984)

GEOMETRICAL OPTICS

CONTENTS

REFRACTION AT PLANE SURFACES

LAWS OF REFRACTION.
REFRACTIVE INDEX

When light is refracted from air to glass across a plane boundary,

(a)   the incident, refracted ray and the normal at the point of incidence all lie in the same plane, and

(b)   the ratio sin i/sin r is a constant, where i and r are the respective angles of incidence and refraction. These are the laws of refraction. They are true for any two media, such as refraction from water to glass.

The ratio sin i/sin r is defined as the **refractive index**, symbol n, between the two media. So for light refracted from a medium 1 to another medium 2 (Fig 6.1).

$$_1 n_2 = \frac{\sin i}{\sin r}$$

As we see later, the phenomenon of refraction is due to the change in speed of light when it travels from one medium to a different medium. Another useful definition of refractive index is given by (Fig 6.1)

$$_1 n_2 = \frac{\text{speed of light in medium 1}}{\text{speed of light in medium 2}}$$

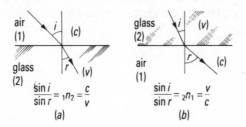

$$\frac{\sin i}{\sin r} = {_1 n_2} = \frac{c}{v}$$

(a)

$$\frac{\sin i}{\sin r} = {_2 n_1} = \frac{v}{c}$$

(b)

Fig 6.1 Refraction of light
and refractive index

Tables of physical constants show the refractive index of glass, water and other media when the incident light is in a vacuum (or air for practical purposes). These are called 'absolute' refractive indices, and the symbol n can then be used for the absolute refractive index. So

$$n = \frac{\text{velocity of light in vacuum, } c}{\text{velocity of light in medium, } v}$$

REFRACTIVE INDEX FOR ANY TWO MEDIA

From the principle of reversibility of light, a ray of light when reversed will travel along its original path. So from *glass to air* (Fig 6.1(*b*)),

(*a*)   the ray is refracted away from the normal, and

(*b*)   refractive index from glass to air, $_gn_a$, is

$$_gn_a = \frac{1}{_an_g}$$

since i is now the angle of incidence in *glass* and r is the angle of refraction in air. (Fig 6.1(*b*)). So if $_an_g = 1.5$, then $_gn_a = 1/1.5 = 2/3$. Similarly, if the refractive index from air to water, $_an_w$, is 4/3, then the refractive index from water to air, $_wn_a$, is 3/4.

The refractive index from water to glass, $_wn_g$, can be found from the velocity formula for refractive index. If v and c represent respectively the velocity of light in a medium and a vacuum (or air), then

$$_wn_g = \frac{v_w}{v_g} = \frac{v_w}{c} \times \frac{c}{v_g} = {}_wn_a \times {}_an_g = \frac{_an_g}{_an_w}$$

Now $_an_g = 1.5 = 3/2$ and $_an_w = 4/3$. So

$$_wn_g = \frac{3/2}{4/3} = \frac{9}{8} = 1.1 \text{ (approx)}$$

n SIN *i* = CONSTANT

Figure 6.2 shows the refraction of light through several media when the plane boundaries are *parallel*. From air to water, $\sin i_a / \sin i_w = n_w$ so

$$\sin i_a = n_w \sin i_w \qquad\qquad (1)$$

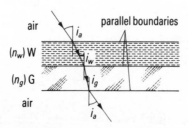

Fig 6.2 Refraction through several media

From glass to air, $\sin i_g / \sin i_a = 1/n_g$ or

$$\sin i_a = n_g \sin i_g \qquad (2)$$

From (1) and (2), we can write, using $n_a = 1$,

$$n_a \sin i_a = n_w \sin i_w = n_g \sin i_g$$

So when light passes through several media with parallel boundaries, or from one medium to another,

$$n \sin i = \text{constant}$$

TOTAL INTERNAL REFLECTION; CRITICAL ANGLE

When a ray of light is incident from air to glass, it is refracted towards the normal. But when ray 1 is incident from glass to air, it is refracted *away* from the normal (Fig 6.3(a)). This is always the case when light travels from one medium to a *denser* medium, for example from glass to water. Note also that a small percentage of the incident light is *reflected* from a glass–air boundary (Fig 6.3(a)). So the refracted light beam is not as intense as the incident beam in this case.

When the angle of incidence in the glass is increased from zero, the angle of refraction in the air increases. During this change, only a small percentage of the light is reflected at the boundary as we mentioned before. Ray 2 has an angle of refraction 90°. Its angle of incidence C is the *critical angle* for the air–glass boundary. So ray 3 is *totally reflected* as shown.

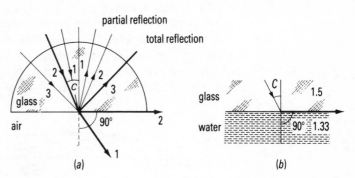

Fig 6.3 Total internal reflection; critical angle

Suppose n is the absolute refractive index of glass (air to glass). Since light here passes from glass to air, then

$$\frac{\sin C}{\sin 90°} = {}_g n_a = \frac{1}{n}$$

Now $\sin 90° = 1$. So

$$\sin C = \frac{1}{n}$$

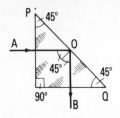

Fig 6.4 Total reflecting prism

Alternatively, using $n \sin i = $ constant for light travelling from glass to air,

$$n_g \sin C = n_a \sin 90°$$

But $n_a = 1$ and $\sin 90° = 1$. So $\sin C = 1/n_g$.

Using n for glass $= 1.5$, $\sin C = 1/1.5 = 2/3 = 0.6667$. So $C = 42°$ (approx). Therefore when light is incident at 45° *in glass* on a glass–air boundary, total internal reflection occurs, that is, *all* the light is reflected back into the glass. Fig 6.4 shows a *total reflecting prism* used in binoculars.

For a glass–water boundary, Fig 6.3(*b*), $\sin C = 1/_w n_g = 0.887$. So $C = 62.5°$.

DISPERSION BY PRISM

Light which contains only one wavelength is called *monochromatic light*. In practice, a source of light contains many wavelengths, which is called the *spectrum* of the source. The visible spectrum of sodium vapour consists of two yellow wavelengths very close to each other. The spectrum of white light or the visible spectrum from the sun consists of many wavelengths, ranging in colour from red (longest wavelength) through orange, yellow, green, blue, indigo to violet (shortest wavelength).

When a ray XO of white light is incident on a glass prism, the different colours or wavelengths travel at different speed through the glass (Fig 6.5). This separates or *disperses* the wavelengths so they travel in different directions after passing through the prism. The red has the fastest speed in glass and violet has the slowest speed. So the red emerges from the prism nearest to its original direction XO in air and violet emerges furthest away from the direction XO, as shown. In a vacuum or air, all wavelengths or colours travel with the same speed. The eye lens brings all the colours to the same place on the retina and the overlapping colours are seen as white.

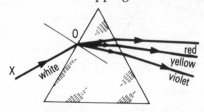

Fig 6.5 Dispersion of white light

Light, an electromagnetic wave (page 231) is dispersed by glass. Radio waves are electromagnetic waves of longer wavelength than light. Radio waves are dispersed when they travel through the ionosphere, a belt or layer of electrons and ions high above the Earth.

OPTICAL FIBRES

Optical fibres can transmit light from one end to the other with only a small loss of light. To transmit information or messages, British Telecom have replaced copper wires in telephone cables by bundles of optical fibres, which are fine strands of glass about one-eighth of a millimetre thick. Telephone signals are converted into light which travels along the fibre. The system is cheaper and more efficient than using copper wire.

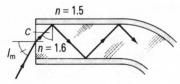

Fig 6.6 Optical fibre

Fig 6.6 shows the principle of the optical fibre. The core (inside) of the glass may have, for example, a refractive index of say 1.6 and is surrounded by a material called 'cladding' of lower refractive index, such as 1.5, to reduce loss of light. At a particular angle of incidence i_m, the ray outside is refracted into the fibre and strikes the fibre-cladding boundary at the critical angle C for these two media. The ray is then totally reflected as shown, and is further totally reflected along the fibre if the incident angle at the core-cladding boundary is C or greater than C. In this way the light is transmitted along the fibre after thousands of reflections at the boundary, from one end to the other. Using the values 1.6 and 1.5 for the refractive index of the fibre and cladding, the critical angle C is given by sin $C=1.5/1.6=0.9375$ and so $C=70°$ approximately.

In Fig 6.6, the angle of incidence i_m is the maximum incident angle in the air which produces internal total reflection inside the fibre. Any angle of incidence greater than i_m will result in an angle of incidence *less* than the critical angle C for the fibre-cladding boundary. Refraction now occurs and some light may then be lost. If i_m is 40°, for example, all the light incident on the end face of the fibre in a cone of semi-angle 40° will pass along the fibre. So i_m is a measure of the so-called 'numerical aperture' of the fibre.

MONOMODE AND MULTIMODE FIBRES AND LIGHT PATHS

Fibres may be *monomode* (light travelling along one path) or *multimode* (light travelling along different paths).

A monomode fibre, with an outside diameter of about 125μm (125×10^{-6}m), has a very narrow central core of about 5μm surrounded by the cladding of smaller refractive index (Fig 6.7(a)). In this case the different wavelengths which make up the pulse of light will all travel along the core.

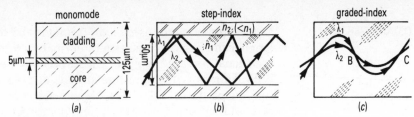

Fig 6.7　(a) Monomode
(b) step-index
(c) graded- index fibres

A multimode fibre may be one of two kinds. The *step-index* type has a relative thick central core of about $50\mu m$ diameter of constant refractive index n_1, which then changes abruptly to cladding of smaller refractive index n_2 (Fig 6.7(b)). The *graded-index* type has also a comparatively large core but the refractive index increases slowly or gradually until the boundary of the fibre is reached (Fig 6.7(c)).

Fig 6.7(b) shows the light paths for two different wavelengths λ_1, λ_2, in a light signal entering the step-index multimode fibre. As the paths have different lengths, the two wavelengths arrive at different times from the other end of the fibre. We met this dispersion effect with a prism on page 182. The signal received is then not clear.

With the graded-index fibre, however, the different wavelengths in the signal are focused at B and C and other points along the fibre, although their paths have different lengths. So the two wavelengths λ_1 and λ_2 arrive at the other end of the fibre with only a very slight time difference, estimated as 10ns (10×10^{-9}s) per kilometre length of fibre. So the graded-index fibre is better than the step-index type.

British Telecom, however, prefer the monomode fibre for transmitting light signals along the central core. Here the time difference for two wavelengths is only about 1ns, or less, per kilometre length of fibre.

The technical details of this system of telecommunications is beyond the scope of this book. Briefly, the audio signal is converted by a microphone into signals which modulate the light from a laser source. The signals travel along the fibre in the form of brief pulses of light which are in coded binary form. At the other end of the fibre cable the light is re-converted by a photo-diode and additional devices into the audio information which was sent in at the transmitting end.

LENSES

Lenses were first used over three hundred years ago. Magnifying glasses were used as simple microscopes about 1650 by Hooke and other pioneers for investigations in human biology. We shall discuss only thin lenses, ones whose thickness is negligible compared with their focal length.

CONVERGING LENSES

Converging lenses are thicker in the middle than at the edges. Parallel light beams converge to a focus after refraction through the lens, so the focus is a *real* one. The focal length f has a positive sign in using optical formulae.

Rays parallel to the principal axis of the lens are brought to a focus at F on the principal axis. This is the *principal focus* of the lens. But a parallel light beam inclined at an angle θ to the principal axis is brought to a 'secondary focus' F_1 which is directly below F as shown in Fig 6.8. An object a long way from the lens, such as the sun, has rays coming from the top T of the object and from the bottom B. If a ray diagram is drawn to show how the lens forms an image of the sun, for example, the rays from B can be drawn parallel to the principal axis to form an image at F as shown. But the rays from the top T must be drawn at an angle θ to the principal axis. The parallel beam is brought to a focus F_1, directly below F. The height h of the image is then represented by the length FF_1. Note that the useful ray through the middle (optical centre) C of a lens should always be drawn first in a ray diagram – it passes straight through the lens without changing direction.

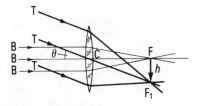

Fig 6.8 Converging lens image of distant object

The sun subtends an angle of about 0.01 radians at the lens. Suppose the lens has a focal length f of 20cm. Then the height h of the sun's image is given by

$$\frac{h}{f} = \tan \theta = \theta \text{ rad when angle } \theta \text{ is very small.}$$

So $h = f\theta = 20 \times 0.01 = 0.2 \text{cm}$

REAL AND VIRTUAL IMAGES; LENS FORMULAE

It is useful to memorize that a converging lens

(*a*)   forms a **real** and inverted image I when the object O is *farther* from the lens than its focus F (Fig 6.9(*a*));

(*b*)   forms a **virtual** upright and magnified image I when the object O is *nearer* to the lens than F (magnifying glass or simple microscope principle) (Fig 6.9(*b*)).

In the 'Real is Positive' sign convention, a real object or image distance is given a *positive* sign before its numerical value; a virtual

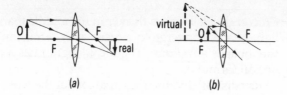

Fig 6.9 Converging lens
images

object and image distance is given a *negative* sign. So if the focal
length of a converging lens (real focus) is 10cm, then $f=+10$cm. The
lens formula in the above convention is

$$1/v+1/u=1/f,$$

where u is the object distance and v is the image distance. If, on
substituting for u and f in the formula, v is found to have a negative
value, then we know that the image is virtual.

The *linear* or *transverse magnification* m produced by a lens is the
ration h_1/h, where h_1 is the image height and h is the object height.
Simple ray diagrams show that, numerically,

$$m=\frac{\text{image distance } v}{\text{object distance } u}$$

If an object O_1 or O_2 is placed in front of the lens, rays from the
object diverge towards the lens (Fig 6.10(a)). This is always the case
for a real object, so the object distance u is positive. If the refracted
rays converge, the image I_1 is real; if they diverge the image I_2 is
virtual (Fig 6.10(a)).

If, however, **convergent** rays are incident on a lens, the point to
which they converge in the absence of the lens is a **virtual object** (Fig
6.10(b)). So the object distance u is negative is this case. In Fig 6.10(b)
the lens converges the incident rays to a real image. So the distance v
is positive in this case.

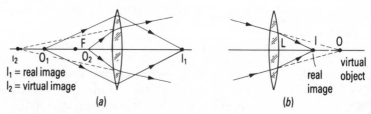

Fig 6.10 Real and virtual
image and object

In the examples which follow, take special care to note how + and
− signs are used when the numerical values of distances are known.

WORKED EXAMPLES ON LENSES

6.1 LENS MAGNIFICATION

An object viewed through a converging lens of focal length 5cm produces an upright magnified image 20cm from the lens. What is the object distance and magnification?

This is the case of the magnifying glass (Fig 6.9(b)). So the image is virtual and v=image distance=−20cm. Since the lens is converging, f=+5cm. Substituting in the lens formula, $1/v+1/u=1/f$, then

$$\frac{1}{-20}+\frac{1}{u}=\frac{1}{+5}$$

So $\frac{1}{-20}+\frac{1}{u}=\frac{1}{5}$, or $\frac{1}{u}=\frac{1}{5}+\frac{1}{20}=\frac{5}{20}$

Hence $u=\frac{20}{5}=4\text{cm}$

Since u is positive in value, the object is real and 4cm from the lens. As a check, the object is less than the focal length distance from the lens and so produces a virtual, upright and magnified image.

The linear or transverse magnification is given numerically by

$$m=\frac{v}{u}=\frac{20}{4}=5$$

6.2 VIRTUAL OBJECT AND LENS

In Fig 6.10(b) the beam incident on the converging lens L of focal length 10cm would pass through O in the absence of the lens, where OL=20cm. Find the image position I after refraction by the lens.

This is the case of a virtual object. So $u=-20$cm. From the lens equation $1/v+1/u=1/f$,

$$\frac{1}{v}+\frac{1}{-20}=\frac{1}{+10}$$

So $\frac{1}{v}-\frac{1}{20}=\frac{1}{10}$, and $\frac{1}{v}=\frac{1}{10}+\frac{1}{20}=\frac{3}{20}$

Hence $v=\frac{20}{3}=6.7\text{cm}$

Since v has a positive value, the image is real as shown in Fig 6.10(b).

OPTICAL INSTRUMENTS

ANGULAR MAGNIFICATION

A person with normal vision can see objects clearly a long distance away. His or her 'far point' is said to be infinity, in which case parallel rays entering the eye are brought to a focus on the retina. With normal vision, the nearest point which can be clearly seen is about 25cm from the eye. At the 'near point' an object produces a diverging beam of rays which are brought to a focus on the retina.

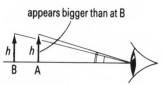

Fig 6.11 Visual angle and object size

The size of the object seen by the eye is proportional to the angle which the object subtends at the eye. A near object at A subtends a bigger angle than when it is some distance away at B, although the physical size h of the object is the same (Fig 6.11). So the near object appears bigger.

The angle subtended at the eye by an object is called the *visual angle*. The purpose of a telescope or a microscope is to increase the visual angle. In general, if α is the visual angle when the object is viewed directly by the eye and α' is the bigger visual angle when using a telescope or microscope, the *angular magnification M* of the instrument is defined as the ratio α'/α:

$$M = \frac{\alpha'}{\alpha}$$

A telescope '$\times 20$' has an angular magnification of 20. A microscope may have an angular magnification of the order of several hundreds. Angular magnification is also called *magnifying power*.

ASTRONOMICAL TELESCOPE; NORMAL ADJUSTMENT

The astronomical telescope has an *objective* converging lens of long focal length f_o and an *eyepiece* converging lens of short focal length f_e. As we shall see later, this arrangement produces a telescope of high angular magnification or magnifying power.

The objective collects the light from the distant object and forms an image I at its focus F_o (Fig 6.12). In the drawing, notice that only rays from the top point T of the distant object are shown at the objective. These are parallel rays. Further, we show them inclined at a small angle α to the principal axis. The rays from the bottom or foot of the

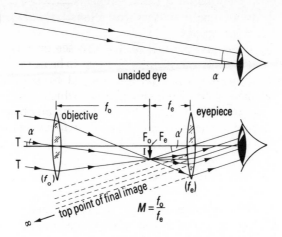

Fig 6.12 Telescope in normal adjustment – final image at infinity

distant object would be drawn parallel to the principal axis at the objective but we omit them for clarity. They form at F_o the foot of the image I. So in Fig 6.12 the angle α is the angle subtended by the distant object at the objective. The image I is inverted.

The eyepiece acts as a magnifying glass. It magnifies the image I and forms a final image which subtends a much greater visual angle α' at the eye than the distant object before the telescope is used. *When the final image is at infinity*, the telescope is said to be in 'normal adjustment'. In this case the rays refracted through the eyepiece are parallel. The image in a converging lens is at infinity only when the image is at a distance from the lens equal to its focal length. So in Fig 6.12 the image I is at a distance f_e from the eyepiece and the principal focus F_e of the eyepiece is here at the same place as F_o.

The length between the objective and eyepiece is therefore $(f_o + f_e)$, *the sum of their focal lengths,* when the telescope is in normal adjustment.

RAY DIAGRAM FOR TELESCOPE ACTION

As we said before, the purpose of the objective is to collect the light from the distant object and form an image I at F_o. In Fig 6.12, the rays which form the top of I go straight on to be refracted by the eyepiece. But before we can show the refracted rays we must first construct the final image formed by the eyepiece.

Fig 6.13 Eyepiece construction lines

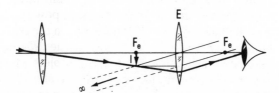

In the diagram, I is an 'object' for the eyepiece E (Fig 6.13). We draw its image for this single lens by taking two construction lines from the top of I:

(a)   a line parallel to the principal axis passes through the principal focus F_e on the other side of the lens,

(b)   a line through the centre of the lens passes straight through.

Now I is at the focus of the eyepiece. So the two refracted construction lines are *parallel*. An eye placed as shown sees the final image at infinity in the direction of the two refracted rays. So to complete the ray diagram we extend the rays from the objective to meet the eyepiece and *then draw rays to the eye which are parallel to the refracted construction lines*. These rays appear to come from the top point of the final image at infinity. In Fig 6.13 only the central ray is drawn entering the eye. Figure 6.12 shows three rays entering the eye but the construction lines should first be drawn as shown in Fig 6.13.

ANGULAR MAGNIFICATION IN NORMAL ADJUSTMENT

We can now find an expression for the angular magnification of the telescope in normal adjustment, that is, with the final image at infinity.

The length of a telescope is usually very small compared with the distance from it of the far object. So the angle α subtended by the object at the objective is practically equal to the small visual angle at the eye of the observer before the telescope is used. From Fig 6.12 we see that $\tan \alpha = h/f_o$ or, since α is very small, $\alpha = h/f_o$ in radians.

If the eye is close to the eyepiece, the angle α' subtended by the final image at the eye if h/f_e in radians. So

$$\text{angular magnification } M = \frac{\alpha'}{\alpha} = \frac{h/f_e}{h/f_o} = \frac{f_o}{f_e}$$

A high value of M is therefore obtained when f_o, the objective focal length, is long and f_e, the eyepiece focal length, is short. If $f_o = 100$cm and $f_e = 5$cm, then $M = 100/5 = 20$. The distance between the lenses $= f_o + f_e = 105$cm.

Examples 6.3 and 6.4 on pages 193 and 194 show how to calculate angular magnification of telescopes in numerical cases.

THE EYE-RING

All the light entering the telescope comes initially from the light passing through the objective. The rim of the lens is a circle. So all the light passing through the eyepiece forms a circular image of the objective lens (Fig 6.14). This ring is called the *eye-ring*. The objective lens, in fact, is an 'object' *whose 'image' in the eyepiece is the eye-ring*. The best position for the eye is the eye-ring. Here all rays entering the objective are seen and so the field of view is wide. If the eye were placed nearer to the eyepiece than the eye-ring, or farther away, fewer rays from the objective would then enter the eye and so the field of view would diminish. For this reason the end of the telescope tube near the eyepiece is lengthened so that the eye, pressing against the end, is practically at the eye-ring position.

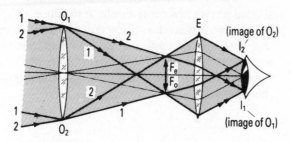

Fig 6.14 Telescope eye-ring

Example 6.3 on page 193 shows how to calculate the position of the eye-ring in a numerical case and its diameter.

MICROSCOPE; SIMPLE MICROSCOPE

Unlike the telescope, the microscope produces angular magnification of objects which are seen initially by the eye at its *near point*. The near point is a distance D from the eye which is 25cm for normal vision. So if the object has a length h, the initial visual angle $\alpha = h/D$ (Fig 6.15(a)).

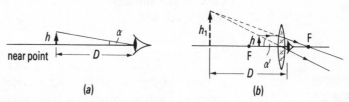

(a) (b)

Fig 6.15 Simple microscope

A magnifying glass is a simple microscope. It has a converging lens and as shown in Fig 6.15(b), it is normally moved until the magnified virtual image seen is a distance D from the eye. In this case the angular magnification M is given by

$$M = \frac{\alpha'}{\alpha} = \frac{h_1/D}{h/D} = \frac{h_1}{h}$$

Now $h_1/h = v/u = D/u =$ the linear magnification produced by the lens. From the lens equation, $v = -D$, where D is the numerical value of the near point from the eye, and $1/v + 1/u = 1/f$. So

$$\frac{1}{-D} + \frac{1}{u} = \frac{1}{f}$$

from which $u = fD/(f+D)$. So, numerically,

$$M = \frac{D}{u} = \frac{f+D}{f} = 1 + \frac{D}{f}$$

If $D = 25$cm, then $M = 1 + D/f = 1 + 25/f$. So a magnifying glass should have a low focal length to produce a high value of M.

An object placed at the focus of a lens forms an image at infinity (Fig 6.16). The image can be seen but not as clearly as when it is formed at the near point. From Fig 6.16, the visual angle α' is h/f when the final image is at infinity, and so the angular magnification M in this case is

$$M = \frac{\alpha'}{\alpha} = \frac{h/f}{h/D} = \frac{D}{f}$$

With $D = 25$cm and $f = 5$cm, then $M = 25/5 = 5$. When the final image is at the near point (Fig 6.15(b)), we have shown that $M = 1 + D/f$. So in the case $M = 1 + 25/5 = 6$, a greater angular magnification than when the final image is at infinity.

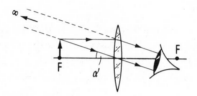

Fig 6.16 Simple microscope
– final image at infinity

COMPOUND MICROSCOPE

A much higher angular magnification is produced by using two lenses to form a compound microscope. Figure 6.17 shows the ray diagram when an object is placed at O slightly *further* from the objective than the focus F_o. The lens L_1 forms a real image at I_1 and this is viewed through the eyepiece L_2, which is adjusted until the big final image I_2 is at the distance D from the eye, 25cm for normal vision. So again the eyepiece acts as a magnifying glass for the real image formed by the objective. Fig 6.17 shows the construction lines for drawing the actual rays which strike the eyepiece and the increased visual angle α'.

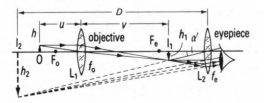

Fig 6.17 Compound microscope

The angular magnification $M = \alpha'/\alpha = h_2/D \div h/D = h_2/h = (h_2/h_1) \times (h_1/h) = m_e \times m_o$, where m_e and m_o are the linear magnifications produced by the eyepiece and objective. From the lens equation, we find $m_e = (1 + D/f_e)$ numerically and $m_o = (v/f_o - 1)$, where v is the distance of the image I_1 from the objective. So

$$M = \left(1 + \frac{D}{f_e}\right)\left(\frac{v}{f_o} - 1\right)$$

From this formula for M we see that high angular magnification is obtained when both the objective and eyepiece have *short* focal lengths.

WORKED EXAMPLES ON TELESCOPES

6.3 MAGNIFYING POWER, EYE-RING

An astronomical telescope has an objective of focal length 100cm and an eyepiece of focal length 5cm.

With the telescope in normal adjustment, calculate the angular magnification (magnifying power), the distance of the eye-ring from the eyepiece and the diameter of the eye-ring if the objective diameter is 50mm.

(a)   Angular magnification $M = \dfrac{f_o}{f_e} = \dfrac{100}{5} = 20$

(b)   The objective distance u from the eyepiece is $(f_o + f_e) = 105$cm. The eye-ring is the image of the objective in the eyepiece. So if v is the eye-ring distance from the eyepiece, then, from $1/v + 1/u = 1/f_e$,

$$\frac{1}{v} + \frac{1}{+105} = \frac{1}{+5}$$

Solving,

$\quad v = 105/20 = 5.25$cm from the eyepiece.

(c)   From $\quad M = \dfrac{\text{diameter of objective}}{\text{diameter of eye-ring } d_e}$

$$20 = \frac{50\text{mm}}{d_e}$$

So $\quad d_e = \dfrac{50\text{mm}}{20} = 2.5$mm

6.4 MAGNIFYING POWER AND MOON

An astronomical telescope has an objective of 100cm focal length and an eyepiece of 5cm focal length. If the image of the moon is seen 25cm from the eye placed close to the eyepiece, calculate the separation of the lenses and the angular magnification.

Assuming the moon subtends an angle of 0.01 rad at the objective, find the diameter of the moon's image formed by the objective.

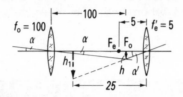

Fig 6.18 Telescopic calculation

Figure 6.18 shows the objective image of height h and the final eyepiece image of height h_1, formed 25cm from the eyepiece. Since this is a virtual image, $v = -25$cm for the eyepiece. The objective image distance from the eyepiece, u, is therefore given, from $1/v + 1/u = 1/f_e$, by

$$\frac{1}{-25} + \frac{1}{u} = \frac{1}{+5}$$

Solving, $u = 4\tfrac{1}{6}$cm

So distance between lenses $= 100 + 4\tfrac{1}{6} = 104.2$cm (approx)

The angular magnification $M = \dfrac{\alpha'}{\alpha} = \dfrac{h/u}{h/f_o} = \dfrac{f_o}{u}$

So $M = \dfrac{100}{4\tfrac{1}{6}} = 24$

Figure 6.18 shows the image formed by the objective. Since $\alpha = 0.01$ rad,

$$\frac{h}{100} = \tan \alpha = \alpha = 0.01$$

So $h = 100 \times 0.01 = 1$cm

1 REFRACTION AT PLANE SURFACES

(a)   Refractive index (absolute) $= \sin i / \sin r = c/v$, where c is the speed of light in a vacuum and v in the medium.

With several media 1, 2 and 3, then $_1n_3 = {_1}n_2 \times {_2}n_3$.

Apparent depth/true depth $= 1/n$ for normal incidence.

(b)   Total internal reflection only occurs when light travels from one medium to an optically *less* dense medium.

If C is the critical angle, $\sin C = 1/n$.

2 DISPERSION

Light of different wavelengths travel at different speeds through glass. So

(i) they are dispersed (separated) by a glass prism and

(ii) they reach the other end of an optical fibre at different times. White light is dispersed into different wavelengths or colours by a glass prism.

3 OPTICAL FIBRES

These fibres have a central glass *core* surrounded by a glass *cladding* of lower refractive index. Light pulses can be transmitted along the fibre by continual *total reflection* at the core-cladding boundary. Critical angle C given by $\sin C = n_2$ (cladding)$/n_1$ (core).

Fibres may be

(a)   *monomode* – light path only along the central core of small diameter or

(b)   *Multimode* – light paths of wavelengths different or

(c)   *graded-index* – light paths different but bunched at points along the fibre. In practice, monomode preferred as dispersion least.

4 LENSES

Lens formulae: $1/v + 1/u = 1/f$ (Real is Positive) and magnification $m = v/u$. f is $+$ ve for a converging lens and $-$ ve for a diverging lens.

Images: Converging lens – object farther than f, then image is real and inverted. Object nearer than f, then image upright, virtual, magnified (magnifying glass).

In lens problems, note that virtual objects have $-ve$ distances for u.

5 OPTICAL INSTRUMENTS

Telescope: Angular magnification $M = \alpha'/\alpha$, where α' is the angle subtended at the eye by the final image and α is the angle subtended at the unaided eye by the object.

Normal adjustment:
(a)   The final image is at infinity.
(b)   The distance between the two lenses $= f_o + f_e$.
(c) $M = f_o/f_e$, so long f_o and short f_e for high M.

Eye-ring = best position for eye (widest field of view) = image of objective in eyepiece. $M =$ objective diameter/eye-ring diameter. Also, high resolving power when objective diameter is large.

Final image at distance D – here the final image is formed at the least distance of distinct vision of the observer, unlike normal adjustment.

Microscope: Simple microscope – for image at the near point distance D from the observer, the object is placed nearer the lens than f. Then $M = (1 + D/f)$. For image at infinity – object placed at focus F. Then $M = D/f$.

Compound microscope – two converging lenses of *short* focal length produce high M.

DIAGRAM SUMMARY

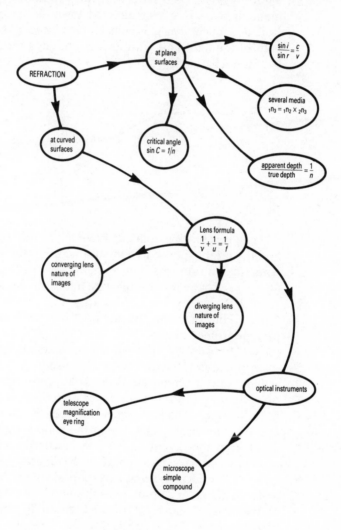

6 QUESTIONS

1   Figure 6.19 shows a narrow parallel horizontal beam of monochromatic light from a laser directed towards the point A on a vertical wall. A semicircular glass block G is placed symmetrically across the path of the light and with its straight edge vertical. The path of the light is unchanged.

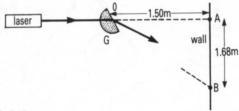

Fig 6.19

The glass block is rotated above the centre, O, of its straight edge and the bright spot where the beam strikes the wall moves from A to B and then disappears. OA = 1.50m, AB = 1.68m.

(a)   Account for the disappearance of the spot of light when it reaches B.

(b)   Find the refractive index of the material of the glass block G for light from the laser.

(c)   Explain whether AB would be longer or shorter if a block of glass of higher refractive index were used.   (L)

2   A lamp and a screen are 80cm apart and a converging lens placed midway between them produces a focused image on the screen. A thin diverging lens is placed 10cm from the lamp, between the lamp and the converging lens. When the lamp is moved back so that it is 30cm from the diverging lens, the focused image reappears on the screen. What is the focal length of the diverging lens?   (L)

3   (a)   A thin converging lens of focal length 0.050m is used to view a disc of diameter 2.0mm, with the eye close to the lens and the image at the least distance of distinct vision (0.25m). Draw a diagram showing how the image is formed. Use the diagram to find the magnifying power of the lens and calculate the diameter of the image.

(b)   By reference to your diagram, state how you would adjust the arrangement so that the image is at infinity. What is then the angle subtended by the image at the eye?   (JMB)

4 (*a*)   Draw a ray diagram showing how a converging lens can be used to form a virtual image of a real object.

(*b*)   An object of length 5.0mm is placed at right angles to the axis of a converging lens of focal length 50mm, and a virtual image is formed 250mm from the lens.

(i) Calculate the angle subtended at the centre of the lens by the object.

(ii) What is the angle subtended at the eye by the object when placed 250mm from an unaided eye?

(iii) Calculate the angular magnification produced by the lens

(*AEB* 1985)

5 (*a*)   Explain what is meant by the magnifying power of a telescope. Show that if f_o and f_e are the focal lengths of the objective and eyepiece respectively, the magnifying power is f_o/f_e when the telescope is in normal adjustment.

(*b*)   For an astronomical telescope, f_o and f_e are 0.90m and 0.10m respectively. When, in normal adjustment, it is used to view a full moon, the image subtends an angle of 0.051 radian at the eye lens. If the distance between the Moon and the Earth is 3.8×10^5km, calculate a value for the diameter of the Moon.   (*JMB*)

6   Calculate the position of the eye ring for an astronomical telescope consisting of two thin converging lenses, an objective of focal length 1.0m and an eyepiece of focal length 20mm, placed 1.02m apart. Explain the advantage of placing the eye at the eye ring position when using the telescope.   (*L*)

7 (*a*)   Draw ray diagrams to show what is meant by the *principal focus* and *focal length* of

(i) a converging (convex) lens and

(ii) a diverging (convex) mirror.

(*b*)   Explain the terms *magnification* and *angular magnification* as applied to optical instruments.

(*c*)   Figure 6.20 shows parts of an astronomical telescope. F_o is one of the principal foci of the objective lens. Parallel rays from the top of a distant object are incident on the objective, and a final image is formed a distance D from the eyepiece lens

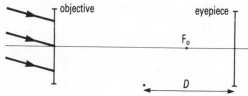

Fig 6.20

(i) Copy the diagram and complete it by showing the paths of the rays through the two lenses. Show clearly the real image of the top of the object formed by the objective lens. Show also the final image produced by the telescope.

(ii) If D is 250mm and the focal lengths of the objective and

eyepiece are 400mm and 50mm respectively, what is the required distance between the two lenses?

(iii) The telescope is used to observe the Moon which subtends an angle of 9.0×10^{-3} rad at the objective. Calculate the angle subtended by the final image at the eyepiece, and determine the angular magnification of the telescope assuming that the telescope is adjusted as in (ii).

(iv) On another occasion the telescope is adjusted to produce a real image of the Sun on the screen placed 500mm beyond the eyepiece. Given that the Sun and Moon subtend equal angles at the Earth, calculate the diameter of the image of the Sun on the screen.   (*AEB* 1986)

WAVES, LIGHT AND SOUND

CONTENTS

TYPES AND SPEEDS OF WAVES

WAVELENGTH, SPEED AND FREQUENCY

The **wavelength**, λ, of a wave is the distance between adjacent crests (or adjacent troughs) in the wave. The **frequency** f, of the wave is the number of wave crests which pass a given point in one second. If, for example, 10 waves pass a point in a second then the frequency is 10 cycles per second or 10 hertz (Hz). The **velocity**, v, of a wave is the distance any wave crest (or trough) travels in one second. If a source produces f waves each second, each of wavelength λ, then the distance the first wave crest moves in one second must be $f\lambda$. Hence

$$v = f\lambda$$

This relation applies to all types of waves.

The **amplitude**, a, of a wave is the maximum displacement from the mean position. As in the case of periodic motion, page 55, the time period, T, of vibration of any point on a wave is related to frequency f by

$$f = \frac{1}{T}$$

TYPES OF WAVES

1 PROGRESSIVE AND STATIONARY WAVES

A **progressive** wave is one whose profile appears to travel along the the medium. Figure 7.1 shows graphs of the displacement in a progressive wave at five different instants, each 1/8th of a time period apart. It can be seen that any two points such as X and Y vibrate with *equal amplitude*, but there is a *phase difference* between the vibration of X and Y. Y reaches zero displacement 1/8th of a cycle after X and thus Y

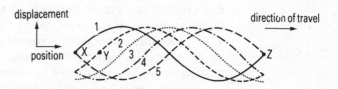

Fig 7.1 Progressive wave

lags in phase behind X by 1/8th of a cycle (or 45° or π/4 radians). In general the phase difference between any particular point and X increases the further along the wave the point is situated. At Z, one wavelength from X, the phase difference is 360° (or 2π radians) and so Z and X vibrate together (in phase).

In a **stationary** or **standing** wave the wave profile does not appear to move, but changes with time as illustrated by the graphs of Fig 7.2. Each graph shows the wave 1/8th of a cycle after the previous graph. In this case it can be seen that X and Y vibrate with *different amplitudes*. As shown, Y vibrates with a larger amplitude than X. However, X and Y vibrate together. So they oscillate *in phase* with one another, unlike the case of the progressive wave in 7.1.

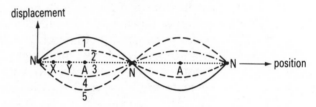

Fig 7.2 Stationary wave

Some points, marked N in Fig 7.2, are permanently at rest and these points are called **nodes**. Midway between nodes there are points, A, which vibrate with maximum amplitude. These points are called **antinodes**. From Fig 7.2, the distance between successive nodes (or successive antinodes) can be seen to be λ/2.

Progressive waves are produced when a source emits waves into an unbounded (unlimited) medium, or one where there are no reflections from any barriers (or walls) surrounding the source. Where waves are produced in a bounded medium, for example in an organ pipe or violin string, then standing waves are produced. *The standing waves are caused by the combined effect of two progressive waves travelling in opposite directions.* In an organ pipe, for example, a progressive wave travelling up the tube is reflected at the open end of the tube and so another progressive wave travels down the tube in the opposite direction. These combine to form the standing wave, discussed on p225.

2 TRANSVERSE AND LONGITUDINAL WAVES

A **transverse** wave is one in which the vibrations are perpendicular to the direction of travel of the wave. Water waves and electromagnetic waves such as light waves are examples of transverse waves.

In a **longitudinal** wave the vibrations are in the same direction as the wave travels. Sound waves are an example of longitudinal waves.

Progressive waves may be transverse or longitudinal. Stationary waves may also be transverse or longitudinal.

VELOCITY OF WAVES

1 TRANSVERSE WAVES ON A STRING

If T is the tension in the string in N, and m is the mass per unit length in kg m^{-1}, then the speed, in m s^{-1}, of the transverse wave is given by

$$v = \sqrt{\frac{T}{m}}$$

2 LONGITUDINAL WAVES IN A SOLID

If E is the Young modulus of the solid material in N m^{-2} and ρ its density in kg m^{-3}, then the speed v in m s^{-1} is given by

$$v = \sqrt{\frac{E}{\rho}}$$

3 ELECTROMAGNETIC WAVES IN A VACUUM

All types of electromagnetic waves (p230) move with the same speed, c, in a vacuum. It can be shown that $c = \sqrt{1/\varepsilon_0 \mu_0}$ where ε_0 is the permittivity of a vacuum (p79) and μ_0 is its permeability (p153). Substituting $\varepsilon_0 = 8.85 \times 10^{-12}$ F m^{-1} and $\mu_0 = 4\pi \times 10^{-7}$ H m^{-1} gives $c = 3 \times 10^8$ m s^{-1} approximately.

4 SOUND WAVES IN A GAS

Here the velocity v, in m s^{-1}, is given by

$$v = \sqrt{\frac{\gamma p}{\rho}}$$

where γ is the ratio of the molar heat capacities of the gas, p is its pressure in N m^{-2} and ρ its density in kg m^{-3}.

The speed of sound varies with temperature and humidity, but not with pressure, as we now show. Since $pV = RT$ for 1 mole of gas and $\rho\frac{M}{V}$, where M is the mass of 1 mole then.

$$v = \sqrt{\frac{\gamma p}{\rho}} = \sqrt{\frac{\gamma p}{M/V}} = \sqrt{\frac{\gamma pV}{M}} = \sqrt{\frac{\gamma RT}{M}}$$

This relationship for v does not involve pressure and so v is *independent* of pressure. Also, we note that the velocity $v \propto \sqrt{T}$ where T is the absolute temperature.

The presence of water vapour in the air reduces its average density. Thus from $v = \sqrt{\gamma p/\rho}$ the speed of sound in moist air is greater than in dry air at the same pressure.

MEASUREMENT OF THE SPEED OF SOUND

Figure 7.3 shows a laboratory method for determining the speed of sound in air. A loudspeaker produces sound waves of frequency of the order of a few kilohertz which are reflected by the hardboard sheet B. Thus standing waves are produced along the line between the loudspeaker and B. A microphone M is moved along this line and the output from M is displayed on an oscilloscope. The average distance d between one antinode and the next is found – each anti-node is located by finding the position of M which gives the greatest amplitude of trace on the oscilloscope. Since $\lambda = 2 \times$ distance between antinodes, then $\lambda = 2d$. The frequency of the signal is found from the scale on the audio oscillator and v is calculated from $v = f\lambda$.

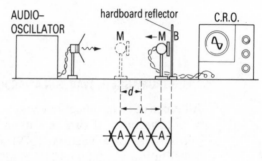

Fig 7.3 Velocity of sound

PROPERTIES OF WAVES

REFLECTION

Waves are reflected when they strike a barrier which they cannot penetrate. A plane wave is reflected from a plane surface according to the law of reflection $i = r$ and the stages in the wave reflection are illustrated in Fig 7.4. Figure 7.5 shows how the law of reflection can be demonstrated for microwaves (3cm wavelength electromagnetic waves).

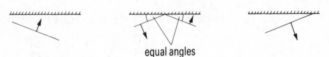

equal angles

Fig 7.4 Reflection of waves

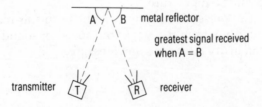

A   B   metal reflector

greatest signal received
when A = B

Fig 7.5 Reflection of microwaves

transmitter   receiver

The reflection of light waves was considered by Huygens who used the idea of *secondary wavelets* to explain the propagation of a wave. Huygens proposed that each point on a wavefront becomes a new source of circular waves (secondary wavelets). A new wavefront can be drawn at a later time by finding the tangent to all the secondary wavelets at that time. Figure 7.6 shows how this idea can explain the law of reflection, that is, the angle of incidence i equals angle of reflection r. When the plane wavefront reaches AC and begins to reflect, A can be considered to emit secondary wavelets. When C has reached B, the tangent to all the secondary wavelets emitted from points between A and B is the line BD, as shown. The secondary wavelets from A travel to D in the same time as the wavefront takes to travel from C to B. Hence CB=AD. It can now easily be shown that triangles ACB and ADB are congruent, and hence $i=r$.

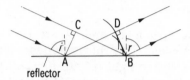

reflector

Fig 7.6 Law of reflection

REFRACTION

Refraction of waves occurs when waves travel into a medium where their *speed* changes. Figure 7.7 shows how refraction can be demonstrated for microwaves. The speed of microwaves changes as they pass into the paraffin and this causes their direction of travel to change. The maximum signal is detected by the receiver in the position shown.

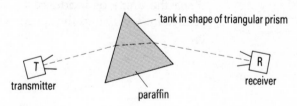

tank in shape of triangular prism

transmitter

receiver

paraffin

Fig 7.7 Refraction of microwaves

The principle of secondary wavelets can be used to explain the law of refraction. Figure 7.8 shows a plane wave AC approaching a boundary XY between two media. BD is the tangent to all the secondary wavelets emitted by A and other points along AC, at the instant when C reaches B. Now the time taken for the wavelet to travel from A to D is equal to the time taken for the wavefront to travel from C to B. Hence, using time = distance/speed

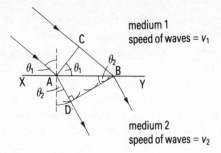

medium 1
speed of waves = v_1

medium 2
speed of waves = v_2

Fig 7.8 Law of refraction

$$\frac{CB}{v_1}=\frac{AD}{v_2} \quad \text{or} \quad \frac{CB}{AD}=\frac{v_1}{v_2} \quad \text{or} \quad \frac{AB \sin \theta_1}{AB \sin \theta_2}=\frac{v_1}{v_2}$$

So $\quad \dfrac{\sin \theta_1}{\sin \theta_2}=\dfrac{v_1}{v_2}$

Since the speeds v_1 and v_2 are constant it follows that $\sin \theta_1/\sin\theta_2 = $ constant. This rule is true for all types of wave and is known as Snell's law in light (p179). If $v_1>v_2$ then $\theta_1>\theta_2$, so the rays are refracted towards the normal. For light waves, v_1/v_2 is the refractive index $_1n_2$ from medium 1 to medium 2. So for light

$$\frac{\sin \theta_1}{\sin \theta_2}= {_1n_2}$$

Before the wave theory of light was proposed, light was believed to be a stream of particles or corpuscles. To explain refraction on the corpuscular theory Newton needed to assume that the speed of light in a medium, such as water, was greater than the speed in air. Measurements made by Foucault showed that the speed of light in water was *less* than that in air. This contradicts the corpuscular theory of Newton but agrees with the wave theory. Here $\sin \theta_1/\sin \theta_2 = v_1/v_2$. Since the angle of incidence θ_1 in air is greater than the angle of refraction θ_2 in water, the speed v_1 in air is greater than that in water, v_2.

DIFFRACTION

Diffraction or spreading of waves occurs when waves pass through apertures (openings or gaps), or around obstacles. Diffraction is appreciable when the wavelength is long compared with the width of the aperture or size of the obstacle. In the case of light, where the wavelength is very short (about 5×10^{-7}m), negligible diffraction occurs with ordinary sized apertures. To observe diffraction with light, very tiny apertures are needed as we shall see later. Sound waves will diffract around doors because of their much greater wavelength (about 1m). An experiment to illustrate diffraction with microwaves is shown in Fig 7.9. Here the width of the aperture (gap) between two metal screens is of the same order as the wavelength (3cm) of the microwaves from T.

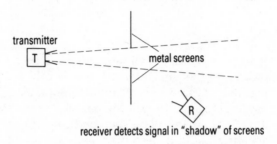

Fig 7.9 Diffraction of microwaves

INTERFERENCE

Interference of waves can occur when two or more waves of equal frequency or wavelength overlap. To produce observable interference effects, the overlapping waves must maintain a *fixed phase relationship* with one another. The waves are then said to be **coherent**.

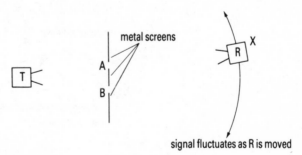

Fig 7.10 Interference of microwaves

Interference of light will be dealt with later in more detail. A double-slit experiment to show interference with microwaves is shown in Fig 7.10. When R is at a position such as X where a *maximum* signal is received by R, then wave crests from A must arrive at X at the same time as wave crests from B. In this case we say that **constructive interference** occurs and the path difference, XB−XA, must be a whole number m of wavelengths, that is

$$XB-XA=m\lambda.$$

If R is at a position, Y, where *minimum* signal is received, then wave crests from A must arrive at Y at the same time as *troughs* from B (and vice versa). In this case we say **destructive interference** occurs. The path difference YB−YA must be $\lambda/2$ or an odd number of half-wavelengths, so generally

$$YB-YA=(m+\tfrac{1}{2})\lambda$$

where m is a whole number.

POLARIZATION

A wave in which the vibrations are confined to a single plane perpendicular to the direction of travel is called a **plane-polarized** wave. Figure 7.11 shows waves on a sting passing through a slit S. Before passing through the slit the particles of the string are vibrating in all directions in the plane perpendicular to the direction of travel as shown at A, and here the waves are said to be unpolarized. After passing through the slit the direction of vibration is vertical and the wave W is confined to a single vertical plane. The wave is now plane-polarized.

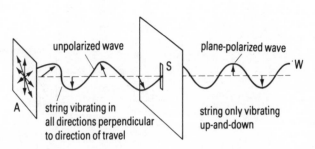

Fig 7.11 Polarization

In a longitudinal wave, the vibrations are in the same direction as the direction of travel. A longitudinal wave cannot show polarization. Thus *if a wave demonstrates polarization effects it must be transverse.* Figure 7.12 shows an experiment which demonstrates polarization with microwaves and so confirms that electromagnetic waves are transverse in nature (see p230).

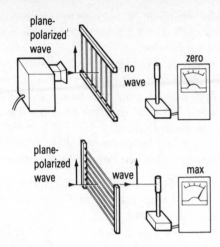

Fig 7.12 Polarization of
microwaves

INTERFERENCE WITH LIGHT WAVES

YOUNG'S FRINGES

An experiment for demonstrating Young's fringes and for measuring the wavelength of light is shown in Fig 7.13. S_1 and S_2 are two narrow parallel slits whose centres are less than one millimetre apart. They are arranged parallel to the single slit S. Light from a sodium lamp passes through S, diffracts into a narrow beam and is incident on S_1 and S_2. These two slits act as *coherent* sources of waves, that is, they have a constant phase difference, since the light spreading from them comes from the same source S.

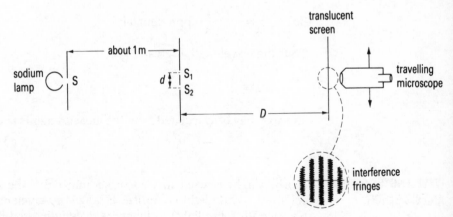

Fig 7.13 Young's fringes

In the region beyond the double slit the light waves from the two sources S_1 and S_2 overlap due to diffraction from each slit. Interference is therefore produced and bright and dark interference fringes can be seen on a translucent screen. In order to measure the wavelength of the light emitted by the sodium lamp, three measurements are needed:

1   The distance x between the centres of adjacent bright fringes. This can be measured using the travelling microscope.

2   The distance, D, from slits to screen, measured with a metre rule. D will need to be at least 1m to make the distance x large enough to measure accurately.

3   The distance d between the slits. This can also be measured with a travelling microscope before the experiment starts.

Theory. In Fig 7.14, X is the central bright fringe and Y is the first bright fringe from the centre. From p210, $S_2Y - S_1Y = S_2M = \lambda =$ the wavelength of light used, where $S_1Y = YM$. Now provided OX is much greater than S_1S_2 and YX, S_1M is very nearly perpendicular to S_2Y. Also, angle S_2S_1M or θ is very nearly equal to angle XOY as shown. Hence triangle S_1S_2M and OYX are almost similar. Thus

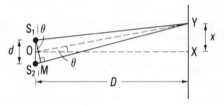

Fig 7.14 Theory of Young's experiment

$$\frac{S_1S_2}{S_2M} = \frac{OY}{YX}$$

So $\dfrac{d}{\lambda} = \dfrac{OY}{x} = \dfrac{D}{x}$ approximately

Thus the wavelength is given by

$$\lambda = \frac{dx}{D}$$

and so λ may be calculated from the measurements taken.

WAVELENGTHS OF RED AND BLUE LIGHT

If white light is used instead of sodium light, the bands seen are coloured. White light comprises a range of wavelengths from short wavelength (blue light) to longer wavelengths (red light). From the last equation it can be seen that the fringe separation, x, is proportrional to the wavelength. Experiment shows that the red components of white light produce wider spaced fringes than the blue components. So red light has a longer wavelength than blue light. Coloured fringes are seen with white light but sodium lamps emit only yellow light. Sources which emit only one wavelength are described as **monochromatic**.

It should be noted that the source slit S in Young's experiment must be narrow and a relatively long way from the two slits S_1 and S_2. If either of these conditions is not satisfied, it is no longer possible to regard the light from S_1 and S_2 as having a constant phase difference and so S_1 and S_2 will not act as coherent sources. The fringes will then become less clear or may even not appear on the screen.

INTERFERENCE IN AN AIR WEDGE

Fringes in a narrow air wedge, formed by using a thin separator such as paper, can be demonstrated by the apparatus shown in Fig 7.15. Light from a sodium lamp is partially reflected down on to the air wedge by the microscope slide A. Light reflected from the air wedge is then observed in the travelling microscope after transmission through A.

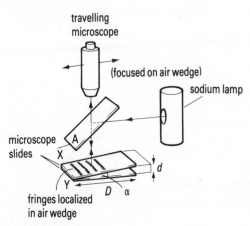

Fig 7.15 Wedge fringes

Theory. In the air wedge, two coherent light waves are derived from the incident light. Light reflected at S from the lower face of the top slide forms one wave, while light transmitted at S and reflected from the upper face of the lower slide forms the other wave. The extra distance travelled by the second beam is $2t$ where t is the thickness of the air wedge at S, Fig 7.16, but an additional $\lambda/2$ optical path difference is introduced because light suffers a phase change by reflection from an optically denser medium (glass) at T. So the net optical difference is $2t + \lambda/2$. So for a bright band or constructive interference,

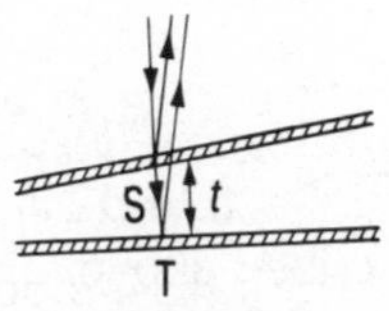

Fig 7.16 Theory of wedge fringes

$$2t + \frac{\lambda}{2} = m\lambda \text{ or } 2t = (m - \tfrac{1}{2})\lambda$$

Similarly, for a dark band or destructive interference,

$$2t + \frac{\lambda}{2} = (m + \tfrac{1}{2})\lambda \text{ or } 2t = m\lambda$$

The bright bands form lines parallel to the edge XY of the slides since t is constant along these lines, see Fig 7.15. Also, at XY there is a *dark* band, although the geometrical path difference between the beams is zero. This is because of the phase change of π radians (180°) on reflection at the lower microscope slide.

ANGLE OF AIR WEDGE AND THICKNESS OF FOIL

The angle, α, of the air wedge can be found as follows. The travelling microscope is used to measure the distance x between the centres of consecutive bright bands. (Measure across 10, say, and divide the total distance by 10). From one bright band to the next path difference must change by λ. So the air thickness here must change by $\lambda/2$, as the second light beam travels down *and* up through the wedge. Hence, from Fig 7.17(a),

$$\tan \alpha = \frac{\lambda/2}{x} \text{ or, since } \alpha \text{ is small, } \alpha = \frac{\lambda}{2x}$$

Knowing λ and x, α can be calculated.

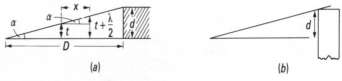

(a)   (b)

Fig 7.17 Angle of air wedge.
Thickness of foil

The thickness d of the separator or foil can now be found. D is measured from the separator to where the slides meet (Fig 7.17(a)). A travelling microscope can be used if high accuracy is required. d is then calculated from

$$\frac{d}{D} = \tan \alpha = \frac{\lambda}{2x}$$

so $d = \lambda D/2x$

If the top slide is supported on a rod as shown in Fig 7.17(b), any expansion of the rod (due to heating, say) can be found. d is measured both before and after expansion, as described above, and the change in the value of d gives the expansion.

CONTOUR FRINGES

In the last section the bright fringes followed lines of constant optical thickness (t) and so are an example of **contour fringes**. Two microscope slides pressed together will have a small air gap between them. When the slides are used to view sodium light by reflection, contour fringes can be seen. Here the pressing together distorts the slides and the bands are irregular, rather like the contours of a map. Observation of the fringes between two optical surfaces in contact can therefore be used to test, to a high degree of precision, whether the surfaces are perfectly plane.

NEWTON'S RINGS

An example of contour fringes is that of Newton's rings which are formed when a lens is placed on a flat piece of glass and illuminated as in Fig 7.18. The interference fringes form concentric rings following the lines of equal thickness of the air gap.

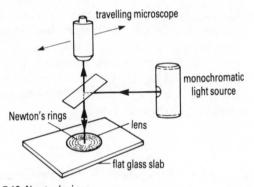

Fig 7.18 Newton's rings

As with the case of wedge fringes, the condition for a bright ring is

$$2t = (m + \tfrac{1}{2})\lambda$$

For a dark ring

$$2t = m\lambda$$

Again, as in the case of wedge fringes, the phase change on reflection produces a dark spot at the centre where the geometrical path difference is zero. Fig 7.19 shows the geometry of the situation, where a is the radius of curvature of the lower face of the lens.

Consider a ring or radius r, where the thickness of the air gap is t. By the intersecting chord theorem

$$r^2 = (2a - t)t = 2at - t^2$$
$$= 2at \text{ approximately since } t \text{ is very much smaller than } r.$$

This gives

$$2t = r^2 a$$

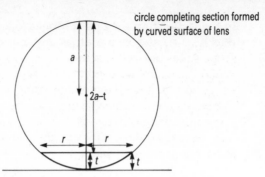

Fig 7.19 Geometry of
Newton's rings

The condition for a bright fringe becomes

$$r^2 a = (m + \tfrac{1}{2})\lambda \qquad\qquad (1)$$

This can be used as a basis for measuring the wavelength of a monochromatic light source provided that the radius of curvature of the lower face of the lens, a, is known. If the radius of the 21st bright ring is measured using the travelling microscope then r and a will be known. m will be 20 bearing in mind the fact that the first bright ring corresponds to $m = 0$. Substituting into equation (1) will enable the wavelength to be calculated.

COLOURS IN THIN FILMS

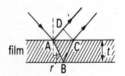

Fig 7.20 Thin film
interference

Figure 7.20 shows light incident on a thin film of some transparent material. Light reflected from A and that passing through the material and reflected from B, act as coherent sources. Interference can occur when these beams are combined, for example by the eye. The thickness of the film must not be too great otherwise DC will be too great for both beams to enter the eye.

The optical path difference between the two beams is $(AB + BC)n - AD$, where n is the refractive index of the material. n must be included because light travels slower (by a factor of $1/n$) in the material than in air. So if a given distance in air occupied m wavelengths, the same distance in the material would occupy nm wavelengths. It can be shown that $(AB + BC)n - AD = 2tn \cos r$, where r is the angle of refraction and t the thickness of the film. Again, however, there is a phase change which occurs at the reflection from A. So there will be constructive interference if

$$2nt \cos r + \frac{\lambda}{2} = m\lambda \quad \text{or } 2nt \cos r = (m - \tfrac{1}{2})\lambda$$

There will be destructive interference if

$$2nt \cos r + \frac{\lambda}{2} = (m + \tfrac{1}{2})\lambda \quad \text{or } 2nt \cos r = m\lambda$$

The thin film will therefore produce contour fringes if viewed from a fixed direction so that r is constant.

Colours are seen in thin films when they are illuminated with white light, and are caused by interference. There will be destructive interference for the colour whose wavelength is given by $2nt \cos r = m\lambda$. The white light will be deficient in that colour and the complementary colour can be seen. The colour can change at different places in the film because

(a) t may change, that is, the film is of varying thickness and
(b) r may change as looking at different points in a large film would mean looking at different angles.

DIFFRACTION OF LIGHT

**DIFFRACTION BY A
SINGLE SLIT**

We have already seen that it is necessary for light to be diffracted to produce the interference effects seen in Young's double slit experiment. If the light from each of the two slits did not diffract, then the beams would not overlap and no interference could be produced.

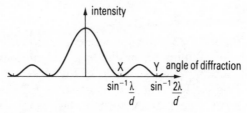

Fig 7.21 Intensity variation
from a narrow slit

If the light from a narrow source illuminates a *single* slit of width d, the intensity pattern seen on a distant screen is as shown by Fig 7.21. The intensity pattern first falls to zero at X where the angle of diffraction is $\sin^{-1}(\lambda/d)$ (approx λ/d in radians). At Y the angle of diffraction is $\sin^{-1}(2\lambda/d)$ (approx $2\lambda/d$ in radians). It can thus be seen that the *narrower* the slit, the wider is the diffraction pattern on the screen. Thus in the double slit experiment it is necessary to have very narrow slits, so that the light causing interference spreads over a fairly wide angle and a large number of interference fringes can then be seen.

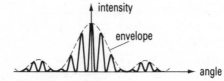

Fig 7.22 Intensity variation
from a double slit

The intensity of the fringes varies with angle as shown in Fig 7.22. The diffraction pattern for *each* single slit (Fig 7.19) forms the envelope to the graph in Fig 7.22, as shown.

RESOLVING POWER

When light enters the aperture or objective lens of a telescope it diffracts as it passes through. The diameter d, of the objective lens is quite large, so the amount of diffraction is small, as we have just seen. The image, however, is not perfectly sharp. The diffraction limits the amount of detail which can be seen.

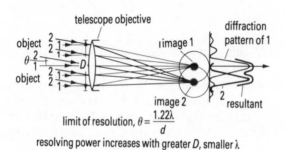

limit of resolution, $\theta = \dfrac{1.22\lambda}{d}$

resolving power increases with greater D, smaller λ

Fig 7.23 Resolving power

    Figure 7.23 shows a telescope receiving light from two objects which subtend an angle θ at the objective. The two images are diffraction patterns formed by the objective. For a circular aperture, it can be shown that if θ is less that 1.22 λ/d, then the two patterns overlap to such an extent that they cannot be distinguished as separate objects. 1.22 λ/d is called the **resolving power** of the optical instrument. Calculation shows that the human eye ($d\approx0.5$cm) can distinguish objects which subtend an angle greater than 10^{-4} radians. A large telescope of diameter 100cm can distinguish objects provided they subtend angles greater than 6×10^{-7} radian. The larger diameter telescope can thus 'resolve' far greater detail than the eye. From $\theta=1.22\lambda/d$, we can see that the telescopes with large diameters have high resolution, that is, they can distinguish between objects closer together. This is one reason why reflector, refractor and radio telescopes have been built with objectives of large diameter.

DIFFRACTION GRATING

A diffraction grating consists of very narrow apertures produced by ruling lines on a transparent surface such as glass. Typically there may be 1000 lines per mm. In practice diffraction gratings may be reproduced photographically.

    Theory. Figure 7.24 shows part of a diffraction grating where the distance between the centres of adjacent slits is d. Each slit is very narrow so light is diffracted from each slit into a widely divergent beam. The diagram shows only the light rays which are diffracted at a

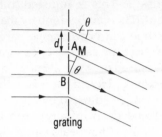

Fig 7.24 Diffraction grating

particular angle θ to the normal of the grating. The waves in this direction will interfere constructively when focused at a point if the path difference, AM, between light from adjacent slits A, B is a whole number of wavelengths. Hence the light diffracted by θ will interfere constructively if

$$AM = m\lambda$$
or $\quad d \sin \theta = m\lambda$

m is called the order of the diffraction image. When $m = 1$ the first order image is formed, $m = 2$ gives the second order.

DIFFRACTION GRATING AND NON-MONOCHROMATIC LIGHT

When light having a range of wavelengths strikes a diffraction grating, the component wavelengths interfere constructively at different angles. Since $d \sin \theta = m\lambda$, in any particular order m the longer wavelengths will interfere constructively at larger angles θ. Thus the light is split up into a *spectrum* in each order m. White light incident on a grating will thus give several spectra, each corresponding to a different value of m. This is unlike the case of a prism where only one spectrum is produced. The colours in the various order spectra may overlap, as illustrated by Example 7.6.

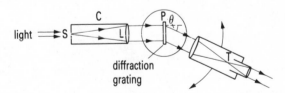

Fig 7.25 Spectrometer

THE SPECTROMETER

A diffraction grating can be used in conjuction with a spectrometer to measure wavelengths to a high degree of precision. Figure 7.25 shows the three essential features of a spectrometer.

1   The *telescope* T. This is adjusted to be focused on infinity before the experiment. This can be done by making distant objects appear clear. The telescope can now receive parallel beams of light from the grating. The telescope can rotate as shown by the arrows and the angle θ can be measured on a vernier scale.

2   The *collimator* C. Here a slit S is at the principal focus of a lens L. The collimator can be adjusted by observing the slit through the telescope (already focused on infinity), and the distance between S and L is varied until the slit appears sharp. The collimator will now produce a parallel beam of light which falls on the grating.

3   The *table* P. This supports the grating and it is equipped with adjusting screws to ensure that the lines in the diffraction are vertical.

     The wavelength is measured by finding the angle between the first diffracted images on either side of the central image. θ is obtained by halving thi angle. λ is then given by $\lambda = d \sin \theta$. If the second order image was used, then $\lambda = \tfrac{1}{2}d \sin \theta$.

POLARIZATION OF LIGHT

POLARIZATION BY POLAROID

A polaroid sheet has the property of producing plane-polarized light from unpolarized light. Light is an electromagnetic wave consisting of electric and magnetic fields, page 230. Polaroid will only allow through those components of the electric field which are in a particular direction. The electric field components in a perpendicular direction are absorbed.

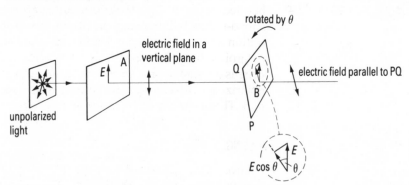

Fig 7.26 Polarization by polaroids

     Figure 7.26 shows the action of two polaroids A and B on an unpolarized incident light beam. The unpolarized beam enters A which only allows through components of electric field in a vertical direction giving a resultant amplitude E as shown. Polaroid B has been rotated through θ, and only allows through components parallel to the edge PQ. The emergent beam therefore has an electric field component of amplitude $E \cos \theta$. As θ is increased the amplitude of the emergent field decreases until no light is allowed through when $\theta = 90°$ and cos $\theta = 0$. Polaroids A and B are then said to be 'crossed'.

USES OF POLAROIDS

1 CONTROL OF INTENSITY

Polaroids can be employed to vary the intensity, I, of a light beam in a controlled manner. The intensity of any wave is proportional to the square of its amplitude. Thus the intensity emerging from polaroid B in Fig 7.26 is proportional to $(E \cos \theta)^2$, that is, $I \propto \cos^2\theta$. Rotating polaroid B by a measured angle θ therefore reduces the intensity by a factor $\cos^2\theta$, which can be calculated.

2 STRAINS IN STRUCTURES

Certain materials, such as perspex, have the property of rotating the plane of polarization of light passing through them in regions where they are stressed. If a model of a structure is made in perspex and placed between crossed (p.221) polaroids, then the stress patterns can be observed. This is often useful in helping an engineer in the design of structures.

3 CONCENTRATION OF SUGAR SOLUTIONS

Some sugar solutions have the property of rotating the plane of polarization of light passing through them. Some sugars give a right-handed rotation (dextro-rotatory), others give a left-handed (laevo-rotatory). The angle of rotation is proportional to the concentration of the solution. If a solution is placed between two crossed polaroids, the rotation, θ, of one polaroid needed to re-extinguish the light is the angle through which the sugar has rotated the plane of polarization. θ can be measured and is proportional to the concentration. This method of testing sugar solutions is called *saccharimetry*.

4 SUNGLASSES

Polaroid sunglasses can reduce the glare of the sun reflected from a surface. This is because light can be polarized by reflection, as we now see.

OTHER METHODS OF POLARIZING LIGHT

1 REFLECTION

Polarized light is obtained when ordinary light is reflected from a non-conductive surface such as unsilvered glass or the surface of water. Brewster showed that when the angle of incidence is given by

$\tan i = n$, where n is the refractive index, the reflected beam is completely plane polarized. This angle i $(= \tan^{-1} n)$ is called the **Brewster angle**. For water, where $n = 1.33$, the Breewster angle is 53°. If polaroid sunglasses are used to look at light reflected from the surface of water at this angle, then no reflected light will be transmitted by the polaroid. At other angles there will be partial polarization – the glare seen through sunglasses will then be reduced but not eliminated.

2 SCATTERING

Light scattered by small particles at 90° to the incident direction is plane polarized. This accounts for the fact that light from the sky is partially plane polarized.

3 DOUBLE-REFRACTION

Crystals of calcite produce two refracted beams when light passes into them. Each beam is plane polarized, the plane of one beam being perpendicular to that of the other. This phenomenon is called *double refraction*.

SOUND WAVES

Sound waves can only travel through a material medium. Sound will not pass through a vacuum. In air, or other medium, the particles vibrate longitudinally as shown in Fig 7.27(a). This shows the position of the particles in the medium at the instant when the graph of displacement against time is that in Fig 7.27(b). Displacements to the right are plotted as positive and to the left as negative.

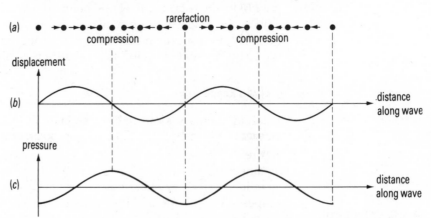

Fig 7.27   Sound wave

It will be seen that at some points the particles are close together and form a region of high pressure, called a **compression**. At other points the particles are further apart than average. Here the pressure is low and this region is called a **rarefaction**. Figure 7.25(c) shows the variation of pressure in the sound wave at the instant considered. It should be noted that this graph is one-quarter of a cycle out of phase with the displacement graph.

As the wave travels the particles vibrate to and fro and the graph profile in Figs 7.27(a) and (b) would travel in the direction of the wave.

CHARACTERISTICS OF NOTES

1 *Pitch* depends on frequency of vibration; a note of higher pitch has a higher frequency. Notes which are one octave apart have frequencies in the ratio 2:1.

2 *Loudness* of a sound depends on *intensity*. The intensity in any wave is measured by the wave energy per second passing through an area of $1m^2$ normal to the direction of travel. The unit of intensity is therefore $W\ m^{-2}$. We have already stated that the intensity in any wave is proportional to the square of its amplitude. There is no simple relation between loudness and intensity since the human ear has different sensitivity at different frequencies. In general, however, at any particular frequency the greater the intensity of the sound, the greater is its loudness.

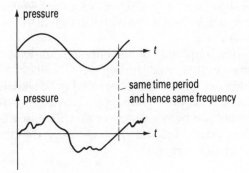

Fig 7.28 Timbre of notes

3 *TIMBRE* or *QUALITY* of a note depends on the particular waveform. Figure 7.28 shows the pressure variations in two sound waves obtained using a microphone connected to an oscilloscope. The notes have identical frequencies but different timbres due to the different shapes of the waveform. This shape is determined by the number and intensity of the overtones present. *Overtones* are notes which may have frequency twice, three times, four times, ..., that of the lowest or *fundamental* note. The upper wave in Fig 7.28 represents a 'pure' note with no overtones.

BEATS

If two notes of nearly equal frequencies, f_1 and f_2, are sounded together, beats are heard, that is, the intensity of the resultant sound fluctuates so that a series of loud and soft sounds are heard alternately. The frequency of the loud sounds = beat frequency = $f_1 - f_2$ if f_1 is greater than f_2. (It is $f_2 - f_1$ if f_2 is greater than f_1.)

To prove this formula, suppose that t is the time between successive loud sounds or beats. In this time one note makes exactly *one* cycle more than the other. Hence $f_2 t - f_1 t = 1$, or $f_2 - f_1 = 1/t$. But $1/t$ is the beat frequency. So beat frequency = $f_2 - f_1$.

Beats can be used to obtain the frequency of a tuning fork X, if another tuning fork Y with a known frequency close to that of X is available. On sounding X and Y together the beat frequency can be measured. To find which fork has the higher frequency each one is loaded in turn with a small piece of Plasticine. The higher frequency fork will give a reduction in beats when so loaded, as its frequency can then be found by adding or subtracting the beat frequency to the known frequency, as required.

RESONANCE

Any mechanical system which can vibrate has a **natural frequency** at which it will vibrate if displaced and released. Such a system may be forced to vibrate at a different frequency called the **forcing frequency**. If, however, the forcing frequency becomes equal to the natural frequency then large vibrations will occur and there is said to be **resonance**.

The amplitude of response at resonance depends on the amount of energy dissipated during each vibration. Systems which dissipate a large amount of energy are said to be heavily *damped*. If little energy is dissipated the damping is small. Figure 7.29 shows how the amplitude of response varies with forcing frequency for a system. Two graphs are shown, one where the damping is small, the other where it is larger.

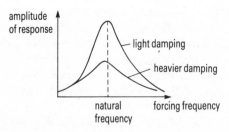

Fig 7.29 Resonance graphs

STANDING WAVES IN PIPES

1 CLOSED PIPE

At the closed end there can be no longitudinal vibration of the air and so there must be a node N at this end. At the open end there is an antinode, A. Figure 7.30 shows the first three possible modes of vibration. The curved lines drawn in the tubes are graphs giving the extreme positions of the longitudinal displacements of the air in the pipes.

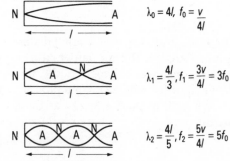

$$\lambda_0 = 4l, \; f_0 = \frac{v}{4l}$$

$$\lambda_1 = \frac{4l}{3}, \; f_1 = \frac{3v}{4l} = 3f_0$$

$$\lambda_2 = \frac{4l}{5}, \; f_2 = \frac{5v}{4l} = 5f_0$$

Fig 7.30 Standing waves in a closed pipe

The fundamental note, one with the lowest frequency, has a wavelength λ_0 which is $2 \times$ distance between nodes $= 4 \times$ distance NA between node and adjacent antinode $= 4l$, where l is the length of the pipe. Hence

$$f_0 = \text{fundamental frequency} = \frac{v}{\lambda_0} = \frac{v}{4l}$$

where v is the speed of sound in the air in the tube. The wavelengths of the first and second overtones are $4l/3$ and $4l/5$ as shown, so the frequencies of these notes are $3f_0$ and $5f_0$. Notice that with a closed pipe, only odd number multiples of f_0 are possible in the series of overtones, which may accompany the fundamental note.

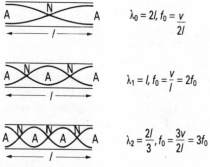

$$\lambda_0 = 2l, \; f_0 = \frac{v}{2l}$$

$$\lambda_1 = l, \; f_0 = \frac{v}{l} = 2f_0$$

$$\lambda_2 = \frac{2l}{3}, \; f_0 = \frac{3v}{2l} = 3f_0$$

Fig 7.31 Standing waves in an open pipe

2 OPEN PIPE

Figure 7.31 shows the first three modes of vibration. The wavelengths can be seen to be $2l$, l and $2l/3$. Hence the frequency of the fundamental, f_o, is given by $v/2l$ and the first and second overtones have frequencies $2f_o$ and $3f_o$. In an open pipe the series of overtones have any whole number multiple of the fundmental frequency. Closed and open pipes therefore have a different series of possible frequencies of overtones. The *quality* of the note emitted by the two types of pipe is therefore different.

END-CORRECTION IN PIPES

In a pipe the antinode is at a small distance c beyond the open end of the pipe. Thus for the fundamental in a pipe *closed* at one end (Fig 7.30), the distance between node and antinode is $(l+c)$. Hence $\lambda = 4(l+c)$ and $f_o = v/4(l+c)$. For a pipe *open* at both ends (Fig 7.31), the effective length of pipe is $(l+2c)$ as there is a correction at each end. Approximately, $c = 0.6r$ where r is the radius of the pipe.

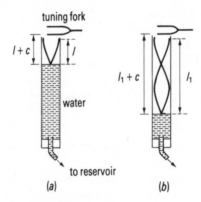

Fig 7.32 Resonance tube experiment

RESONANCE TUBE EXPERIMENT

Figure 7.32 shows an apparatus suitable for demonstrating vibrations in pipes. The water level in the tube starts near the top of the tube and is gradually lowered while a tuning fork is sounded over the top. At some point a loud sound is heard. Here the forcing frequency of the fork is equal to the natural frequency of the tube and *resonance* occurs (page 224). The first position of resonance occurs when the tube vibrates as in Fig 7.32(a). Here $l+c = \lambda/4$. A second position of resonance can be found as in Fig 7.32(b) where $l_2 + c = 3\lambda/4$. From these two equations

$$3(l+c) = l_1 + c$$

giving $\qquad c = (l_1 - 3l)/2 \qquad\qquad\qquad (1)$

Also by subtration

$$l_2 - l = \lambda/2 \qquad (2)$$

USES

1 The end-correction, c, can be found from equation (1)
2 The wavelength, λ, of the wave can be found from equation (2).
If the speed of sound is known, then the frequency of the fork can be calculated from $f = v/\lambda$. On the other hand if the frequency of the fork is known, then v can be found. To achieve greater accuracy in the measurement of velocity it is better to use a range of tuning forks and measure the length of the first resonance for each. We know $l + c = \lambda/4 = v/4f$. Hence a graph of l against l/f will have a slope of $v/4$ and the negative intercept on the l/f axis gives c.

MODES OF VIBRATION ON A STRETCHED STRING

Figure 7.33 shows the first three modes of vibration of a stretched string fixed between two points such as a violin or guitar string. The two fixed ends are both nodes, NN, so $NN = \lambda/2 = l$. Hence the wavelength of the fundamental is $2l$ and so the fundamental frequency f_0 is given by

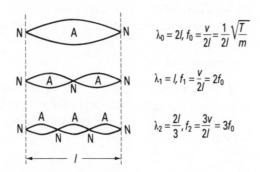

$$\lambda_0 = 2l, \; f_0 = \frac{v}{2l} = \frac{1}{2l}\sqrt{\frac{T}{m}}$$

$$\lambda_1 = l, \; f_1 = \frac{v}{2l} = 2f_0$$

$$\lambda_2 = \frac{2l}{3}, \; f_2 = \frac{3v}{2l} = 3f_0$$

Fig 7.33 Standing waves on a string

$$f_0 = \frac{v}{\lambda_0} = \frac{v}{2l} = \frac{1}{2l}\sqrt{\frac{T}{m}}$$

where T is the tension in the string and m is its mass per unit length ($v = \sqrt{T/m}$, from p205). The frequencies of the first and second overtones are $v/l = 2f_0$ and $3v/2l = 3f_0$. All whole number multiples of f_0 are therefore possible in the series of overtones.

THE SONOMETER

Figure 7.34 shows a sonometer, consisting of a stretched wire S whose vibrating length *l* can be varied by movable bridges X and Y. S is fastened to a sounding board B, so that vibrations from S are transmitted to B which has a large surface area in contact with the air. In this way a much louder sound is heard when S vibrates than otherwise would be the case. The tension *T* in the wire can be varied by means of weights *W* which hang over the pulley.

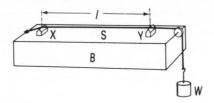

Fig 7.34 Sonometer

The wire can be tuned to give the same note as the frequency of a tuning fork by ear if one has a good sense of pitch. Otherwise, the wire is first tuned approximately and then a small paper rider is placed on the centre of S. when the fork and wire have the same frequency, resonance occurs when the lower end of the sounding tuning fork is placed on B and there is a large vibration of the wire which throws the paper rider off.

The following experiments are possible with a sonometer.

1 TO SHOW $f \propto 1/l$

Use five different tuning forks of known frequency and for each measure the length *l* of S when the tuning fork and S are in tune. Plot *f* against $1/l$, when a straight line passing through the origin is obtained.

TO SHOW $f \propto \sqrt{T}$

The wire is kept of constant length. For each of five tuning forks the weight *W* is adjusted until the wire and tuning fork are in tune. *f* is then plotted against $\sqrt{T}$ when a straight line through the origin is obtained.

3 TO MEASURE THE FREQUENCY OF AN A.C. CURRENT

The wire S is connected to the a.c. supply so that an alternating current flows through the wire. A horseshoe magnet is placed with its poles either side of the wire. The length of the wire is adjusted until the wire is seen to vibrate with maximum amplitude. The forcing oscillation caused by the current flowing in the magnetic field is then equal to the natural frequency of the wire and resonance occurs. In this case the frequency of the a.c. is given by

$$f = \text{natural frequency of wire} = \frac{1}{2l}\sqrt{\frac{T}{m}}$$

f can therefore be calculated from measurements of l, T and m. f is in Hz if l is in m, T in N and m in kg m^{-1}

DOPPLER EFFECT

This is the name given to the apparent change of frequency when there is *relative motion* between a source of sound S and the observer, O.

CASE 1. SOURCE MOVES

Here, if S moves towards O an observer at O will receive sound of a reduced wavelength λ' and increased frequency f'. From Fig 7.35, $\lambda' = $ true wavelength $\lambda - $ distance source moves in time of one cycle. Hence

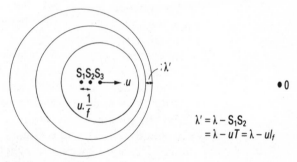

$$\lambda' = \lambda - S_1S_2$$
$$= \lambda - uT = \lambda - u/f$$

Fig 7.35 Doppler effect

$$\lambda' = \lambda - u_s T = \lambda - u_s/f$$

where u_s is the speed of the source, T is the time period and f the frequency of the emitted sound. Thus

$$\lambda' = \frac{v}{f'} = \frac{v}{f} - \frac{u_s}{f} = \frac{v - u_s}{f}$$

So $$f' = \frac{v}{v - u_s} \cdot f$$

When S approaches O the frequency is raised. When S moves away from O, u_s is negative and f' is decreased.

CASE 2. OBSERVER MOVES

In this case the speed of sound relative to the observer changes. If the observer approaches S with a speed u_o then the speed of sound relative to the observer is $v + u_o$. Hence

$$f' = \frac{\text{speed}}{\lambda} = \frac{v + u_o}{\lambda} = \left(\frac{v + u_o}{v}\right) f$$

CASE 3. GENERAL CASE

If both source and observer move the apparent frequency f' can be calculated from

$$f' = \left(\frac{v + u_o}{v - u_s}\right) f$$

In this formula u_s is positive if the movement of S is towards O, and u_o is positive if the movement of O is towards S.

DOPPLER EFFECT IN LIGHT

For light there is a similar effect to sound, but it is only the relative speed between source and observer which matters. Provided, however, the speed of movement is much less than the speed of light, the formulae derived above will give answers to high accuracy. (If speeds of movement become comparable with the speed of light, then the full relativistic formula derived by Einstein must be used.)

When a star moves away from an observer on the earth with the velocity v, its apparent wavelength λ' measured by the observer is longer than its true wavelength λ. This is the so-called 'red-shift' in wavelength ($\lambda' - \lambda$), and is equal – for non-relativistic speeds – to $v\lambda/c$. Hence v can be measured when the shift is known. If the star approaches the earth the wavelengths appear shorter and 'blue-shift' is observed.

ELECTROMEGNETIC WAVES

NATURE OF ELECTROMAGNETIC WAVES

All electromagnetic waves move with a speed c of $3 \times 10^8 \text{m s}^{-1}$ approximately in a vacuum (page 231). These waves all contain electric and magnetic fields which oscillate perpendicular to the direction of travel. Figure 7.36 shows diagrammatically a plane-polarized electro-magnetic wave. The electric field oscillations are confined to a vertical plane and the magnetic field oscillations to a horizontal plane. The essential differences between types of electromagnetic wave are due to their differences in wavelength.

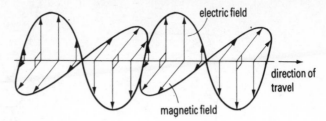

Fig 7.36 Electromagnetic wave

ELECTROMAGNETIC SPECTRUM

The whole range of wavelength of electromagnetic waves in called the **electromagnetic spectrum**. Below the spectrum is shown diagrammatically together with the names of the waves and some facts about them. Light waves are electromagnetic waves which have wavelength about 1000 times longer than X-rays. Radio waves are the longest electromagnetic waves and γ-rays are shortest.

γ-rays	X-rays	Ultraviolet	Visible light	Infrared	Micro and radio
Emitted from radio-active sources (p312)	Produced when fast electrons stopped by metals (p307)	In sunlight. From quartz lamps. Produce suntan	Affects eye	Heat radiation from hot objects	Radiated by aerials
Wavelength: 10^{-11}m	10^{-10}m	10^{-8}m	10^{-7}m	10^{-6}m	10^{-2}m to ~km

SPEED OF ELECTROMAGNETC WAVES

The speed of electromagnetic waves in the form of light may be measured using a ROTATING MIRROR method. Fig 7.37 shows the apparatus used by Michelson in 1927.

Light from slit source S falls on to one face of an octagonal mirror R. It is then reflected via M_1 and M_2 to the large curved mirror C_1. The focal length of C_1 is such that the image of the slit in M_2 is at the focus of C_1. Light from C_1 is thus transmitted in a parallel beam to distant curved mirror C_2. From here it is reflected to a small plane mirror M_3 placed at the focus of C_2. The light thus travels back to C_2 and then in a parallel beam back to C_1. From here the light goes *via* M_4, M_5, to the opposite face of R and finally to the eyepiece E. Mirror M_2 is just above the plane of the diagram and mirror M_4 is just below the plane.

In this experiment the rotating mirror part of the apparatus was at Mount Wilson observatory, whilst C_2 was about 22 miles away. Once

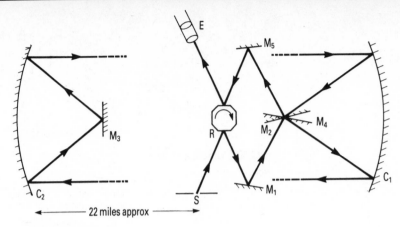

Fig 7.37 Michelson's
rotating mirror experiment

the arrangement is set up the mirror R is rotated. When the light arrives back at R the mirror will be in a different position from that shown in the diagram and so the light will not be reflected to E. If, however, the speed is sufficiently fast for the mirror to turn through one eighth of a revolution while the light is on its journey, then the diagram is again correct and the light once again becomes visible in E.

To do this it was necessary to rotate the mirror at 528 revolutions per second. The time for one eighth of a revolution is $1/8 \times 528$ s. The distance travelled in this time was about 44 miles, and was measured exactly by Michelson to be 7.08×10^4m. the speed of light is therefore given by

$$c = \text{distance/time} = 7.08 \times 10^4 \times 8 \times 528$$
$$= 3.00 \times 10^8 \text{m s}^{-1}$$

In fact the measurements were more precise than is given here and they gave the result $299\,798 \pm 4$km s^{-1}.

WORKED EXAMPLES

7.1 VELOCITY OF SOUND   A signal generator emits sound waves of frequency 2kHz. The velocity of sound is 331.5m s^{-1} at 0°C. Calculate the wavelength of the waves at 20°C.

(**Overview** Use $v \propto \sqrt{T}$ to find the new velocity, then use $v = f\lambda$.)

The velocity of sound $\propto \sqrt{T}$, hence

$$\frac{v_0}{v_{20}} = \sqrt{\frac{273}{293}}$$

So $\qquad v_{20} = 331.5 \times \sqrt{\dfrac{293}{273}}$

$\qquad\qquad = 343.4\text{m s}^{-1}$

$\therefore \qquad\qquad \lambda = \dfrac{v}{f} = \dfrac{343.4}{2000} = 0.172\text{m} = 17.2\text{cm}$

7.2 STANDING WAVES

A ship sails in a straight line between two transmitting stations A and B which emit radio waves of the same frequency. The signal fluctuates in intensity and passes through a minimum each time the ship travels 100m. Calculate the frequency of the radio signals. (Speed of radio waves, $c = 3 \times 10^8\text{m s}^{-1}$.)

(**Overview** The two radio waves from A and B travel in opposite directions and set up a standing wave on the line joining them.)

The distance between the minimum signals is the distance between nodes, which is 100m. So the wavelength $\lambda = 2 \times 100\text{m} = 200\text{m}$. Using

$$c = f\lambda$$

we obtain $\qquad f = \dfrac{c}{\lambda} = \dfrac{3 \times 10^8}{200} = 1.5 \times 10^6 = 1.5\text{MHz}$

7.3 WEDGE FRINGES

Two glass slides and a wire of diameter 0.04mm are used to produce wedge fringes. If the wavelength of the light used is 500nm, find approximately how many fringes can be produced.

(**Overview** Use $2t = m\lambda$)

At the point where the wire is situated the thickness of the film is $0.04\text{mm} = 0.04 \times 10^{-3}\text{m}$. Hence the number, p, of the nearest dark band is given by

$$2t = p\lambda$$

or $\qquad 2 \times 0.04 \times 10^{-3} = p \times 500 \times 10^{-9}$

So $\qquad p = \dfrac{2 \times 0.04 \times 10^{-3}}{500 \times 10^{-9}} = \dfrac{800}{5} = 160$

Hence 160 bands can be seen in the wedge.

7.4 NEWTON'S RINGS

In an arrangement to observe Newton's rings light of wavelength 500nm is used. When the lens is carefully lifted from the glass plate

120 fringes appear at the centre and move away into the fringe system. By how much was the lens raised?

(**Overview** To produce one extra fringe the lens must be lifted by half a wavelength, so that an extra whole wavelength is introduced into the path difference.)

To produce 120 extra fringes the distance moved must be 60 wavelengths, i.e. $60 \times 500 \times 10^{-9}$m $= 3 \times 10^{-5}$m
$$= 3 \times 10^{-2}\text{mm}$$

7.5 DIFFRACTION GRATING

A grating with 1000 lines per mm is illuminated by monochromatic light of wavelength 4×10^{-7}m. How many diffraction images are produced?

(**Overview** There can be no value of sin θ greater than 1 and this limits the value of m in $d \sin \theta = m\lambda$.)

We have $d = 10^{-3}$mm $= 10^{-6}$m. Hence for $m = 1$ we have

$$d \sin \theta = \lambda$$
so $10^{-6} \sin \theta = 4 \times 10^{-7}$,
giving $\sin \theta = 0.4$ and $\theta = 23.6°$

The second order image is formed where

$$d \sin \theta = 2\lambda$$
giving $\sin \theta = 0.8$ and $\theta = 53.1°$

There can be no third order image as this would make sin θ greater than 1, which is impossible. So we have the undiffracted image ($m = 0$) and two diffracted images on either side of this image, giving a total of five images.

7.6 DIFFRACTION GRATING WITH TWO COLOURS

Light of two wavelengths 4×10^{-7}m (violet) and 6×10^{-7}m (yellow) is incident on a diffraction grating with 500 lines per mm. At what angle (or angles) do the two colours of light interfere constructively together?

(**Overview** Each colour has a different wavelength so it may be possible for a lower order diffraction of the long wavelength to be at the same angle as a high order diffraction for the short wavelength.)

$d = \dfrac{1}{500}$mm $= 2 \times 10^{-6}$m. Suppose the pth order for the 4×10^{-7}m wavelength overlaps with the qth order for the 6×10^{-7}m wavelength. Then $d \sin \theta = p \times 4 \times 10^{-7} = q \times 6 \times 10^{-7}$ which gives $p/q = 1.5$. Hence possible pairs of values of p and q are $p=3$, $q=2$ or $p=6$, $q=4$, and so on. Now $q=4$ gives sin θ greater than 1 and so it is not possible to obtain this image. Thus only one angle exists where the two wavelengths overlap and this is given by

$$d \sin \theta = 3 \times 4 \times 10^{-7} \text{ (or } 2 \times 6 \times 10^{-7})$$

So $\quad\quad 2 \times 10^{-6} \times \sin \theta = 12 \times 10^{-7}$

and $\quad\quad\quad\quad\quad \sin \theta = 12/20$, from which $\theta = 36.9°$

7.7 DIFFRACTION GRATING WITH TWO WAVELENGTHS

Light consisting of two wavelengths which differ by 160nm passes through a diffraction grating with 2.5×10^5 lines per metre. In the diffracted light the third order of one wavelength coincides with the fourth order of the other. What are the two wavelengths and at what angle of diffraction does this coincidence occur?

(**Overview** As previous example.)

Let λ be the shorter wavelength. Then $(\lambda + 160 \times 10^{-9}\text{m})$ is the longer wavelength. Also $d = 1/(2.5 \times 10^5) = 4 \times 10^{-6}\text{m}$.

We have $\quad\quad d \sin \theta = 3(\lambda + 160 \times 10^{-9}) = 4\lambda$ at coincidence

hence $\quad 3\lambda + 480 \times 10^{-9} = 4\lambda$ or $\lambda = 480 \times 10^{-9}\text{m} = 480\text{nm}$

and the other wavelength is $480 + 160 = 640\text{nm}$.

Since $\quad\quad\quad\quad d \sin \theta = 4\lambda$ at coincidence

$$4 \times 10^{-6} \sin \theta = 4 \times 480 \times 10^{-9}$$

So $\quad\quad\quad\quad\quad \sin \theta = 0.48$ or $\theta = 28.7°$

7.8 USE OF POLAROIDS

Light from source A passes through two uncrossed (parallel) polaroids and falls on a screen S_1. Light from a second source B falls on a screen S_2. the screens S_1 and S_2 are equally illuminated. B now moves twice as far from S_2 as it was originally. Through what angle must one of the polaroids be rotated so that the screens are again equally illuminated?

The intensity of light from a source obeys an inverse-square law with distance. Hence the illumination on S_2 decreases by a factor of $\frac{1}{4}$. Since rotating a polaroid by θ reduces the intensity by $\cos^2 \theta$, then the illumination on S_1 will be equal to that on S_2 if $\cos^2 \theta = \frac{1}{4}$. So $\cos \theta = \frac{1}{2}$ and hence $\theta = 60°$.

7.9 RESONANCE TUBE

A tube 1m long closed at one end has its lowest resonance frequency at 86.2Hz. With the same tube open at both ends the first resonance occurs at 171Hz. Calculate the speed of sound in the column of air and the end correction for the tube.

(Overview Use the appropriate equation for each instance given. This gives two equations in the two unknowns which can then be solved simultaneously.)

For the tube open at one end the wavelength of the fundamental λ is given by

$$l+c=\lambda/4=v/4f$$

Hence
$$1+c=v/(4\times 86.2) \tag{1}$$

For the tube open at both ends we have

$$l+2c=\lambda/2=v/2f$$

Hence
$$1+2c=v/(2\times 171) \tag{2}$$

Eliminating c between (1) and (2)

$$1+2\left(\frac{v}{4\times 86.2}-1\right)=\frac{v}{2\times 171}$$

Simplifying

$$v\left(\frac{1}{172.4}-\frac{1}{342}\right)=1$$

and
$$v=348\text{m s}^{-1}$$

Substituting in (1), we obtain

$$c=\frac{348}{4\times 86.2}-1$$
$$=9.3\times 10^{-3}\text{m}=9.3\text{mm}.$$

7.10 BEATS

Two open-ended organ pipes are sounded together and 8 beats per second are heard. If the shorter pipe is of length 0.80m what is the length of the other pipe? You may ignore end corrections. (Speed of sound in air $=320\text{m s}^{-1}$.)

The wavelength in the shorter pipe is 1.6m and hence its frequency of vibration is $v/\lambda=320/1.6=200$Hz. The longer pipe will have a lower frequency and since the difference in frequency is 8Hz, the longer pipe has a frequency of 192Hz. Hence the wavelength is $v/f=1.67$m. The length of the tube is $\lambda/2=0.83$m.

7.11 VIBRATING STRINGS

A string fixed at both ends is vibrating in the lowest mode of vibration for which a point a quarter of its length from one end is a point of maximum vibration. The note emitted has a frequency of 100Hz.

What will be the frequency emitted when it vibrates in the next mode such that this point is again a point of maximum vibration?

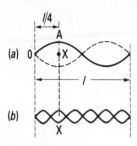

Fig 7.38 Worked example

If a point one quarter way along the wire is an antinode, then the lowest mode of vibration which produces this is shown in Fig 7.38(a). Here $\lambda = l$. The next mode of vibration for which this happens is shown in Fig 7.38(b). In this case $\lambda = l/3$. Thus the wavelength is $\frac{1}{3}$rd of the previous value and so the frequency is three time greater than the previous value, that is 300Hz.

7.12 THE SONOMETER

The wire of a sonometer, of mass per unit length 1.0g m^{-1}, is stretched over the two bridges by a load of 40 N. When the wire is struck at its centre point so that it executes its fundamental vibration, and at the same time a tuning fork of frequency 264Hz is sounded, beats are heard and found to have a frequency of 3Hz. If the load is slightly increased the beat frequency is lowered. Calculate the separation of the bridges.

By what amount must the load be increased to produce a beat frequency of 10Hz if the same tuning fork is used?

(Overview Use beats to find the frequency of the wire. Use the formula for the frequency of stretched wire to calculate the length of the stretched wire. Finally this formula can be used again to find the new tension needed to produce the increase in pitch.)

When the load is increased the frequency of the wire is raised. This reduces the beat frequency and so the wire must have *lower* frequency than the fork. Hence the frequency of the wire is $264 - 3 = 261$Hz. But

$$\text{fundamental frequency} = \frac{1}{2l}\sqrt{\frac{T}{m}}$$

Hence
$$261 = \frac{1}{2l}\sqrt{\frac{10}{10^{-3}}}$$

as $m = 1\text{g m}^{-1} = 10^{-3}\text{kg m}^{-1}$. This equation gives $1/l = 2.61$ and hence $l = 0.38$m.

To produce a beat frequency of 10Hz by *increasing* the frequency of the wire, the tension must be raised so that the frequency of the wire is $264 + 10 = 274$Hz. Hence the tension is given by

$$f = \frac{1}{2l}\sqrt{\frac{T}{m}}$$

So
$$274 = \frac{1}{2 \times 0.38}\sqrt{\frac{T}{10^{-3}}}$$

which gives $T = 44$ N

Thus the tension must be increased by 4 N.

7.13 THE DOPPLER EFFECT

A train approaches a tunnel in a cliff face at a speed of 2m s^{-1} and sounds its whistle at a frequency of 500Hz. Calculate the number of beats heard between the whistle and the reflection of the sound in the cliff face. (Assume speed of sound $= 330$m s^{-1}.)

(**Overview** The sound 'image' of the train's whistle can be considered as a source of sound an equal distance behind the cliff face. This 'virtual' source therefore approaches the train with a speed of 2m s^{-1}.)

With the sign convention of p230, $u_s = +2$m s^{-1} as the source approaches the observer, and $u_o = +2$m s^{-1} as the observer approaches the source. Hence

$$f' = f\left(\frac{v + u_o}{v - u_s}\right)$$

$$= 500\left(\frac{330 + 2}{330 - 2}\right) = 500 \times \frac{332}{328} = 506\text{Hz}$$

Hence 6 beats per second would be heard between the whistle and the echo from the cliff face.

7.14 THE DOPPLER EFFECT

A source of sound S, $f = 1000$Hz approaches a stationary observer at 20m s^{-1}. Calculate the frequency heard by O. Also calculate the frequency heard by O if O recedes from S at 20m s^{-1}. (Assume speed of sound $= 330$m s^{-1}.)

In the first case $u_s = +20$m s^{-1}, $u_o = 0$

Hence
$$f' = f\left(\frac{v + u_o}{v - u_s}\right)$$

$$= 1000\left(\frac{330}{330 - 20}\right) = 1064\text{Hz}$$

The frequency has thus been raised by 64Hz.

In the second case O is moving away from S so that $u_o = -20\text{m s}^{-1}$ and $u_s = 0$. Hence

$$f' = f\left(\frac{v + u_o}{v - u_s}\right)$$

$$= 1000\left(\frac{330 - 20}{330}\right) = 939\text{Hz}$$

Here the frequency has been lowered by 61Hz.

1 TYPES OF SPEED OF WAVES

Speed v, wavelength λ, frequency $f: v = f\lambda$.

Waves can be
(a)   progressive or stationary,
(b)   transverse or longitudinal.

Standing (stationary) waves have regions of zero amplitude (nodes) and regions of maximum amplitude (antinodes).

Speeds: String $\sqrt{T/m}$; longitudinal waves in a solid $\sqrt{E/p}$; e-m waves in a vacuum $\sqrt{1/\mu_o\varepsilon_o}$; sound $\sqrt{\gamma p/\rho}$; speed of sound $\propto \sqrt{T}$ but independent of pressure.

2 WAVE PROPERTIES

Reflection.

Refraction: $\sin \theta_1/\sin \theta_2 = v_1/v_2$.

Diffraction: large when wavelength is long compared to gap width.

Interference constructive for path difference $m\lambda$, destructive for path difference $(m + \frac{1}{2})\lambda$.

Polarization: Occurs with transverse waves only and used as a test for transverse waves.

3 LIGHT WAVES

Young's fringes: $\lambda = dx/D$. Interference fringes can be used to measure wavelength. Coloured fringes with white light source.

Air wedge: Bright fringe when $2t = (m + \frac{1}{2})\lambda$. Used to find thickness of foil or small expansions.

Newton's rings formed in air space between lens and flat glass surface – $r^2/a = (m + \frac{1}{2})\lambda$ – for a bright fringe.

Interference produces colours in thin films.
Diffraction limits resolution of optical instruments. Smallest angle resolved $= 1.22\lambda/d$ radians.

DIAGRAM SUMMARY

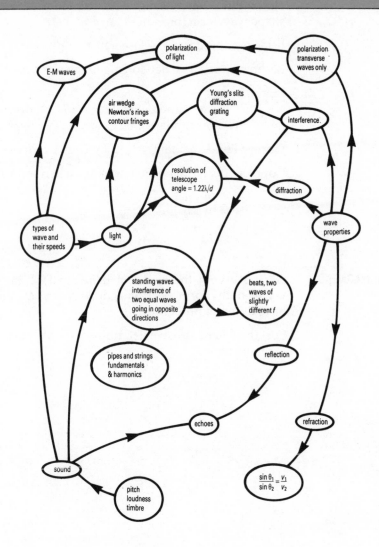

Diffraction grating: $d \sin \theta = m\lambda$ (m = order of spectrum).

Polarization: Produced by polaroids, reflection, scattering, double refraction. Uncrossed polaroids turned through θ reduces intensity by $\cos^2 \theta$ factor.

---

4 SOUND WAVES

Sound is a longitudinal compression wave.

Notes: Pitch depends on frequency, loudness on intensity, timbre on shape waveform (presence of overtones).

Beats produced by two close notes. $F = f_1 - f_2$.

Resonance occurs when forcing frequency = natural frequency.

Pipes:
(a)   Closed – fundamental $f_o = v/4(l+c)$, overtones = $3f_o$, $5f_o$, etc
(b)   Open – fundamental $f_o = v/2(l+2c)$, overtones = $2f_o$, $3f_o$, etc.

Strings: Fundamental $f_o = (1/2l)(\sqrt{T/m})$, overtones = $2f_o$, $3f_o$ etc.

Doppler effect: Change in frequency when source or observer moves,

$$f' = \left(\frac{v+u_o}{v-u_s}\right) f.$$

---

5 ELECTROMAGNETIC WAVES

All move with the speed of light, $c = 3 \times 10^8$m s^{-1}. Wide range of wavelengths form electromagnetic spectrum, different wavelengths have different properties.

Speed of light can be determined by Michelson's Rotating Mirror experiment.

7 QUESTIONS

1 State the conditions required for interference to be observed between waves from two sources.

Describe how you would demonstrate the interference between sound waves from two sources. Give appropriate values for the variables involved and show how you could use your measurements to find a value for the wavelength of the sound used.

Both speakers of stereo hi-fi system are fed from the same mono orchestral recording of music. Explain why you cannot find a place in the room at which no sound is heard.

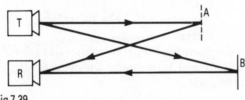

Fig 7.39

An ultrasonic burglar alarm operates at 40kHz. The transmitter T is set up as shown (Fig 7.39), 900mm from a curtain A covering a window and 1000mm from the wall B. The receiver R is alongside the transmitter.

Calculate the path difference between the beam reflected from the wall and that reflected from the curtain. Express this distance in terms of the wavelength of sound used.

The window is left open by mistake when the burglar alarm is switched on. As a result the curtain moves. Describe quantitatively, including a sketch graph, how the output of the receiver varies with time when the curtain is moving towards the receiver at a steady speed of 0.1m s^{-1}.

(Speed of sound in air $= 340 \text{m s}^{-1}$.)   (O&C)

2 (a)   A soap film is formed on a vertical wire frame and viewed in reflected white light. Initially a number of horizontal coloured fringes are seen. As the soap film gradually drains the fringe pattern changes until eventually, just before the film breaks, the top part appears black.

(i) Explain how the coloured fringes are formed.

(ii) Describe how the appearance of the fringes changes as the film drains.

(iii) Explain why part of the film appears black just before it breaks.

(b)   The diagram (Fig 7.40) shows an experimental arrangement for demonstrating the diffraction of light through a narrow horizontal slit. The apparatus is used in a darkened room.

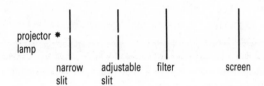

projector ✱
lamp

narrow   adjustable   filter   screen
slit   slit

Fig 7.40

(i) Sketch a graph to show the intensity pattern you would see on the screen when the apparatus is correctly adjusted.

(ii) The width of the adjustable slit is 0.50mm, the filter transmits yellow light of wavelength 6.0×10^{-7}m and the screen is 1.0m away from the adjustable slit. Calculate the positions of the first and second minima in the diffraction pattern.

(iii) What would be the effect on the pattern of removing the filter?   (JMB)

3   The diagram Fig 7.41 (not drawn to scale) shows the apparatus used in an attempt to measure the wavelength of light using double slit interference.

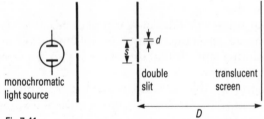

monochromatic
light source

double
slit

translucent
screen

D

Fig 7.41

(a)   Explain how the apparatus produces double slit interference fringes on the translucent screen.

(b)   If measurable fringes are to be seen on the screen, then
(i) the slit separation s must be small;
(ii) the distance D between the slits and the screen must be large.
Explain each of these conditions.

(c)   The light from the monochromatic source has a wavelength of 5.0×10^{-7}m and it is required to produce a fringe separation of 5.0mm. Suggest suitable values for s and D justifying your answer.
Describe the important features of the pattern you would hope to obtain.

(d)   Describe and explain any changes in the pattern which would be brought about by each of the following changes made separately:

(i) one slit is covered with an opaque material;

(ii) the slits are made narrower although their separation remains the same.   (AEB 1984)

4   (a)   Explain with the aid of diagrams what is meant by (1) diffraction (2) constructive interference of waves.

Describe the parts played by each of these effects in the production of spectra by an optical diffraction grating.

(b)   Describe how you would use a grating and a spectrometer to determine the wavelength of light from a sodium vapour lamp as precisely as possible. Identify the readings you would take and state how you would use them to calculate an accurate result. (You may assume that

(i) the spectrometer has already been correctly set up with the telescope and collimator correctly adjusted and

(ii) the grating spacing is accurately known.)

(c)   Sodium light has two principal components, of wavelength 589.0nm and 589.6nm respectively. Given that the focal lengths of both the collimator lens and the objective lens of the telescope are 200nm, and the grating has 600 lines per millimetre, calculate:

(i) the angular separation of the two beams emerging from the grating in the second order;

(ii) the greatest width of slit which can be used if the two lines seen through the telescope in the second order can be seen as separate (explaining clearly, with the aid of a diagram, the principles of your calculation).   (O)

5   The wavelengths of the two yellow sodium lines are 589.0nm and 589.6nm. The eye, with the aid of the spectrometer telescope, can resolve an angle of 0.2 minutes. How many lines per metre must there be in a diffraction grating for the two sodium lines to be seen as just separate in the first order in the spectrometer? Assume the diffraction angle is small so that $\sin \theta \simeq \theta$ in radians.   (W)

6   A beam of electromagnetic waves of wavelength 3.0cm is directed normally at a grid of metal rods, parallel to each other and arranged vertically about 2.0cm apart. Behind the grid is a receiver to detect the waves. It is found that when the grid is in this position, the receiver detects a strong signal but that when the grid is rotated in a vertical plane through 90°, the detected signal falls to zero. What property of the wave gives rise to this effect? Account briefly in general terms for the effect described above.   (L)

7   (a)   Describe the contrasting features of progressive (travelling) and stationary (standing) waves.

With the aid of a sketch, outline an experiment to demonstrate standing waves. Briefly describe the apparatus used and any adjustment that has to be made to obtain a satisfactory result.

(b)   A woodwind player in an orchestra tunes his instrument to a fundamental frequency of 256.0Hz in a cold dressing-room where the mean air temperature in the instrument is 285K. During the performance in a warm concert hall his instrument, playing what is intended to be the same note, produces a beat frequency 2.5Hz with a similar instrument which was tuned in the warmer air of the hall to 256.0Hz. (Assume that the dimensions of the instrument do not change significantly, and that $c \propto \sqrt{T}$, where c is the speed of sound in air and T is the thermodynamic (absolute) temperature.)

    (i) Explain whether the first instrument produces a note of higher or lower frequency when played in the warmer air of the hall.

    (ii) Estimate the mean temperature of the air in the first instrument when played in the concert hall.

(c)   (i) Explain what is meant by plane-polarized light.

    (ii) Describe how you would produce a parallel beam of plane-polarized light, starting with a tungsten-filament lamp.

    (iii) How would you determine by a different technique the plane of polarization of the beam?

    (iv) With the aid of a diagram, explain how a photographer makes use of a Polaroid filter in front of his camera lens to photograph objects seen through a shop window from a brightly-lit street.   (O)

8   (a)   Describe the apparatus you would use to measure the speed of sound in *free* air (not air in a tube) in the laboratory.
Give the measurements you would make and show how you would use them to calculate the speed of sound.
How does an everyday observation show that the speed of sound in air is the same for all audible frequencies?

(b)   A resonance tube is held vertically in water and can be raised or lowered. A tuning fork of frequency 384Hz is struck and held above the open end of the tube on a day when the speed of sound in air is 344m s^{-1}. The shortest tube length at which resonance occurs is 21.6cm and the corresponding length when the tube is filled with carbon dioxide is 16.7cm.
Calculate

    (i) the end correction for the tube,

    (ii) the speed of sound in carbon dioxide.

(c)   The sirens of two police cars, A and B, have fixed frequencies f_A and f_B, the frequency of car A being the higher. With both cars standing still in the road and with their sirens sounding, a bystander hears a single note with a beating rate of 20Hz.
Car A drives off and accelerates uniformly till it reaches a steady speed v, which equals $c/10$, where c is the speed of sound in air at the time. The bystander notices the beating rate decrease, reach zero and increase again to 20Hz by the time the speed of A is steady. The siren of car B is switched off and the bystander notices that the apparent frequency of the siren of car A has fallen to f_A'.

    (i) Calculate f_A'/f_A.

    (ii) Calculate f_A.   (L)

9   (a)   State **two** of the principal differences between progressive waves and stationary (standing) waves.

(*b*)   The diagram (Fig 7.42) shows the apparatus used in an experiment to investigate the vibrations of a length of copper wire. End A of the wire is attached to a vibrator which moves up and down at the frequency of the signal generator. A load attached to end B keeps the wire in tension. When the signal generator is adjusted to aparticular frequency a stationary wave is obtained as shown.

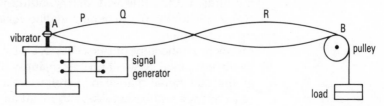

Fig 7.42

(i) Explain why the mid-point of the wire is stationary.
(ii) Compare and contrast the motion of the wire at points P, Q and R.
(iii) The length AB of the copper wire used in the experiment is 1.5m and its cross sectional area is 0.059mm^2; the tension in the wire is 2.0 N. If the density of copper is 8.9×10^3kg m^{-3}, calculate the mass per unit length of the wire, and show that the lowest frequency at which you would expect a stationary wave is about 20Hz. Draw a diagram of the wave.

(*c*)   In another experiment the leads to the signal generator are removed from the vibrator and connected to the wire near to points A and B so that an alternating current flows in the wire. A horseshoe magnet is arranged to produce a horizontal magnetic field at right angles to the wire near the mid-point.

(i) Describe the additional force now experienced by the wire and explain how it arises.
(ii) State and explain what you would expect to observe if the frequency of the alternating current were gradually increased from zero to about 70Hz.
(iii) How could you use this experiment, suitably adapted, to measure the frequency of an alternating voltage?

(*AEB* 1985)

10   Describe the Doppler effect and derive an equation for the apparent change in frequency when
(i)   the source is moving and the observer is stationary and
(ii)   the source is stationary and the observer is moving.
When do the equations not apply to electromagnetic waves?
A line of wavelength 590nm in the sun's spectrum is observed first at one edge then at the opposite edge of the sun's disc on an equatorial diameter of the sun. A difference in wavelength of 0.0074nm is found. The rotational period of the sun in 27 days. Calculate the sun's diameter.
(Speed of light $= 3.00 \times 10^8$m s^{-1}.)   (*W*)

11   Describe a direct terrestrial method by which the speed of light may
be determined. What aspects of the method require careful design?
Place the radio, visible, infra-red and ultra-violet regions of the electromagnetic spectrum in order of increasing wavelength and give a
typical wavelength for each region.
Electromagnetic standing waves may be set up in a cavity with reflecting ends in the same way that standing waves can be set up on a
string. Write down a general equation for a standing wave to be set
up in the cavity.
Monochromatic light of frequency 5.1×10^{14} Hz resonates in a certain
gas-filled cavity. A small further quantity of gas leaks into the cavity
giving a fractional change in the refractive index of 10^{-5}. What change
of frequency will be necessary to maintain the standing wave in the
cavity in the same mode?   (C)

ELECTRON PHYSICS AND ELECTRONICS

CONTENTS

PRODUCTION AND PROPERTIES OF ELECTRONS

THERMIONIC EMISSION

In a metal there are free electrons which can move throughout the volume of the metal. When the temperature of the metal is increased the free electrons move with greater speed and so they may be able to escape from the metal. This process is called **thermionic emission**. The least energy needed to remove an electron from the metal is called the **work function** of the metal. The lower the work function, the greater will be the thermionic emission at any given temperature. Coating metals with barium or strontium oxide has the effect of lowering the work functions and so facilitating electron emission. As we shall see later, the work function also governs the **photoelectric effect**, page 303.

ACCELERATION OF ELECTRONS – ELECTRON GUN

Normally, in experiments with electrons, very high electron velocities are required. This can be achieved by accelerating the electrons by an electric field. Figure 8.1 shows an **electron gun**. A power supply S, which may be a.c. or d.c., is connected to a heating coil H which is used to heat the metal cathode C. Electrons emitted from C by thermionic emission are accelerated towards the metal anode A by the electric field set up between C and A by the high voltage V, as shown. Some fast moving electrons pass through an opening in the anode A and enter the field-free region beyond A. Here they continue with a steady speed v.

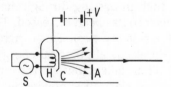

Fig 8.1 Electron gun

The kinetic energy gained by an electron=potential energy change in moving through V volts

Thus $\frac{1}{2}mv^2 = eV$

where m is the mass of an electron and e its charge, assuming zero initial kinetic energy for the electron. The potential energy $= eV$ from p81. As we shall now show, electrons can reach very high speeds when they are accelerated by electric fields.

DEFLECTION IN UNIFORM ELECTRIC FIELD

Consider an electron entering a uniform electric field of intensity E, with a velocity v at right angles to the direction of E (Fig 8.2). The force F on the electron, charge e, due to the field $= Ee = eV/d$, since $E =$ potential gradient $= V/d$, page 83. F acts in the direction BA, which is opposite in direction to the field as the electron charge is negative. Hence from $F = ma$, the acceleration $a = eV/md$.

This acceleration is vertical and along BA in Fig 8.2. The horizontal component of the electron's velocity does not therefore change. Thus the time t for which the electron is within the field is D/v, where D is the length of the plates. In this time the vertical speed gained,

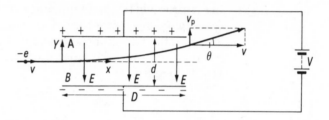

Fig 8.2 Deflection in an electric field

$$v_p = a \times t = \frac{eV}{md} \times \frac{D}{v}.$$

The electron emerges at an angle θ to the plates shown, where $\tan \theta = v_p/v$.

So $\tan \theta = \dfrac{eVD}{mdv^2}$

Note that, in passing through the electric field, the kinetic energy of the electron has been increased. The velocity now $= \sqrt{v_p^2 + v^2}$ and so the k.e. $= \frac{1}{2}m(v_p^2 + v^2)$. This increase in kinetic energy comes from the loss in electrical potential energy which the electron has undergone, as it has 'fallen' towards the top (positive) plate.

Whilst the electron is moving within the field, we can show that the path is a *parabola*. The horizontal distance, x covered in time t is vt, whilst the vertical distance, y, is $\frac{1}{2}at^2$, where the acceleration $a = eV/md$.

So $x = vt$ or $t = x/v$

and hence $y = \dfrac{1}{2} \cdot \dfrac{eV}{md} \cdot t^2 = \dfrac{1}{2mdv^2} \cdot x^2$

Since $y \propto x^2$, this is the equation of a parabola.

Clearly if y becomes equal to $d/2$ then the electrons will have hit the positive plate. Otherwise the electrons leave the field as shown in Fig 8.2 and then continue in a *straight line* path. Note that gravity affects electrons, like all other objects, and in a vacuum they fall at about 10m s^{-2}. However, electron speeds are usually so high (see Example 11.1) that the time of flight within any apparatus is very short. Thus the distance they fall under gravity is far too small to be measurable.

DEFLECTION IN UNIFORM MAGNETIC FIELD

Suppose an electron enters a uniform magnetic field B with a velocity v at right angles to B (Fig 8.3). As we have seen (p140), the force F on the electron is then Bev, and F is directed always perpendicular to the field and the direction of motion. Since F is constant, it produces a *circular* path as shown. The centripetal force towards the centre, O, of the circle is provided by F, and since the acceleration is v^2/R where R is the circle radius (p44),

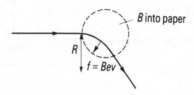

Fig 8.3 Deflection in a magnetic field

$$Bev = ma = mv^2/R$$

So $\qquad R = \dfrac{mv}{Be}$

We see that the radius of the orbit is proportional to the *momentum*, mv, of the electron.

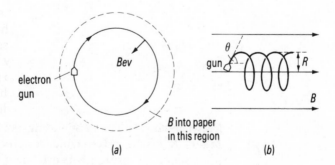

Fig 8.4 Circular and helical orbits

If an electron gun is within the field region, and the magnetic field B is sufficiently strong, the electrons can move in a complete circle (Fig 8.4(a)). If the electrons are not fired at right angles to B, but at an angle θ to B, then a *helical* path is produced as shown in Fig 8.4(b). This is because the initial velocity has two components,

(a) $v \cos \theta$ along the field direction, and

(b) $v \sin \theta$ perpendicular to the field.

$v \cos \theta$ produces no magnetic force and so the electrons drift along the field direction with this speed. $v \sin \theta$ is the component perpendicular to the magnetic field and which produces the circular motion. The radius, R, of the helix will therefore be given by $R.mv \sin \theta/Be$.

Note carefully that a *magnetic field cannot change the energy of the electrons*. The force is always perpendicular to the motion, and so no work is done by the magnetic force. Thus the kinetic energy of the electrons cannot change.

MEASUREMENT OF e/m, SPECIFIC CHARGE OF ELECTRON

Figure 8.5(a) shows a *fine beam tube*. An electron gun emits electrons from the anode A, and the tube is situated in a uniform magnetic field, B, produced by two current carrying *Helmholz* coils. These coils are spaced at a separation equal to their radius, which produces a uniform field over a large volume within the coils, as shown in Fig 8.5(b).

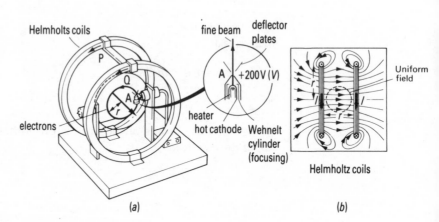

(a)   (b)

Fig 8.5 e/m and Helmholtz coils

Measurements: (1) The accelerating voltage on the anode, V. (2) The radius R of the orbit, which can be measured using a travelling microscope. Measurements are taken when the beam is seen at each end of a diameter, and hence the radius is calculated. (3) The value of the current I in the Helmholz coils. (4) The number of turns N and radius r of each Helmholz coil.

Theory. The value of B produced by the two coils can be calculated from

$$B = \left(\frac{4}{5}\right)^{3/2} \mu_o \frac{NI}{r}$$

From the formula governing the acceleration of electrons in the gun,

$$\tfrac{1}{2}mv^2 = eV$$

so $\quad \dfrac{m}{e}v^2 = 2V \qquad\qquad (1)$

From the magnetic field deflection

$$Bev = \frac{mv^2}{R}$$

or $\quad \dfrac{m}{e}v = BR \qquad\qquad (2)$

Dividing (1) by (2) we get $v = 2V/BR$. Substituting for v in $m/e = BR/v$ we obtain $e/m = 2V/B^2R^2$. From these two equations, v and e/m may be calculated. An approximate value for e/m is 1.8×10^{11} C kg^{-1}.

COMPARISON WITH e/M FOR PROTON

In the electrolysis of water it is found that 1g (10^{-3} kg) of hydrogen is liberated by about 96 500 coulombs. Now if e is the charge of each hydrogen ion (that is, of each proton) and M is its mass, then if N is the number of ions liberated,

$$\frac{\text{total charge}}{\text{total mass}} = \frac{Ne}{NM} = \frac{96\,500}{10^{-3}}$$

So $\quad \dfrac{e}{M} = 9.6 \times 10^7$ C kg^{-1}

This is a much smaller *charge/mass* ratio value than that for an electron, showing that the proton has a much greater mass. In fact, from the two results we can calculate the ratio of the mass M of the proton to the mass m of the electron. Thus

$$\frac{M}{e}\cdot\frac{e}{m} = \frac{M}{m} = \frac{1.8 \times 10^{11}}{9.6 \times 10^2} = 1870$$

Hence the proton has a mass nearly 2000 times that of an electron.

MEASUREMENT OF e– MILLIKAN'S EXPERIMENT

Millikan measured the charge on an electron using an apparatus whose principle is shown in Fig 8.6(a). Two metal plates A and B, a few cm in diameter, are situated a distance d, about 0.5cm, apart. Plate A has a small central hole through which tiny oil drops, produced by an atomiser, enter the region between A and B. A variable voltage supply V is connected between A and B as shown in Fig 8.6(a) and the oil drops can be observed by a microscope due to light which they reflect from a light source (Fig 8.6(b)). Because of their very small size, the drops have a very small final or terminal velocity as they fall through the air.

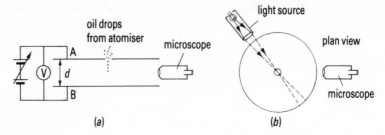

Fig 8.6 Millikan's experiment

WEIGHT OF DROP

When the terminal velocity is reached, we have, from p404,

$$\text{wt of oil drop} - \text{upthrust} = 6\pi a \eta v,$$

where a is the radius of the drop, η is the viscosity of air and v is the terminal velocity.

so
$$\frac{4}{3}\pi a^3 \rho g - \frac{4}{3}\pi a^3 \sigma g = \frac{4}{3}\pi a^3 (\rho - \sigma) g = 6\pi a \eta v,$$

where ρ is the density of the oil and σ the density of the air.

Rearranging,
$$a = \sqrt{\frac{9\eta v}{2(\rho - \sigma)g}}$$

from which the radius of the drop can be calculated. The terminal velocity of the drop is measured by timing the time to fall through a certain distance measured on a graduated scale in the eyepiece of the microscope. Knowing η, ρ and σ from tables, a is then calculated from the last equation, and the weight W of the oil drop can be found from $W = \frac{4}{3}\pi a^3 \rho g$.

METHOD FOR e

The main part of the experiment can now begin. Here the weight of the drop is balanced by an upward electric force on the drop, which is produced when it becomes charged. The stages are as follows:

1 Introduce α-radiation (or X-rays) into the space between the two plates A and B. This produces ionization and some ions or electrons may settle on the selected drop and give it a charge q.

2 Switch on the power supply V. If the drop falls more quickly, then reverse V. This ensures there is an upward force on the drop.

3   Adjust the value of V until the drop is *stationary* and then measure the value of V

4   Change the charge on the drop (or try to) by repeating stage 1. Then repeat 2 and 3.

    Operation 4 is repeated many times to obtain a series of voltages, V, which can cause the drop to be stationary.

Theory. When the drop is at rest, $W = qE$, where E is the electric intensity between the plates. But $E = V/d$. So

$$W = \frac{qV}{d} \text{ or } q = \frac{Wd}{V}$$

From the values of W, V and d, we can calculate q.

CALCULATION A series of values of q are obtained corresponding to the values of V. It is observed that:

(*a*) q has a common factor, that is, q is always a multiple of some basic number. This shows that a charge cannot be obtained in any amount but only *in multiples of some basic quantity.*

(*b*)   The basic common factor is 1.6×10^{-19} coulomb. This is considered to be the charge on *one* electron. In Millikan's experiment the oil drops have a charge which is a multiple of this charge. If, for example, $q = -6.4 \times 10^{-19}$ coulomb, then the oil drop would have had 4 electrons, or 4 negative ions, collected on it. If $q = +4.8 \times 10^{-19}$ coulombs then 3 positive ions would have collected on the oil drop.

    In practice Millikan had to make allowance for two factors:

(i)   he did not know accurately the viscosity of air at the particular temperature of the experiment and

(ii)   Stokes' law is not accurately obeyed for such small drops.

Their size is comparable with the mean free path of air molecules and this means that the air cannot be regarded here as a 'smooth' fluid through which the oil drop falls.

THE CATHODE-RAY OSCILLOSCOPE

STRUCTURE OF OSCILLOSCOPE; BRIGHTNESS AND FOCUS

Figure 8.7(*a*) shows a schematic diagram of the basic features of a cathode-ray tube. A heated cathode C emits electrons which are accelerated by an anode A_1. A small negative potential can be applied between the grid and the cathode. As this p.d. increases, fewer electrons are able to pass G, and so controlling the potential of G controls the *brightness* of the spot.

    Focusing of the spot is achieved by varying the p.d. between two metal anode cylinders A_1 and A_2. This difference of potential sets up electric fields between the electrodes which can focus the electrons. Figure 8.7(*b*) shows the pattern of equipotential surfaces between A_1, which is at 600V, and A_2 at 1200V. If an electron approaches the gap

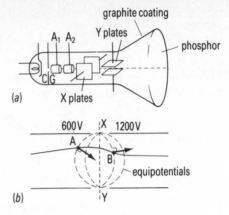

Fig 8.7 Oscilloscope and
electron lens

between the electrodes at an angle, then there will be a component of
force perpendicular to its motion. At A, for example, the electric field,
which is perpendicular to the equipotential surface, gives a force as
shown in Fig 8.7(*b*). The region to the left of the equipotential surface
XY therefore acts as a 'converging lens'. To the right of XY the region
acts as a 'diverging lens' as illustrated by the arrow showing the
direction of force B. The speed of the electrons is, however, now
greater due to the accelerating p.d. between A_1 and A_2. Hence the
diverging effect to the right of XY is not as great as the converging
effect to the left of XY and overall the arrangement acts as a converging *electron lens*. This can produce a focused spot on the screen.

The beam now passes through two pairs of plates. Voltages
applied to the *X-plates* deflect the beam horizontally. The p.d. applied
to the *Y-plates* deflect the beam vertically. A varying p.d. from a
potentiometer circuit also can be applied to one of the X-plates. This
allows the spot to be *shifted* so that a given trace on the screen can be
positioned as needed, and the control is called the *X-shift*. There is
also a similar *Y-shift* control.

The screen is coated with a phosphor which glows when the
electrons strike it. The inside of the tube is coated with a conducting
graphite paint and is earthed, together with the anode A_2. This
screens the electron beam for other electrostatic effects as it travels to
the screen. The cathode is at a high negative potential, usually a few
kV, with respect to the earth, and is therefore dangerous to touch.

TIME-BASE

A circuit inside the oscilloscope can be used to apply a *time-base* to the
X-plates. A p.d., which varies as shown in Fig 8.8 is generated
electrically and is connected to the X-plates. The spot is then driven at
a uniform rate horizontally across the screen, and then, during the
flyback, the spot returns swiftly to the left hand side of the screen and

repeats the motion. The rate at which this happens can be controlled by a multi-position switch on the front panel of the oscilloscope. This switch is labelled with the time taken for the spot to cross each centimetre of the screen, and is called the *time-base speed* or the time/cm switch. If the switch is set in the position 5ms/cm, then it takes 5ms or 5×10^{-3} s for the spot to cross each centimetre of the screen.

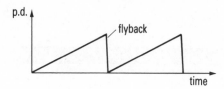

Fig 8.8 Time-base voltage

Y-AMPLIFIER

Signals from the input connected to the Y-plates go *via* the Y-amplifier. This is an accurate amplifier whose *gain* is set precisely and which can be controlled from the front panel by means of a switch. The switch is labelled with the number of volts/cm. Thus a setting of 5 volts/cm means that it takes a p.d. of 5 volts at the input to give a vertical deflection of 1cm.

X-INPUT

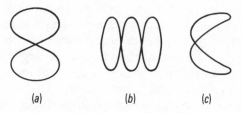

(a)   (b)   (c)

Fig 8.9 Lissajous' figures

It is possible to connect external signals to the X-plates without using the time-base. This is done to display *Lissajous' figures*, which can give accurate comparison of frequencies where the frequencies are in simple whole number ratios. In Fig 8.9(a), a.c. signals are connected to the X- and Y-plates respectively. It can be seen that the spot moves side to side twice as often as it moves up and down. Hence the frequency of the signal applied to the X-plates is exactly double that applied to the Y-plates. In Fig 8.9(b) the spot moves up and down three times whilst it moves from side to side once. Here the frequency applied to the Y-plates is three times the frequency applied to the X-plates.

It should be noted that the precise shape of the figure depends on the phase between the two signals. Figure 8.9(*c*) shows the same frequencies as 8.9(*a*) but with a different phase relation. When the X- and Y-voltages are both of the same frequency and are in phase, a slanted *line* is obtained. If the voltages are 90° out of phase a *circle* or an *ellipse* is produced when the voltages vary sinusoidally.

ELECTRONICS

SEMICONDUCTORS – THE TRANSISTOR

CONDUCTION IN PURE SEMICONDUCTORS

Semiconductors are a class of materials with a conductivity between that of metals and that of insulators. Silicon and germanium are the two most commonly used semiconductor materials. At absolute zero (0 K) pure semiconductors behave as insulators and there are no free charge carriers. As the semiconductor temperature increases, more charge carriers are produced and the conductivity increases as we now explain.

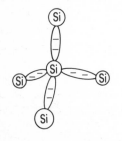

Fig 8.10 Structure of silicon

Silicon and germanium are tetravalent and have four electrons which form covalent bonds with neighbouring atoms in a tetrahedral fashion, as shown in Fig 8.10. At absolute zero there are no thermal vibrations to disturb the electrons so they remain within the bonds. As the temperature rises, however, the thermal energy of vibration is sufficient to displace some electrons and give them enough energy to become free. The electrons attached to atoms are said to be in the **valence band** of energy. Free electrons are in the **conduction band**, which is a band of higher energy values. There is an energy gap between these bands and the thermal vibrations can give some electrons in the valence band enough energy to cross the gap.

If an electron, charge $-e$, leaves an atom, then that atom is left with a net positive charge, $+e$. This region of positive charge can move as units $+e$ of positive charge and are called **holes**. In a semiconductor the electric current will be carried by positive holes moving in the conventional current direction, and electrons moving in the opposite direction. In general, however, the electrons are much more mobile and move with a greater speed when a p.d. is connected across the semiconductor. For this reason the Hall voltage of a pure semiconductor shows that the predominant charge carriers are negative, although holes and electrons are present in equal numbers.

EFFECT OF TEMPERATURE RISE

When the temperature of a semiconductor increases, the thermal vibrations cause many more **electron-hole pairs** to be formed. Thus the conductivity increases (or the resistivity decreases) rapidly as the temperature rises due to the greatly increased concentration of charge carriers. In a metal the same number of conduction electrons are free at all temperatures, and so temperature does not affect the charge carrier concentration. In contrast to the pure semiconductor, the conductivity of a pure metal *decreases* with temperature due to the lattice vibrations impeding the electron flow, that is, the electrical resistance of a pure metal increases with temperature.

DOPED SEMICONDUCTORS

Semiconductors can be **doped** by the addition of impurities – typically 1 part per million. Arsenic has five valence electrons and if this is added to germanium, the arsenic atom will take the place of a germanium atom in the crystal lattice. There is, however, one valence electron spare and this becomes free to move. Thus *each arsenic atom contributes a free electron*. There is also a *fixed* positive ion at that lattice site. This positive charge is not free to move and so is not a hole. The doping with arsenic thus produces a majority of negative charge carriers and this is called an **n-type** semiconductor. There are still electrons and holes produced thermally, as we described above, but these are comparatively few in number. The electrons are called **majority** carriers and the holes **minority** carriers. Doped semiconductors are described as **extrinsic**, while pure semiconductors are described as **intrinsic**. The arsenic atoms which provide the electrons are called **donors**.

Boron or indium are trivalent elements. If these are used to dope the germanium than there is one valence electron too few at impurity sites. Thus a positive hole is formed which is free to migrate. At the impurity lattice site there is a *fixed* negative ion. Impurity atoms in this case are called **acceptors** as they accept electrons to form the negative ion. In this type of doped semiconductor the majority carriers are positive holes and so the material is called **p-type**.

p-n JUNCTION; JUNCTION DIODE

Figure 8.11(*a*) shows diagrammatically a **p-n junction diode** which is manufactured by fusing together p- and n-type semiconductors so that a very thin junction is formed between them. Immediately the junction is formed some positive holes on the left of the junction will drift (or diffuse) across the junction into the n-type region and recombine with electrons. This means that there is a net positive charge in the region immediately to the right of the junction since that region was previously neutral. Similarly, some electrons will diffuse to the left and recombine with holes, making that region electrically negative (Fig 8.11(*b*)). Thus a p.d. is set up across the junction which prevents further diffusion of charge carriers. This p.d. is called a **barrier p.d.** and is about 0.6 volts in value.

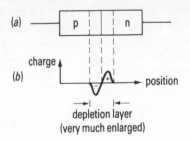

Fig 8.11 p-n junction

Since recombination has taken place either side of the junction, this area is deficient in charge carriers and is called the **depletion layer** or **depletion region**. Its thickness is very small and of the order of 10^{-3}mm.

JUNCTION DIODE AS RECTIFIER

Suppose a p.d. is applied to the p-n junction as in Fig 8.12(a). Provided the p.d. applied is greater than the barrier p.d., good conduction takes place and the diode is said to be **forward-biased**. The conduction takes place as follows. Electrons from the wire flow into the n-type material at C. These are urged towards the depletion layer at B. At A, electrons flow out of the p-type material and so holes move to the right. In the depletion layer holes and electrons recombine, so the depletion layer decreases in width. The net effect is to produce a current in the circuit, which is carried by holes along AB and electrons along CB.

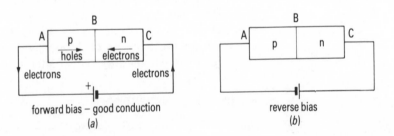

Fig 8.12 p-n junction diode

If the p.d. is applied as in Fig 8.12(b) the junction is said to be **reverse-biased**. In this case the holes in the p-type material and the electrons in the n-type region are pulled away from the junction. The depletion layer thus increases in width and the conduction ceases since charge cannot now cross the depletion layer. The only conduction which takes place is by the minority charge carriers in each semiconductor region. Thus the current is very low.

The p-n junction therefore acts as a rectifier and the semiconductor used is called a **junction diode**. The symbol is shown in Fig 8.13(*a*) and a graph of current *I* against p.d. *V* is shown in Fig 8.13(*b*).

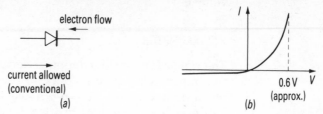

Fig 8.13 Diode symbol and characteristic

RECTIFIER CIRCUITS

Figure 8.14(*a*) shows a simple **half-wave** rectifier circuit. A graph of p.d. against time *t* is illustrated inset. The diode only conducts when A is positive with respect to B, that is for half a cycle. In Fig 8.14(*b*) the d.c. across the load is *smoothed* by use of a suitable large capacitor C. The capacitor charges during one half-cycle, and remains partly charged during the next half cycle when the diode does not conduct. Hence the p.d. across the load will fluctuate less than the arrangement of Fig 8.14(*a*).

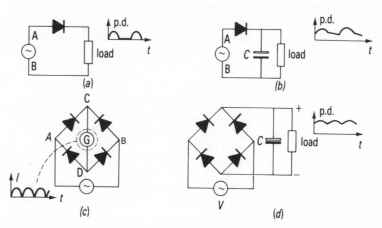

Fig 8.14 Rectifier circuit

Figure 8.14(*c*) shows a **bridge rectifier**, connected so that a d.c. meter, G, can be used for a.c. When A is positive with respect to B the current flows from A to C to D to B. No current can flow from A to D or from C to B. When A is negative with respect to B the current flows from B to C to D to A. So the current through the meter is always in the same direction and flows during *both* half cycles. The variation of current *I* with time *t* through G is shown. A bridge rectifier is often used in a power supply as in 8.14(*d*). Here a suitable capacitor C is also connected across the output to produce smoothing of the voltage.

THE TRANSISTOR

The **transistor** is a three terminal device which consists essentially of two p-n junctions placed back to back. It exists in two forms, the n-p-n transistor and the p-n-p transistor. The arrangement of the p- and n-type semiconductor materials together with their circuit symbol are shown in Fig 8.15(*a*) and (*b*). The three regions of semiconductor used are called respectively, the **emitter** E, **base** B, and **collector** C, as shown.

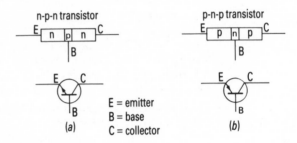

Fig 8.15 Transistor
construction and symbols

In use the emitter-base junction is *forward-biased* and the collector-base junction is *reverse-biased* as shown in Fig 8.16.

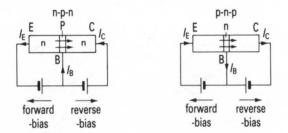

Fig 8.16 Transistor
operation

In the n-p-n transistor, electrons from the emitter flow into the base region. If the base region were wide, these electrons would recombine with holes in the p-type base and the emitter current I_E would be equal to the base current I_B. However, the base is deliberately made *very thin*, much smaller than the average distance travelled by an electron before recombining with a hole. Thus only a small number of electrons combine with holes in the base region and so I_B is much less than I_E. Most electrons travel on into the n-type material of the collector and so give rise to the collector current I_C. From this discussion, it is clear that

$$I_E = I_B + I_C$$

The amplifying property of the transistor stems from the fact that I_C is much greater than I_B. A roughly constant fraction of the electrons from the emitter recombine in the base. So if I_B changes by a small amount such as 0.02mA there is a large change, such as 2mA, in I_E (and I_C). This represents a current amplification of 100 times.

In the p-n-p transistor circuit of Fig 8.16(b), holes from the p-type emitter only partly recombine in the n-type base. Most pass into the p-type collector and produce the collector current I_C. In other respects the action is similar to an n-p-n transistor, but the polarities of the supply battery must be reversed.

TRANSISTOR CHARACTERISTICS

There are three main characteristic graphs for a transistor.

1 *The input characteristic.* Here the input current is plotted against the input voltage for constant output voltage. The reciprocal of the gradient of the graph at any point is therefore the *input resistance* at that point.

2 *The output characteristic.* Here the output current is plotted against the output voltage for a constant input current. Normally a family of curves is plotted for various input currents. The reciprocal of the gradient of the graph is the *output resistance*.

3 *The transfer characteristic.* Here the output current is plotted against the input current for a constant output voltage. The slope of the graph is the *current amplification factor* or *current gain*.

Figures 8.17–8.20 shows the characteristics obtained for a common-base and common-emitter configuration. The figures on the graphs are typical of a small n-p-n transistor. As we shall see shortly, the common-emitter circuit is widely used as an amplifier circuit.

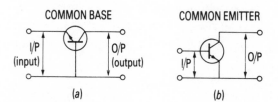

Fig 8.17 Transistor configurations

The input resistance in the first diagram varies since the graph is curved. It is about 20Ω to 100Ω.

The input resistance in the second diagram varies as the graph is curved. It is about 500Ω to $1k\Omega$

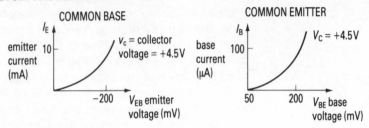

Fig 8.18 Input characteristic

Output characteristic

The output resistance in the first diagram is very high since the graphs are almost parallel to the horizontal axis.

The output resistance in the second diagram is high although not as high as in the common base configuration.

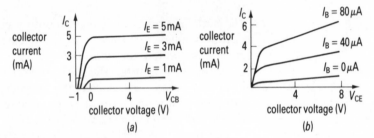

Fig 8.19 Output
characteristic

Transfer characteristic

Current gain, α, is slightly less than unity in common-base configuration.
In common-emitter configuration the current gain, β, is about 50.

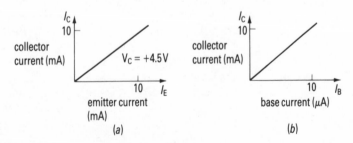

Fig 8.20 Transfer
characteristic

RELATIONSHIP BETWEEN α AND β

We have seen that

$$I_E = I_C + I_B$$

so $$\Delta I_E = \Delta I_C + \Delta I_B$$

where Δ denotes the changes in the various quantities.

Thus $$\frac{\Delta I_E}{\Delta I_C} = 1 + \frac{\Delta IB}{\Delta I_C}$$

so $$\frac{1}{\alpha} = 1 + \frac{1}{\beta} \text{ or } \beta = \frac{\alpha}{1-\alpha}$$

If $\beta = 50$, then $50 - 50\alpha = \alpha$, so $\alpha = \frac{50}{51} = 0.98$.

TRANSISTOR AS A SWITCH

In normal use the base-emitter voltage V_{BE} of the transistor is less than 0.7V. Once the base-emitter voltage exceeds 0.7V, the barrier potential across the base-emitter junction is overcome and large changes in base current can occur for negligible change in base-emitter voltage.

In Fig 8.21, if the input voltage, V_{in}, is zero, then $I_B = 0$ and so $I_C = 0$. In this case the voltage V_1 across the load R is zero, and so the output voltage, $V_{out} = V_{CC}$, the supply voltage. In this condition the transistor is not conducting and is said to be *off*.

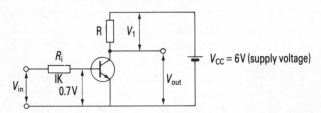

Fig 8.21 The transistor as a switch

If the input voltage V_{in} rises to 0.9V say, then the p.d. across the input resistor $R_i = 0.9 - 0.7 = 0.2$V. The base current $I_B = 0.2/R_i = 0.2/1000 = 0.2$mA. The collector current $I_C = \beta I_B = 50 \times 0.2 = 10$mA, if $\beta = 50$. Suppose the load resistor $R = 500\Omega$. Then the p.d. V_1 across $R = IR = 10 \times 10^{-3} \times 500 = 5$V. Hence the output voltage, $V_{out} = 1$V. Thus the small rise in input voltage has caused the output voltage to fall considerably. If the input voltage increases to 0.94V a similar calculation shows that $I_C = 12$mA and $V_1 = 6$V, so V_{out} is now theoretically zero. Any further increase in V_{in} cannot increase I_C and the transistor is said to be *saturated* or *bottomed*. In this case the current is limited by V_{cc} and R. The transistor is now said to be *on*.

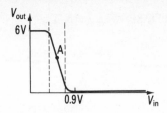

Fig 8.22 Input/output p.d. for
a switch

Figure 8.22 shows the variation of the output voltage with input
voltage. The transistor can thus switch from 'off' to 'on'.

COMMON-EMITTER AMPLIFIER

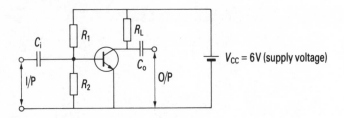

Fig 8.23 Common-emitter
amplifier circuit

Figure 8.23 shows a circuit diagram of a simple amplifier. As was
stated on p264 in the n-p-n transistor the emitter-base junction is
forward-biased and so the base must be maintained at a positive
potential with respect to the emitter. The collector-base junction is
reverse-biased, so the base must be negative with respect to the
emitter. This biasing, which sets the suitable operating point on the
characteristic curves, is provided by the potential divider arrangement R_1, R_2. If the current through R_1 and R_2 is much larger than I_B,
then the base-emitter p.d. is $R_2/(R_1 + R_2) \times V_{CC}$. This p.d. is usually
set at a few-tenths of a volt, and the amplifier operates at the point A
shown in Fig 8.22. It is clear that if the a.c. signal applied at the input
causes V_{in} to move outside the range shown by the dotted lines, then
distortion of the waveform will occur.

The load resistor R_L in Fig 8.23 is needed so that changes in
collector current produce changes in the output voltage.

Suppose a small a.c. signal is applied to the input, I/P. The capacitor C_i isolates any d.c. component in the input signal and so the
biasing cannot be affected. If, at some instant, the base current
increases by $\triangle I_B$, then the collector current increases by $\beta \triangle I_B$. This
causes the p.d. across R_L to *rise* by $\beta \triangle I_B.R_L$. Thus the output voltage
falls by $\beta \triangle I_B.R_L$. The amplifier thus produces a 180° phase shift.

$$\text{The voltage gain} = \frac{\text{change in O/P voltage}}{\text{change in I/P voltage}}$$
$$= \frac{\beta \triangle I_B.R_L}{\triangle V_i}$$

But $\triangle V_i = \triangle I_B.r_i$ where r_i is the input resistance, which can be found from the reciprocal of the gradient of the input characteristic at the operating point. Hence

$$\text{voltage gain} = \frac{\beta R_L}{r_i}$$

If, for example $\beta = 50$, $r_i = 1000\Omega$ and $R_L = 2.2k\Omega$, then the voltage gain $= 50 \times 2200/1000 = 110$.

DIGITAL AND ANALOGUE CIRCUITS

A **digital circuit** is one where the voltage levels are either high or low – no other level is permitted. Values in digital circuits can be represented by a number of levels which form a binary number. **Analogue circuits**, however, have voltages which can vary continuously. Analogue circuits include amplifiers, where the output voltage varies in proportion to the input voltage, or sine wave oscillators, where the output varies in a smooth sine curve way. We deal first with some aspects of digital electronics.

DIGITAL ELECTRONICS – LOGIC GATES

VOLTAGE LEVELS

In digital electronics voltages can be either HIGH (ON) or LOW (OFF). Only these two voltage states are possible. It is usual to denote the high voltage by a so-called logic state '1', and a low voltage by logic state '0'. The following discussion of a NOT and other gates illustrates these ideas. Gates have one or more inputs and one output.

NOT GATE

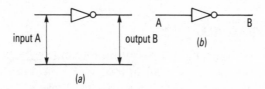

input A   output B

(a)

A   B

(b)

Fig 8.24 The NOT gate

A NOT gate is simply an inverter, that is, a high voltage input is converted to a low voltage output and vice-versa. The circuit symbol of a NOT gate is shown in Fig 8.24(*a*). Usually the 'earth' or zero potential line is omitted as in Fig 8.24(*b*).

INPUT		OUTPUT	
high	1	low	0
low	0	high	1

Fig 8.25 Truth table for a
NOT gate

The behaviour of the gate can be described by a *truth table* (Fig 8.25). Usually the truth table is shown only in terms of '1's and '0's.

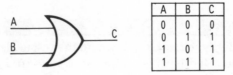

A	B	C
0	0	0
0	1	1
1	0	1
1	1	1

Fig 8.26 The OR gate

OR GATE

An OR gate has two inputs A, B and a single output C. The output C is high if either A or B (or both) is high. The circuit symbol is shown in Fig 8.26(*a*) and the truth table in Fig 8.26(*b*).

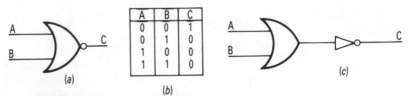

A	B	C
0	0	1
0	1	0
1	0	0
1	1	0

(a)

(b)

(c)

Fig 8.27 The NOR gate

If the output of the OR gate is inverted, the resulting gate is called a NOR gate. The symbol is given in Fig 8.27(*a*), the truth table in Fig 8.27(*b*), and an equivalent circuit comprising an OR gate followed by an inverter in Fig 8.27(*c*).

AND AND NAND GATES

An AND gate gives an output only if inputs A and B are both high. A NAND gates gives an output which is the inverse of AND. Fig 8.28 illustrates the circuit symbols for these gates and shows their truth tables.

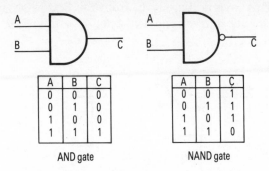

A	B	C
0	0	0
0	1	0
1	0	0
1	1	1

AND gate

A	B	C
0	0	1
0	1	1
1	0	1
1	1	0

NAND gate

Fig 8.28 The AND & NAND gates

EXCLUSIVE-OR GATE

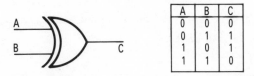

A	B	C
0	0	0
0	1	1
1	0	1
1	1	0

Fig 8.29 The EXCLUSIVE-OR gate

This gate is often useful. It gives a high output if either of the inputs A or B is high, but not both. (Compare this carefully with the OR gate.) The symbol and truth table are shown in Fig 8.29.

GATE INTERCHANGEABILITY

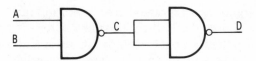

Fig 8.30 An AND gate from two NAND gates

When gates are purchased they are normally packaged so that several come on one chip. It might therefore be the case that, for example, several NAND gates are unused but an AND gate is required. Fig 8.30 shows how this can be done. In this circuit if C is low then both inputs of the second NAND gate will be low, and from the truth table this will give a high output at D. If C is high then both inputs of the NAND gate will be high and so the output will be low. Thus the second NAND gate with its inputs connected together behaves as an inverter.

This inverts the output from the first NAND gate and so the whole circuit behaves as an AND gate.

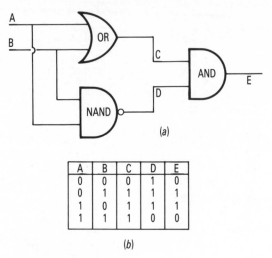

(a)

A	B	C	D	E
0	0	0	1	0
0	1	1	1	1
1	0	1	1	1
1	1	1	0	0

(b)

Fig 8.31 EXCLUSIVE-OR from other gates

The above circuit shows how an exclusive OR gate can be constructed from a combination of other gates. Fig 8.31(a) shows the circuit and Fig 8.31(b) shows the truth table.

Looking at the truth table it can be seen that the column for C is obtained by ORing the columns for A and B. The column for D is obtained by NANDing the columns for A and B. Since C and D are then connected to an AND gate, the final output E is obtained by ANDing the C and D columns.

ADDERS

A	B	C	S
0	0	0	0
1	0	0	1
0	1	0	1
1	1	1	0

Fig 8.32 Truth table for a half-adder

In computers it is often necessary to add two binary digits, A and B. If A is 0 and B is 0, then the sum S is zero. If either A or B=1 (but not both), then S must be 1. If, however, A and B are both 1 then the sum in binary is 10. In electronic terms this is handled by making the sum S=0 and generating a 'carry', C, to indicate the 1 in the next place to the left. The binary addition process can be summarized in the truth table shown in Fig 8.32. S represents the sum of the two digits, and C indicates if there is a carry. It is easy to see how this can be generated electronically. The carry column is related to A and B by an AND operation. The sum column is related to A and B by an EXCLUSIVE-OR operation. The resulting circuit is called a **half-adder** and is shown in Fig 8.33.

The problem with the half-adder is that it generates a carry if A and B are both 1, but it does not add in a carry from a previous stage. This would be necessary if we were adding binary numbers several

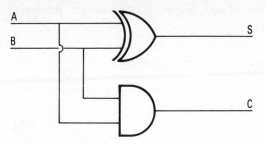

Fig 8.33 A binary half-adder

digits long. To add carries previously generatred a **full adder** is needed. Essentially it has three binary digits to add, the two currently being added, together with any previous carry. Fig 8.34 shows the truth table. C_{in} is the carry in and C_{out} is the carry out. S is the sum as before.

C_{in}	A	B	C_{out}	S
0	0	0	0	0
0	1	0	0	1
0	0	1	0	1
0	1	1	1	0
1	0	0	0	1
1	1	0	1	0
1	0	1	1	0
1	1	1	1	1

To make a full adder electronically two half-adders are used. First A and B are added to give a sum. This sum is then added to the carry in to give the final sum. A carry is generated from the full adder if one or other of the two half-adders generates a carry. The carries from each half-adder are therefore 'OR'ed together. The reader should verify that it is impossible for both half-adders to generate a carry, and that

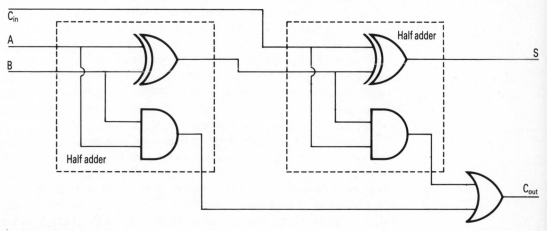

Fig 8.35 A binary full adder

the circuit of Fig 8.35 does correctly produce the truth table for the full adder.

BISTABLE CIRCUITS; FLIP-FLOPS

A bistable circuit has two electronically stable states. It is important in memory circuits of computers since it can 'remember' the last state it was put into even when the signal putting it into that state has been removed. Bistables are also used in binary counting circuits as we shall see.

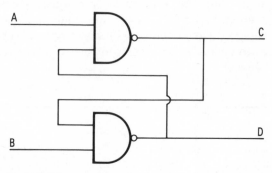

Fig 8.36 A simple set-reset
flip-flop

The circuit of Fig 8.36 is a simple form of bistable circuit constructed from two NAND gates. Suppose for the moment that both inputs A and B are high, and that C is high and D is low. The output C is connected to the input of the lower NAND gate. Hence the inputs of this gate are both high and so D is low. D in turn is connected to the input of the upper NAND gate. This gate therefore has one high and one low input and so the output C is high. Thus our initial assumptions about the circuit are possible. The reader should verify that with A and B both high it is equally possible to have C low and D high. Hence the circuit has two stable states.

Now suppose that while C is high and D is low, input B goes low. The inputs to the lower NAND gate are now dissimilar and so D goes high. Since D is connected to the upper gate both inputs are now high and so C goes low. The bistable has therefore switched to its other state. As soon as this has happened B can go high again and the circuit will remain in its new state. The reader can verify that to switch the circuit again it is necessary to use input A.

This circuit is called set-reset flip flop, or RS flip-flop for short. The outputs are normally given the symbols Q and $\bar{Q}$, where $\bar{Q}$ denotes the opposite state to Q. The flip-flop is always used with the outputs in opposite states. The flip-flop is said to be set when Q is high, and reset when Q is low. As we have seen the upper input is used to make Q become high and so this is the 'set' input. It is denoted by $\bar{S}$.

The bar on the S indicates that the input must go low in order to switch the circuit. Similarly, the lower input is given the symbol $\bar{R}$.

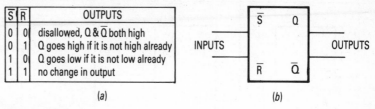

$\bar{S}$	$\bar{R}$	OUTPUTS
0	0	disallowed, Q & $\bar{Q}$ both high
0	1	Q goes high if it is not high already
1	0	Q goes low if it is not low already
1	1	no change in output

(a)

INPUTS   OUTPUTS

(b)

Fig 8.37 The RS flip-flop

Fig 8.37(a) gives the behaviour of the RS flip-flop in tabular form and Fig 8.37(b) gives the circuit symbol. Note that it is not allowed to make the S and R inputs both zero. If this were done both the Q and the $\bar{Q}$ output would be high and the flip-flop is never used in this state.

BINARY COUNTERS

An RS flip-flop is not suitable for all purposes because the state cannot be changed continuously using only one input. A more complex bistable called a JK flip-flop can overcome these difficulties. The behaviour of this device is more complicated in general but one particular use can be discussed here. If the J and K inputs are both high then the third 'clock' input C behaves in a simple manner. Each time the clock input falls from high to low the state of the bistable switches. Fig 8.38(a) shows the circuit symbol of the flip-flop, and Fig 8.38(b) shows how the output varies if a square wave signal is applied at the input.

In Fig 8.38(b), the frequency of the output is half that of the input

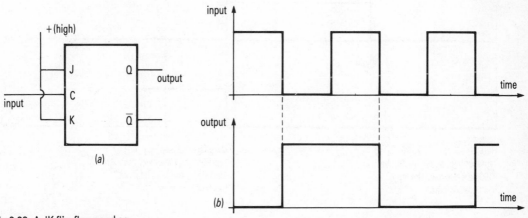

Fig 8.38 A JK flip-flop used as a binary divider

since the bistable switches only when the input falls. If a series of
such flip-flops are arranged in a row then the result is called a **binary**

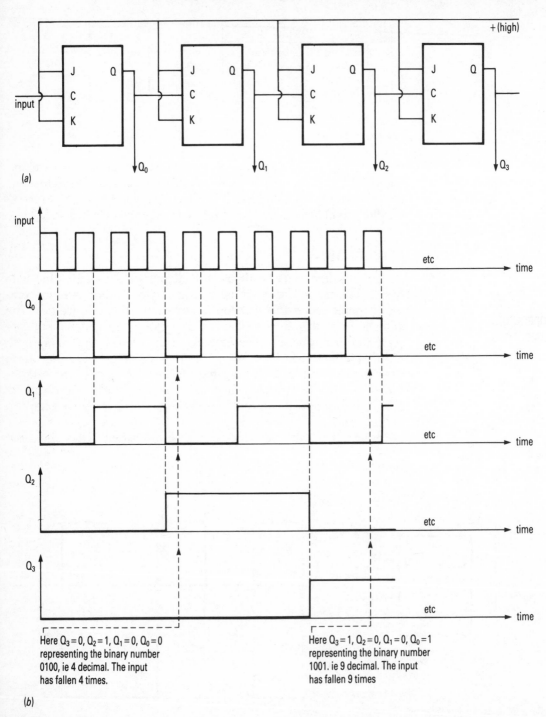

(a)

Here $Q_3 = 0$, $Q_2 = 1$, $Q_1 = 0$, $Q_0 = 0$
representing the binary number
0100, ie 4 decimal. The input
has fallen 4 times.

Here $Q_3 = 1$, $Q_2 = 0$, $Q_1 = 0$, $Q_0 = 1$
representing the binary number
1001. ie 9 decimal. The input
has fallen 9 times

(b)

Fig 8.39 A binary counter
with waveforms

counter. Fig 8.39(*a*) shows the circuit and Fig 8.39(*b*) shows the waveforms at the various outputs. The circuit is called a counter because as the pulses arrive the data at Q_0 to Q_3 behaves as a 4 digit binary number. To start with these outputs are all zero and then go through the following sequence:

Q_3	Q_2	Q_1	Q_0
0	0	0	1
0	0	1	0
0	0	1	1
0	1	0	0
....	....	....	....
1	1	1	0
1	1	1	1

This sequence generates the binary numbers 1–15. Note from Fig 8.39(*b*) that what is being counted is the number of falling edges in the input waveform.

We can now move from digital circuits to analogue circuits and consider the operational amplifier.

OPERATIONAL AMPLIFIERS

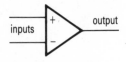

Fig 8.40 Opamp circuit symbol

Fig 8.40 shows the basic circuit symbol of an operational amplifier. It has two inputs, the **inverting** input ($-$) and the **non-inverting** input ($+$), together with one output. The amplifier has a very high gain such as 10^6, and amplifies the **difference** in p.d. between the inverting and the non-inverting inputs. The output is positive if the non-inverting input is more positive than the inverting input. The output goes negative when the non-inverting input is at the higher potential. There is no need for the non-inverting input to be $+$ve and the inverting input to be negative. It is only the relative potential that matters. For example -3.21V on the ($+$) input and -3.22V on the ($-$) input would give a $+$ve output since the ($+$) input is at the higher potential.

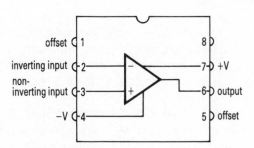

Fig 8.41 Opamp 'pin-outs'

In practice, the operational amplifier must have a **split** power supply, usually $+15$V and -15V with respect to the 0V line for inputs and outputs. Fig 8.41 shows the pin connections to the 741, a common op-amp used in many practical circuits. Pin 8 is not used.

In addition to the power supply terminals and input and output terminals there are two **offset-null** terminals. These are connected to some external circuitry to make sure that the output of the amplifier is precisely zero when the p.d. between the (+) and (−) input terminals is zero. Without this circuit there would be some output even if the input terminals were connected together.

The gain of the amplifier is so high that a p.d. of only a fraction of a millivolt between the inputs causes the output to **saturate**, that is, to become as large as it can. This will be very nearly the supply voltage and will be positive if the (+) input is at the higher potential, and negative if the (−) input is at the higher potential.

It is very important to note that the input current drawn by the opamp is very small, about 0.1μA for a 741 and much less for FET opamps.

The saturating behaviour of opamps means that they can be used effectively for switching and as voltage comparators.

THE OPAMP AS A SWITCH AND VOLTAGE COMPARATOR

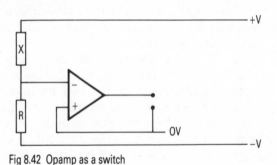

Fig 8.42 Opamp as a switch

Fig 8.42 shows the circuit diagram of an opamp used as a **switch**. In this and all other opamp circuits the power supply and off-set circuits are not shown.

In this circuit the (+) input is held at 0V by a direct connection. The (−) input will be at 0V only when X is exactly equal to R, as in this case there will be 15V across both X and R. Suppose, for example, X is a light dependent resistor (LDR). In the dark its resistance will be very high, much greater than R. In this case the (−) input is negative because the potential drop across the LDR is now greater than 15V. Hence the output is positive. When light shines on X its resistance may drop to below that of R, if the value of R is chosen correctly. The (−) input will now be +ve, and so the output will be negative. Because of the high gain, the switch in output from −ve to +ve will occur very suddenly at a particular illumination of the light dependent resistor. The reader should also consider the behaviour of the above circuit if X and R were interchanged, or if the (−) input were connected to 0V and the (+) connected to the join of X and R. Circuits such as this could be used to switch on the lights in an unattended house when it got dark. The output of the opamp would be connected to a circuit which operates a relay when the output switches.

The operation of Fig 8.43, a **voltage comparator**, is very similar. In this circuit the (+) input is connected to a potential divider and is thus at a fixed potential. If the (−) input is higher than this the output will be negative, and vice-versa. The circuit will therefore switch as the potential of the (−) input changes from just above to just below the fixed potential of the (+) input.

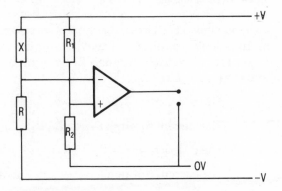

Fig 8.43 Voltage comparator

A circuit based on Fig 8.43 can be used to control temperature. X would then be a thermistor with R a variable resistor. The value of R would define the particular temperature required. When the temperature reached its required value the opamp would switch and the change in output could be used to cut off the input of heat energy (via some type of relay). When the temperature fell, the output would switch back and the heat supply be re-established.

FEEDBACK

An operational amplifier is also used in amplifier circuits with **feedback**, that is, with some component connecting the output and input. The behaviour of opamp circuits with feedback can be found, to a good degree of approximation, by using the following rules.

1 The inputs to the opamp itself draw no current. In practice this is usually less than $0.1\mu A$ which is generally small enough to ignore.

2 The output will adjust itself so that the feedback makes the difference in p.d. between the inputs as small as possible.

INVERTING AMPLIFIER

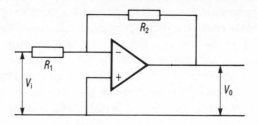

Fig 8.44 Inverting amplifier

Fig 8.44 shows the circuit diagram of an inverting amplifier. The $(+)$ input is at 0V, so rule 2 says that the $(-)$ input will also be at 0V. Rule 1 says that the current through R_1 flows on through R_2 as none flows into the $(-)$ input.

The p.d. across $R_1 = V_i - 0 = V_i$, as the $(-)$ input is at 0V.

The current through $R_1 = I = V_i/R_1$.

The p.d. across $R_2 = 0 - V_0$ as the $(-)$ input is at 0V.

The current through $R_2 = -V_0/R_2$ and this is also I.

Hence $V_i/R_1 = -V_0/R_2$

The gain of the amplifier is defined as V_0/V_i and so

$$\text{GAIN} = V_0/V_i = -R_2/R_1.$$

From this it can be seen that V_0 is opposite in sign to V_i and hence the name **inverting** amplifier. The gain is numerically R_2/R_1.

If, for example, $R_2 = 100k$ and $R_1 = 10k$ then the gain would be 10. The above analysis holds for a.c. or d.c. In the former case V_0 will alternate out of phase with V_i and will have peak values R_2/R_1 times as large.

Notes

1   A fraction of the signal is fed back to the $(-)$ input. The increasing positive output causes the $(-)$ input to become more positive and this reduces the change in output. This is called negative feedback. Feedback to the $(+)$ input would have the opposite effect and the output would immediately saturate.

2   Since the gain is $-R_2/R_1$ the gain is NOT determined by the amplifier itself but by the feedback network. This greatly improves the stability of the amplifier since the very high gain of the opamp itself is subject to wide variation as the conditions and temperature of use change.

3   The energy supplied at the output comes from the power supply.

4   The input resistance of the amplifier is R_1. This means that the input 'sees' a resistance of R_1 when connected to the circuit. This is because a current of V_i/R_1 flows.

5   The output resistance of the circuit is very low – a fraction of an ohm.

This means that the output of the opamp can be regarded as a source with very low internal resistance. The output voltage is therefore practically independent of any load to which the opamp is connected. This is a considerable advantage.

NON-INVERTING AMPLIFIER

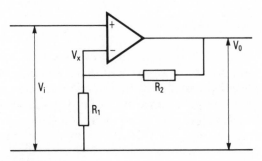

Fig 8.45 Non-inverting amplifier

Fig 8.45 shows the circuit diagram of a non-inverting amplifier. From Rule 2

$$V_i = V_x$$

From Rule 1 no current flows into the inverting input, so that I, the current through $R_1 =$ current through R_2. Now

$$I = V_0/(R_1 + R_2) = V_x/R_1 = V_i/R_1$$

Thus

$$V_0 = V_i (R_1 + R_2)/R_1 = V_i (1 + R_2/R_1)$$

The gain is given by

$$gain = V_0/V_i = 1 + R_2/R_1$$

Note that here again the feedback is negative since the feedback network is connected to the $(-)$ input.

For $R_1 = 10k$ and $R_2 = 100k$ the gain is 11.

Apart from the fact that this amplifier does not invert there is another important difference compared with the inverting amplifier. Here the input is connected directly to the input of the opamp, and so by rule 2, a very small current flows. The input resistance is therefore very high. This is often useful since many sources of p.d. cannot provide a large current without severely affecting their operation.

The output impedance is still very low.

BUFFER

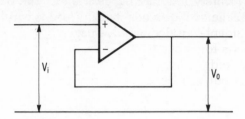

Fig 8.46 Opamp buffer

In Fig 8.46 the $(-)$ input is connected directly to the output. By rule 2 the potential of each input is the same so $V_1 = V_0$. This may seem a pointless circuit, but it is in fact very useful. The device has a very high input resistance but its output resistance is very low. The device can therefore take its input from a source capable of providing only a small current, and deliver at its output the same p.d. with higher current providing capability. Since this circuit protects the sensitive p.d. source from circuits which draw heavy current, it is called a 'buffer'.

OSCILLATORS

SQUARE WAVE OSCILLATOR

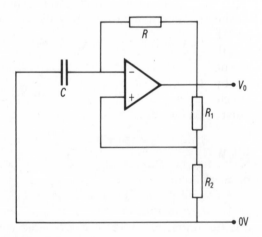

Fig 8.47 Square-wave
oscillator

Consider the circuit of Fig 8.47. Note firstly that there is positive feedback since the output is connected *via* R_1 and R_2 to the $(+)$ input. In this case the output will saturate. Suppose the output is positive somewhere near $+V$, the positive supply voltage. Also let $R_1 = R_2$

In this case the $(+)$ input will be at $+V/2$. If initially C is uncharged there is no p.d. across it. The $(-)$ input is therefore at 0V. This is consistent with the output being $+V$. C will now charge up through R from this positive output. When the p.d. across C is greater than $V/2$ the $(-)$ input will have the higher potential and the output will suddenly switch and saturate so that $V_0 = -V$. The capacitor now starts to discharge and then charge with the opposite polarity until the p.d. across it drops below the $-V/2$ potential of the $(+)$ input. The output then suddenly swings to $+V$ and so the cycle continues.

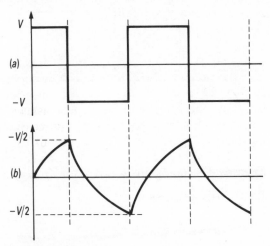

Fig 8.48 Voltages in square-wave oscillator

The graph of p.d. against time for the output is shown in Fig 8.48 (a). The p.d. variation at the $(-)$ input is shown in Fig 8.48(b). Note that this is also the p.d. across C.

The time constant for the charge and discharge of the capacitor is set by RC. The period of oscillation is related to this but is also affected by R_1 and R_2 since they decide the p.d. which the $(-)$ input must reach before switching occurs.

SINE WAVE OUTPUT

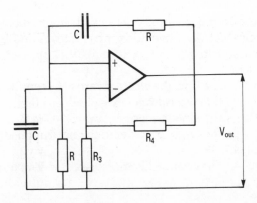

Fig 8.49 Sine-wave oscillator

To provide a sine wave oscillation a more complex circuit is needed which is shown in Fig 8.49.

Here a non-inverting amplifier of gain $1+R_4/R_3$ has an extra C-R feedback circuit connected to the $(+)$ input. There is thus both negative and positive feedback and to create sinusoidal oscillations these must be in the correct balance. Analysis beyond the scope of this book shows that this occurs when the gain of the non-inverting amplifier is precisely 3. Thus $R_4 = 2R_3$. In this instance the sinusoidal oscillations have a frequency of $1/2\pi RC$.

WORKED EXAMPLES

8.1 ACCELERATION OF ELECTRONS

Calculate the velocity of an electron accelerated by a p.d. of 1000V. Assume the charge to mass ratio of an electron $(e/m) = 1.8 \times 10^{11}$ C kg^{-1}.

(**Overview:** Use k.e. gained $= eV$ so $mv^2/2 = eV$.)

From the last equation,

$$v^2 = 2.\frac{e}{m}.V = 2 \times 1.8 \times 10^{11} \times 10^3 = 3.6 \times 10^{14}$$

so $v = \sqrt{3.6 \times 10^{14}} = 1.9 \times 10^7 \text{m s}^{-1}$

This speed is about 6% of the speed of light c in a vacuum (3×10^8 m s^{-1}).

8.2 ELECTRIC DEFLECTION OF ELECTRONS

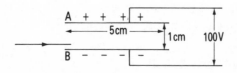

Fig 8.50 Worked example

An electron beam accelerated through 1000V enters horizontally a region between two horizontal plates A and B 5cm long separated by 1cm and to which a p.d. of 100V is applied, as shown in Fig 8.50. Calculate
(a)   the time the electrons are between the plates,
(b)   the vertical velocity gained
(c)   the kinetic energy gained,
(d)   the angle of travel to the horizontal after emergence from the plates,
(e)   the vertical distance moved whilst between the plates,

(f)   the potential energy lost by movement in the electric field.
Take $e/m = 1.8 \times 10^{11}$ C kg^{-1}, and $m = 9 \times 10^{-31}$kg.

(**Overview** Calculate speed of electrons as in 8.1. Find the acceleration
from $F = ma = eV/d$. Once this has been found the normal formulae for
constant acceleration can be used to calculate the motion parallel to
the field. The motion horizontally is with constant speed.)

(a)   For the acceleration of the electrons from zero to a velocity v
by the 1000V p.d.,

$$\tfrac{1}{2}mv^2 = eV$$

So $v = \sqrt{\dfrac{2eV}{m}} = \sqrt{2 \times 1.8 \times 10^{11} \times 10^3}$

$$= 1.9 \times 10^7 \text{m s}^{-1}$$

This horizontal speed is unchanged as it passes through the plates.

Thus $t = \dfrac{\text{distance}}{\text{velocity}} = \dfrac{0.05}{19 \times 10^6} = 2.6 \times 10^{-9} \text{s}$

(b)   Between the plates, the vertical acceleration, $a = \dfrac{eV}{md}$, so

$$a = \frac{1.8 \times 10^{11} \times 100}{0.01} = 1.8 \times 10^{15} \text{m s}^{-2}$$

The velocity gained $v_1 = at = 1.8 \times 10^{15} \times 2.6 \times 10^{-9}$
$$= 4.7 \times 10^6 \text{m s}^{-1}$$

(c)   The kinetic energy gained $= \tfrac{1}{2}mv_1^2$
$$= \tfrac{1}{2} \times 9 \times 10^{-31} \times (4.7 \times 10^6)^2$$
$$= 9.9 \times 10^{-18} \text{J}$$

(d)   The angle of travel to the horizontal is given by

$$\tan \theta = \frac{v_1}{v} = \frac{4.7 \times 10^6}{10 \times 10^6}, \text{ from which } \theta = 13.9°.$$

(e)   The vertical distance travelled $= \tfrac{1}{2}at^2$
$$= \tfrac{1}{2} \times 1.8 \times 10^{15} \times (2.6 \times 10^{-9})^2$$
$$= 0.0061\text{m} = 6.1\text{mm}$$

(f)   Potential energy lost $=$ electric force $\times$ vertical distance
$$= \frac{eV}{d} \times 0.0061$$

$$= \frac{e}{m} \times m \times \frac{V}{d} \times 0.0061 = 1.8 \times 10^{11} \times 9 \times 10^{-31} \times \frac{100}{0.01} \times 0.0061$$

$$= 9.9 \times 10^{-18} \text{J}.$$

Note that this is equal to (c) as we would expect.

8.3 MAGNETIC DEFLECTION OF ELECTRONS

An electron beam, accelerated through 400V, passes into a magnetic field of flux density 0.005T whose direction is perpendicular to the electron beam. Calculate

(a)   the radius of the electron orbit,
(b)   the time taken to complete one orbit.

Repeat the calculation when the accelerating voltage is reduced to 100V and comment on the results. Take $e/m = 1.8 \times 10^{11} C\ kg^{-1}$.

(**Overview**: Calculate the speed of the electrons as in 8.1. Then use the fact that the necessary centripetal force mv^2/R is provided by the magnetic force, Bev.)

(a)   Using $\frac{1}{2}mv^2 = eV$,

$$v = \sqrt{2eV/m} = \sqrt{2 \times 1.8 \times 10^{11} \times 400}$$
$$= 1.2 \times 10^7 m\ s^{-1}$$

For the magnetic deflection into a circular orbit of radius R,

$$Bev = \frac{mv^2}{R}$$

So
$$R = mv/Be = v(B \times e/m)$$
$$= 12 \times 10^6/(0.005 \times 1.8 \times 10^{11})$$
$$= 1.3 \times 10^{-2}m = 1.3cm$$

(b) The time of orbit $= 2\pi R/v = \dfrac{2 \times \pi \times 1.3 \times 10^{-2}}{12 \times 10^6}$
$$= 7.0 \times 10^{-9} s.$$

When the p.d. is reduced to 100V which is one-quarter of the original voltage, we have

$$v = \sqrt{2eV/m} = 6 \times 10^6 m\ s^{-1}$$

(Note: v is halved as $v \propto \sqrt{V}$.)

Similarly, R is now halved to 6.5cm. The time of flight is *unaltered* since $2\pi R$ is halved and v is halved, that is, the electrons have half as far to go with half the speed. The time of orbit is therefore *independent* of the speed of the charged particle. (This fact is used in the cyclotron, a device which can accelerate protons to high energies.)

8.4 MILLIKAN'S EXPERIMENT

In a Millikan-type experiment a particular oil drop weighs $2 \times 10^{-14}N$ and is between plates 0.5mm apart. Calculate the voltage needed to hold the drop stationary if it carries

(a)   3 electronic charges
(b)   5 electronic charges.
(Take $e = 1.6 \times 10^{-19}C$.)

(**Overview**: Use the idea that the weight is balanced by the upward electrical force.)

Here, weight $W =$ force qE due to electric field $= \dfrac{qV}{d}$

so, $V = Wd/q$

(a) For 3 electronic charges, $q = 4.8 \times 10^{-19}$C
So $V = 2 \times 10^{-14} \times 0.005/4.8 \times 10^{-19} = 208$V
(b) For 5 electronic charges, $q = 8.0 \times 10^{-19}$C
So $V = 2 \times 10^{-14} \times 0.005/8.0 \times 10^{-19} = 125$V

8.5 MOTION OF OIL DROPS

An oil drop, mass 3.25×10^{-15}kg, falls vertically with uniform velocity between two parallel vertical plates 2cm apart. When a p.d. of 1000V is applied to the plates, the path of the drop is then inclined at 45° to the vertical. Calculate the charge on the drop. Calculate the new charge on the drop when its path changes to 26.5° to the vertical. (Assume $g = 9.8$N kg^{-1})

(**Overview** The oil drop moves with its terminal velocity at 45° to the vertical. In this case the horizontal and vertical forces are equal.)

The force on the drop due to an applied p.d. between the plates is horizontal. Hence the vertical component of its velocity is unaltered. With a p.d. of 1000V, the potential gradient $= 1000/2 \times 10^{-2} =$ intensity E.

So force on charge q on drop $= \dfrac{1000}{2 \times 10^{-2}} \times q$ newton

Since the path is 45° to vertical,
$$\text{force} = \text{weight of drop}$$
$$= 3.25 \times 10^{-15} \times 9.8\text{N}$$

So $\dfrac{1000 \times q}{2 \times 10^{-2}} = 3.25 \times 10^{-15} \times 9.8$

and $q = \dfrac{3.25 \times 10^{-15} \times 9.8 \times 2 \times 10^{-2}}{1000}$
$$= 6.4 \times 10^{-19} \ (= 4e)$$

When the angle is 26°30′ to the vertical, the charge has decreased to q_1 say. Then, if E is the intensity between the vertical plates,

$$\tan 26°30' = \frac{Eq_1}{mg} = \frac{Eq_1}{Eq} = \frac{q_1}{q} = \frac{q_1}{4e}$$

So $q_1 = \tan 26°30' \times 4e = 0.5 \times 4e = 2e.$

8.6 OSCILLOSCOPE MEASUREMENTS

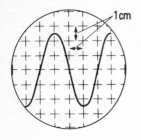

Fig 8.51 Worked example

Figure 8.51 shows the trace obtained on an oscilloscope screen when the time/cm switch is set at $2\mu s$/cm, and the volts/cm switch is set at 10V/cm. Calculate the frequency of the signal and its r.m.s. voltage.

(**Overview** The problem can be solved using the facts that each cm vertically represents a p.d. of 10V, and each cm horizontally represents a time of $2\mu s$.)

One cycle on the screen takes 4cm. Thus the time for one cycle is $4\times2=8\mu s$. Hence the frequency $=1$/time period $=1/(8\times10^{-6})=$ 125 000Hz = 125Hz.
On the screen the peak voltage measures 2.5cm. Hence the value of the peak voltage is $2.5\times10=25$V. The r.m.s. voltage for the sinusoidal variation is, from p157, (peak voltage)/$\sqrt{2}$. Hence r.m.s. voltage $=25/\sqrt{2}=17.4$V.

8.7 COMBINATION OF GATES

Draw a truth table for the circuit in Fig 8.52 which is constructed of NAND gates. Show that the circuit behaves as an exclusive-OR gate.

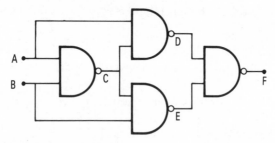

Fig 8.52 Worked example

(**Overview** C is related to A and B by a NAND rule. Find C first. Then use the fact that C and A produce D via a NAND rule. Similarly C and B produce E. Finally D and E can be 'NAND'ed together to get F. It is a good idea to have the NAND truth table to hand while following this example.)

A	B	C	D	E	F
0	0	1	1	1	0
0	1	1	1	0	1
1	0	1	0	1	1
1	1	0	1	1	0

It can be seen from the truth table that F relates to A and B by an exclusive-OR operation.

**8.8 TRAFFIC LIGHT
SEQUENCE**

Use an astable, a J-K flip-flop and any other gates you require to design a system that operates three lights, red, orange and green in traffic light sequence.

(**Overview** The sequence of lights is illustrated by Fig 8.53.)

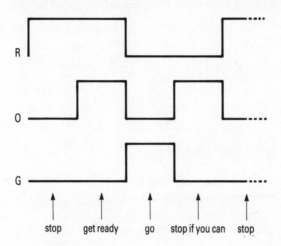

Fig 8.53 Worked example

Note
(*a*)   The red switches state whenever the orange falls from on to off.
(*b*)   The green is on only when the orange and red are both off.

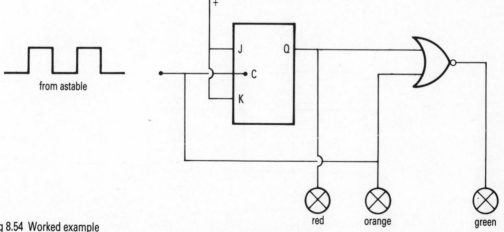

Fig 8.54 Worked example

The astable can be connected straight to the orange lamp. If the astable is also connected to the clock input of the flip-flop then the output of the flip-flop will light the red lamp correctly – see note (*a*) above. From note (*b*) the green lamp is related to the red and orange lamps by a NOR operation. Hence the circuit in Fig 8.54 solves the problem.

8.9 PULSE GENERATOR

In Fig 8.55 C is initially uncharged and V_i is at logic '0'. Describe what happens to V_0 if V_i is switched from logic '0' to logic '1' and held there.

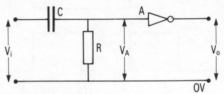

Fig 8.55 Worked example

(Overview The output of the gate is always high or low. As the capacitor charges the p.d. at the input of the gate changes gradually, but the output switches suddenly. The precise input voltage which switches the gate depends on the particular type of gate used.)

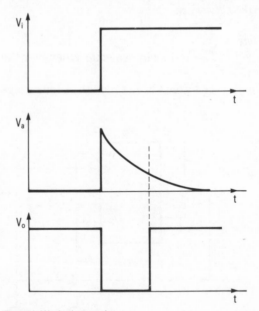

Fig 8.56 Worked example

Fig 8.56 shows the variation of p.d. with time for V_i, V_A, and V_0.

When V_i switches to logic high, V_A will instantaneously follow this since there is initially no p.d. across the capacitor. V_A is the p.d. across R, and so a current will now flow to charge up C. As the current falls exponentially, so will V_A which is just equal to IR. Thus V_A

varies as shown. V_0 is initially high since V_A is low. When V_A switches to high, V_0 falls to low. Then as V_A falls there will be a particular value where V_A is small enough to switch the output of the NOT gate back to high.

Thus a 'negative going' pulse is produced at the output. The duration of the pulse depends on C, R and the precise value of V_A at which the NOT gate switches.

8.10 AN ASTABLE CIRCUIT

Describe how an astable can be made using two NOT gates.

(**Overview** Two pulse producers are used, each of which triggers the other.)

Figure 8.57 shows a suitable circuit. When B goes from logic low to high, A will produce a pulse going from high to low and back again. This in turn produces a pulse at B, and so on.

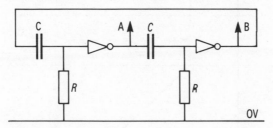

Fig 8.57 Worked example

8.11 OPAMP WITH CAPACITOR

Describe what happens to V_0 in the circuit of Fig 8.58 when the input voltage is raised suddenly from 0 to 1.0V and remains at that voltage.

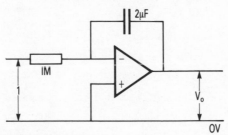

Fig 8.58 Worked example

(**Overview** A constant current will flow through the 1M resistor which will charge the capacitor uniformly until the output voltage saturates.)

In this circuit a capacitor provides feedback. From rule 2, p279, the $(-)$ input is at 0V. Hence the current through the 1M resistor is $1/10^6 A = 10^{-6}$A.

By rule 1, p279, no current flows into the opamp, so this current must flow to charge up C. After t seconds the charge on C is $10^{-6} \times t$ coulombs. The p.d. V_C across C is given by

$$V_C = Q/C = (10^{-6} \times t)/(2 \times 10^{-6})$$
$$= 0.5t$$

But $V_0 = -V_C$ and hence $V_0 = -0.5t$. The output voltage therefore falls linearly with time. After $t = 30$s the output voltage will reach -15V and will then saturate. Fig 8.59 illustrates the variation of V_i and V_0 with time.

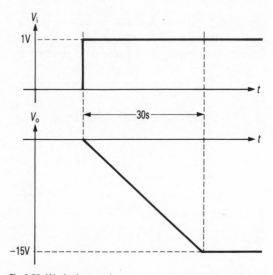

Fig 8.59 Worked example

VERBAL SUMMARY

1 THERMIONIC EMISSION

Electrons are emitted from hot metals.

2 ELECTRONS IN FIELDS

Acceleration of electrons from rest to velocity v by p.d. V, $mv^2/2 = eV$.

Uniform electric field: Force $= eV/d$, $a = eV/md$. Use $v = at$, $s = at^2/2$ to calculate motion. Path is parabolic; there is energy gain in the field.

Uniform magnetic field: Path circular, radius $= R = mv/eB$. No change of speed or energy in magnetic field, only deflection.

e/m for electron measured by fine beam tube.

e measured by Millikan's experiment. Charge on oil drops always a multiple of 1.6×10^{-19} coulomb $=$ charge on one electron.

3 CATHODE-RAY OSCILLOSCOPE

Contains grid to adjust brightness, anodes to accelerate electrons and also to produce an electron lens to focus beam. X- and Y-plates deflect beam. Shift controls centre trace. Time-base circuit drives spot across screen at uniform speed.

4 SEMICONDUCTORS

Pure (intrinsic) semiconductors contain equal numbers of electrons $(-e)$ and holes $(+e)$.

Electrons more mobile. Temperature rise greatly increases number of charge carriers, so resistance decreases.

Doped (extrinsic) semiconductors. Five-valency atom doping gives n-type semiconductor. Three-valency atom doping gives p-type semiconductor.

Junction diode. p-n junction has a *barrier* p.d. (about 0.6V) across the *depletion layer*. When forward-biased (greater than 0.6V) good conduction takes place. Reverse-biased gives very low current.

Transistor. Three electrodes, base, emitter, collector. Base is very thin. n-p-n type has p-type base, p-n-p has an n-type base.

Transistor can be used as a switch where current is limited by collector resistor.

Transistor can be used as a common-emitter amplifier. With correct base bias, small changes of base voltage cause large changes of output voltage. Gain $= \beta R_L/r_i$.

5 DIGITAL ELECTRONICS

Voltages can be 'on' or 'off', represented by logic 0 and 1. NOT, OR, NOR, AND, NAND, EXCLUSIVE-OR gates are represented by their truth tables.

Gates can be combined. The overall effect can be calculated by combining their truth tables.

Half-adder adds to binary digits producing a sum and a carry. A full-adder adds in any previous carry.

A flip-flop has two electrically stable states. A J-K flip-flop can be used as a binary divider.

6 ANALOGUE ELECTRONICS

An opamp has a very high gain amplifying the p.d. between the inverting and non-inverting inputs. It also has a very high input impedance. Without feedback it can be used as a switch or as a voltage comparator.

Feedback enables the opamp to be used as an inverting amplifier (gain $= -R_2/R_1$), or a non-inverting amplifier (gain $= 1 + R_2/R_1$).

Opamps can be used as an oscillator to produce a square or a sine wave output.

DIAGRAM SUMMARY

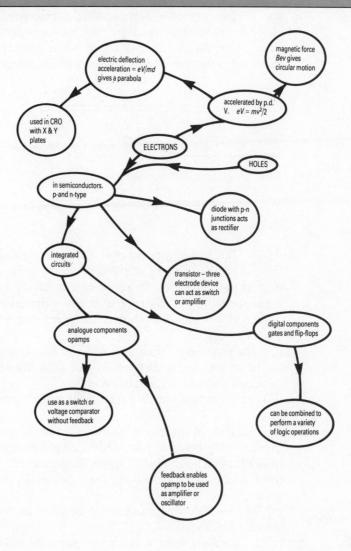

electric deflection
acceleration = eV/md
gives a parabola

magnetic force
Bev gives
circular motion

accelerated by p.d.
V. $eV = mv^2/2$

used in CRO
with X & Y
plates

ELECTRONS

HOLES

in semiconductors.
p- and n-type

diode with p-n
junctions acts
as rectifier

integrated
circuits

transistor – three
electrode device
can act as switch
or amplifier

analogue components
opamps

digital components
gates and flip-flops

use as a switch or
voltage comparator
without feedback

can be combined to
perform a variety
of logic operations

feedback enables
opamp to be used
as amplifier or
oscillator

1 (a)   Describe an experiment to determine the charge to mass ratio of electrons, indicating clearly the measurements made. How is the value of e/m calculated from them? What information from the experiment indicates that electrons are *negatively* charged?

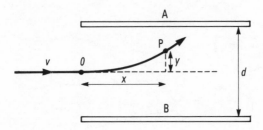

Fig 8.60

(b)   In Fig 8.60 a beam of electrons, travelling with a velocity v in the x-direction, enters at point O ($x=0$, $y=0$) a region of uniform electric field provided by applying a voltage V between plates A and B, separated by a distance d in the y-direction. The electrons are deflected towards A as shown, the point P (x, y) being a point along the trajectory.

(i) Is the potential of plate A positive or negative with respect to B?

(ii) In terms of the distance x, calculate the angle between the x-direction and the electron beam at P.

(iii) Prove that the path is parabolic, namely $y = ax^2$, and find the value of a.

The position of the electron source is moved so that the direction of the incoming beam at point O is now at an angle θ to the x-direction towards plate B. The initial speed is unchanged.

(iv) Find the distance L along the x-axis at which the beam again has $y = 0$.

(v) Explain how this effect could be used as the basis for a velocity selector for electrons.   (O&C)

2 (a)   Explain how a beam of electrons may be produced in a vacuum tube, and describe an arrangement by which the beam may be deflected by a magnetic field. Draw a diagram showing clearly the directions of the beam, the field and the deflection.

(b)   An electron moving at velocity v passes simultaneously through a magnetic field of uniform flux density B and an electric field of uniform intensity E. It emerges with direction and speed unaltered.

(i) Explain how the fields are arranged to achieve this result.

(ii) Derive the relationship between v, B and E

(c)　　　An electron is travelling at 2.0×10^6m s^{-1} at right angles to a magnetic field of flux density 1.2×10^{-5} T; its path is a circle. Uniform circular motion of an electron is accompanied by the emission of electromagnetic radiation of the same frequency as that of the circular motion.

[Take the specific charge of the electron e/m_e to be 1.8×10^{11}C kg^{-1}, and the speed of electromagnetic waves in air c to be 3.0×10^8m s^{-1}.]

(i) Explain why the path of the electron is a circle.

(ii) Calculate the radius of the circle.

(iii) Calculate the frequency of the circular motion of the electron.

(iv) Calculate the wavelength of the electromagnetic radiation emitted and identify in which part of the electromagnetic spectrum this radiation lies.

(v) How would this wavelength be affected by a decrease in the speed of the electron?　　　　　　　　　　　　　　　　　　　(O)

3　(a)　　　A charged oil drop falls at constant speed in the Millikan oil drop experiment when there is no p.d. between the plates. Explain this.

(b)　　　Such an oil drop, of mass 4.0×10^{-15}kg, is held stationary when an electric field is applied between the two horizontal plates. If the drop carries 6 electric charges each of value 1.6×10^{-19}C, calculate the value of the electric field strength.　　　　　　　　　(L)

4　(a)　　　(i) Describe the layout of a cathode-ray tube that utilizes electrical deflection and focusing. Explain how the focusing and the brightness of the spot on the screen are controlled.

(ii) What is the function of the linear time-base in an oscilloscope?

(iii) Why is the tube of a cathode-ray oscilloscope always mounted inside a metal screen of very high magnetic permeability? Explain how the screen achieves its purpose.

(b)　　　Explain in detail how you would use an oscilloscope:

(i) to observe the waveform of the musical note emitted by a clarinet;

(ii) to check the frequency calibration of an audio sine-wave oscillator at any one chosen frequency (you may assume that the frequency of the mains supply is 50Hz exactly);

(iii) to measure the r.m.s. value of sinusoidal alternating current of about 0.2 A flowing in a 5.0 Ω resistor.　　　　　　　　　　(O)

5　(a)　　　Why does the electrical conductivity of an intrinsic semiconductor increase as the temperature rises?

(b)　　　Introducing certain impurities into semiconducting material also increases its electrical conductivity. Describe briefly *either* an n-type *or* a p-type semiconductor and explain why this increase in conductivity occurs.　　　　　　　　　　　　　　　　　　(L)

6　(a)　　　(i) Explain the origin of holes in intrinsic semiconducting materials. What is the process by which holes participate in current flow?

(ii) Explain how the presence of the donor impurities in an n-type semiconducting material raises the number of free electrons per unit volume without increasing the number of mobile holes per unit volume.

(b)   The diagram Fig 8.61 shows two kinds of semiconducting material, p-type and n-type, in contact. What is the important charac-teristic which distinguishes the depletion layer from the rest of the assembly?

Explain the effect on the depletion layer of applying a *small* potential difference (about 0.1V) across XY

(i) if X becomes negative with respect to Y,

(ii) if X becomes positive with respect to Y.

Hence explain the rectifying action of a p-n junction.

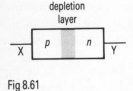

Fig 8.61

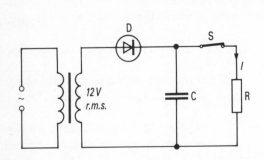

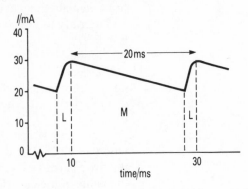

Fig 8.62

(c)   The diagram (Fig 8.62) shows a half-wave rectifying circuit connected to a resistor R through a switch S. The graph shows how the current I in resistor R varies with time.

Write down the source of the current in R

(i) during the period L,

(ii) during the period M.

Calculate the maximum reverse bias potential difference the diode, D, must be capable of withstanding when the switch, S, is open.   (L)

7   Draw the circuit symbol for a two-input AND gate.

An EXCLUSIVE-OR gate is one where the output is high only when one or other but not both of the two inputs is high. Write out a truth table for this gate.   (C)

8   (a)   A light emitting diode (LED) which is connected to the output of a logic gate is required to light when the output is logic 1 ($+5V$). For this to be achieved a forward current of 10mA together with a forward p.d. of 2.0V is required. Draw a circuit diagram to show how the LED is connected and calculate the value of any additional com-ponent required.

(b)   The symbol shown in Fig 8.63 represents a T or toggle flip-flop circuit. The small circle on the clock input indicates that toggling occurs when an input pulse goes from logic 1 to logic 0. The output Q

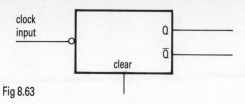

Fig 8.63

can be set to logic 0, i.e. cleared, by momentarily connecting the clear input to a zero voltage line. With the aid of as truth table for input and outputs, explain what is meant by the toggling action of this circuit.

Fig 8.64

The clock pulses shown in Fig 8.64 are applied to the flip-flop circuit. Using the same time scale as Fig 8.64, sketch the outputs from Q to $\overline{Q}$. (c)Draw a diagram to show how four of the circuits shown in Fig 8.63(*b*) can connect together to make a four-bit binary counter to count upwards from zero. Show on your diagram where you would connect four LEDs to indicate the binary count, labelling them A to D with A being the least significant bit.

The clock pulses shown in Fig 8.64 are fed into the input of the four-bit counter after its outputs have been cleared. Using a common time scale, show by means of graphs how the output signal from **each** of the four flip-flops varies between logic 0 and logic 1 for nine input pulses. (*JMB*)

9 The figure (Fig 8.65) shows a circuit for an electronic thermometer. The thermocouple generates an e.m.f. between the terminals X and Y of $50\mu V$ per K temperature difference between its junctions. The thermocouple is connected to the non-inverting input of an operational amplifier and feedback is provided by the resistor of resistance R_f to the inverting input.

(*a*)   State what is meant by an *inverting input* and *a non-inverting input*. State the type of feedback used in this circuit and explain how it functions.

(*b*)   The voltage gain of this amplifier is given by

$$\frac{V_{out}}{V_{in}} = 1 + \frac{R_f}{R_a}.$$

If $R_f = 100k\Omega$ and $R_a = k\Omega$, calculate the reading on the voltmeter when there is a temperature difference of 100K between the junctions of the thermocouple.

(*c*)   The thermometer is to be used to measure body temperatures over the range 35°C to 45°C, the voltmeter giving zero deflection at 35°C and full-scale deflection of 1.0V at 45°C. State how the ther-

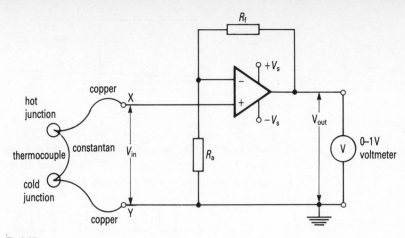

Fig 8.65

mocouple can be set up so that the voltmeter reads zero when the temperature of the hot junction is 35°C.

If $R_a = 1k\Omega$, calculate the value of R_f which will produce a full-scale deflection when measuring a temperature of 45°C

Explain why you would expect the value of R_f to be very much greater than the value of R_a in this application.   (JMB)

ATOMIC AND NUCLEAR PHYSICS

CONTENTS

QUANTA OF ENERGY

Energy is emitted by atoms in discrete amounts called **quanta**. One of the first pieces of evidence to show that light consists of quanta of energy is the photoelectric effect, which we now discuss.

PHOTOELECTRIC EFFECT

When light falls on a metal, some free electrons in the metal may be liberated. This is called the **photoelectric effect**. It is found experimentally that increasing the intensity of the light does *not* increase the energy of the emitted electrons. This cannot be explained on the wave theory of light. In this model light is an electromagnetic wave, and the electric field in the light wave produces a force on the free electrons. A greater intensity light wave has larger electric fields. So the force on any electron, and hence the energy of those liberated, would be larger with a more intense light beam. This is not the case.

To explain the photoelectric effect Einstein proposed that light was made up of particles called **photons**, and that the energy of *each* photon was hf, where f is the frequency of the light and h is the *Planck constant*. h is about 6.6×10^{-34}J s and was first introduced by Planck in his theory of radiation. Since velocity $c = f\lambda$, where λ is the wavelength, a high frequency implies a short wavelength. So high frequency, or short wavelength, gives large photon energy. The photoelectric effect is therefore more likely to be noticed with light of short wavelength, which is at the violet end of the spectrum.

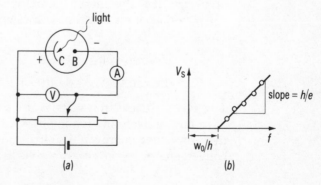

(a) (b)

Fig 9.1 Photoelectric effect

The work function, ω_0 of a metal is defined as the least energy to remove an electron from the metal. Since hf is the energy of the incident photon, we have, by the conservation of energy:

Max kinetic energy (k.e.) of emitted electrons $= hf - \omega_0$

or $hf = \omega_0 + \text{max k.e.}$

An apparatus for studying photoelectricity is shown in Fig 9.1(a). A is a very sensitive current meter, such as a d.c. amplifier, and the potentiometer provides a varying p.d. V to the photocell. B is made *negative* in potential relative to C, so that the electrons emitted from the metal C are repelled. The potentiometer is adjusted until A *just* reads zero. Then the kinetic energy of the emitted electrons is just not sufficient for them to overcome the retarding p.d. V_s and reach B. In this case

max kinetic energy of electrons $= eV_s$

V_s is called the **stopping potential** and is measured on the voltmeter. Hence,

$$hf = \omega_0 + eV_s, \text{ or } V_s = \frac{h}{e}f - \frac{\omega_0}{e} \qquad (1)$$

V_s can be measured for varying frequencies, f, of incident light and a graph plotted. This is a straight line graph as shown in Fig 9.1(b). From equation (1) we see that the slope of the graph is h/e. Hence h can be determined if e is known. Its value is 6×10^{-34} J s (joule second), approximately.

Until f exceeds a certain value, no electrons are emitted. In these cases the photon energy is less than ω_0 and so is insufficient to liberate electrons. The critical frequency, f_c, below which electron emission does not take place is given by $hf_c = \omega_0$. Hence the intercept on the f-axis is ω_0/h, the critical frequency.

THE ELECTRON-VOLT

In atomic physics it is often convenient to use the *electron-volt* as a unit of energy. It is the energy gained when a particle, whose charge is equal to that of an electron, is accelerated through a p.d. of 1 volt. Thus

1 electron-volt $= e \times V$

$= 1.6 \times 10^{-19} \times 1 = 1.6 \times 10^{-19}$ J

The MeV ($= 10^6$ electron-volts) is also used as a larger unit of energy.

In any problem in atomic physics the student must be careful to use consistent energy units. If in any doubt it is always safer to convert energies from electron-volts to joules.

**ENERGY LEVELS IN
ATOMS**

We have just seen that light can be considered to be particles or bundles of energy called photons. When light is emitted from atoms it is also emitted in certain amounts of energy called *quanta*.

In the process of emission an atom moves between two *energy levels*. In fact, the electron structure within an atom only allows the atom to exist, at any instant, in one of a number of definite energy levels characteristic of that atom. When an atom has the lowest possible energy it is said to be in the **ground state**. It may be *excited* to higher energy levels by (*a*) striking the atom with other particles such as electrons in a discharge tube or (*b*) shining light of suitable frequency on the atom.

Once an atom is in a higher energy level (an **excited state**) it will fall back to the ground state and the energy released is emitted as light. For example, consider an atom with four energy levels as shown in Fig 9.2. If the atom is struck by an electron of less energy than E_1, then nothing will happen. The atom will absorb none of the electron's energy and the collision will be **elastic** (see page 24). Only if the energy is greater than E_1 can the atom be excited. Suppose that the atom is struck and raised to the third excited state E_3. It can then fall back to the ground state in various ways as shown by the arrows. Possible energy releases are E_3-E_2, E_2-E_1, E_1-E_0, E_3-E_1, E_2-E_0, E_3-E_0. For each of these possible changes of energy a particular wavelength of light (or other electromagnetic radiation) will be emitted. If the emitted frequency is f the energy of the photon is hf, where h is the Planck constant. Hence the possible frequencies f_1, etc., of the radiation emitted are given by:

$$hf_1 = E_3 - E_2, \; hf_2 = E_2 - E_1, \text{ etc.}$$

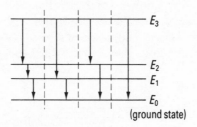

E_3

E_2
E_1

E_0
(ground state)

Fig 9.2 Energy levels

This means that the light emitted from excited atoms can only have certain definite frequencies (or wavelengths). The spectra obtained in such circumstances are called **line** or **emission** spectra, and the frequencies emitted can provide experimental values for the energy levels of the atoms. Note carefully however that *one* particular frequency corresponds to a difference between *two* energy levels.

Other evidence for the existence of energy levels within atoms comes from electron collision experiments such as those performed by Franck and Hertz. Here it was shown that the energy absorbed by atoms during collisions with electrons could only be one of a definite series of values. The potential differences through which an electron must be accelerated just to produce excitation are called **excitation potentials**.

ENERGY LEVELS OF HYDROGEN – IONIZATION

The energy levels of hydrogen obey a simple rule. If the highest energy level is counted as the 'zero' of potential energy, then the ground state energy is -21.8×10^{-19}J. The third energy level has an energy -5.45×10^{-19}J $= -\dfrac{21.8}{2^2} \times 10^{-19}$J. The third energy level has an energy -2.42×10^{-19}J which is $-\dfrac{21.8}{3^2} \times 10^{-19}$J. In general the nth energy level has a value $-\dfrac{21.8}{n^2} \times 10^{-19}$J. This simple rule was discovered by Balmer in 1885 and later found theoretically by Bohr.

In hydrogen, electron transitions to the ground state ($n=1$) must emit at least $(21.8-5.45) \times 10^{-19} = 16.35 \times 10^{-19}$J. The wavelength of a photon of this energy is given by

$$hf = \frac{hc}{\lambda} = 16.35 \times 10^{-19}$$

or $\qquad \lambda = \dfrac{6.6 \times 10^{-34} \times 3 \times 10^8}{16.35 \times 10^{-19}} = 1.2 \times 10^{-7} \text{m}$

using $h = 6.6 \times 10^{-34}$J s.

This is a wavelength in the ultra-violet region of the spectrum. The family of spectral lines in the ultra-violet produced by transitions to the ground state are called the **Lyman** series, after the discoverer (Fig 9.3). Transitions down to the second energy level ($n=2$) give wavelengths in the visible region of the spectrum and are called the **Balmer** series. The **Paschen** series is in the infra-red region and occurs with transitions down to the third energy level ($n=3$).

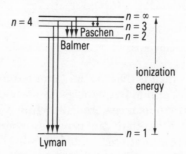

Fig 9.3 Hydrogen energy levels

If a hydrogen atom in its ground state is struck by an electron with *more* energy than 21.8×10^{-19}J, the hydrogen atom may become **ionized**. The electron in the hydrogen atom has been given enough energy to escape from the atom. 21.8×10^{-19}J is called the **ionization energy** of hydrogen (Fig 9.3). To ionize a hydrogen atom the electron striking the atom must have been accelerated through a p.d. whose minimum value V, is given by

$$eV = 21.8 \times 10^{-19}$$

or $\quad V = 21.8 \times 10^{-19} / 1.6 \times 10^{-19} = 13.6$ volts.

13.6V is called the **ionization potential** of hydrogen.

X-RAYS

An X-ray tube is shown diagrammatically in Fig 9.4(*a*). Electrons from a heated cathode are accelerated across a vacuum by a very high p.d. supply, E.H.T., of many kV. When they hit the tungsten (or other metal) anode they rapidly decelerate and some of their kinetic energy is emitted as X-rays. (The bulk of the electron kinetic energy heats the metal anode which often needs artificial cooling aids.) The *intensity* of the X-ray beam is controlled by the current of electrons, which in turn is controlled by the heater temperature.

*PIPES FILLED WITH OIL etc...

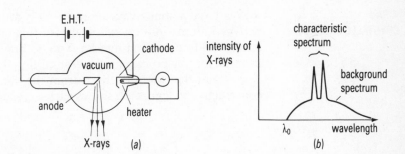

Fig 9.4 X-ray tube and spectrum

X-rays are emitted over a range of wavelengths as shown by the graph in Fig 9.4(*b*). There are two distinct parts of this spectrum.

1 THE BACKGROUND SPECTRUM

This does *not* depend on the nature of the metal target anode. A range of wavelengths is emitted but none shorter than a certain critical wavelength λ_0 (Fig 9.4(*b*)). This means that there is a certain *maximum* frequency f_0, which occurs when the *whole* of the kinetic energy of an incident electron is transferred into an X-ray photon of the same energy. In this case,

$$hf_0 = eV, \text{ where } V \text{ is the accelerating p.d.}$$

So $\quad \dfrac{hc}{\lambda_0}=eV$, or $\lambda_0=\dfrac{hc}{eV}$

Experiments with different voltages V show that the shortest wavelength measured, λ_0, agrees with this formula, thus verifying that X-rays may be considered as photons.

2 THE CHARACTERISTIC SPECTRUM

This depends on the nature of the actual metal used as the target. The high velocity incident electrons eject an electron near the nucleus of the metal atoms. Another electron then falls into this 'space' and in so doing emits a particular amount of energy in the form of an X-ray photon. The energy to remove an electron from near the nucleus is likely to depend on the nuclear charge, Ze, of the target element, where Z is the proton number. Hence the X-ray frequency is also likely to depend on Z. This was studied by Moseley who found, by using different elements as targets, that

$$\sqrt{f}=a(Z-b)$$

where a and b are constants and f is the X-ray frequency.

X-RAY CRYSTALLOGRAPHY

X-rays have a short wavelength ($\sim10^{-10}$m) and the regular spacing between atoms in a crystal is also of this order. Thus crystals form natural diffraction gratings for X-rays. Bragg worked out the basic formula for this process on the assumption that layers of atoms reflect X-rays and that the reflections from the various layers can interfere constructively. Figure 9.5 shows such a crystal.

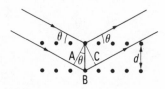

Fig 9.5 Bragg's law

The path difference between the two beams is AB+BC=2AB. But AB=d sin θ where d is the separation of the atom layers, so the path difference is $2d$ sin θ. If this is a multiple n of the wavelength, then constructive interference will occur. Thus for a strong X-ray beam

$\quad 2d$ sin $\theta=n\lambda$

This is called **Bragg's law**.

In any crystal there are various sets of planes with differing values of d. In practice, therefore, a crystal will produce many diffracted beams, provided that it is oriented correctly to the X-ray beam.

WAVE-PARTICLE DUALITY

In the last section we have considered that electromagnetic radiation can behave as **particles**. For example, in the photoelectric effect the explanation was based on the idea that light consists of photons each with energy hf. Also, the lower limit of the wavelength in X-ray spectra was explained in terms of the idea that all of the k.e. of an electron was transferred to an X-ray photon. At the same time, however, we have talked about the frequency and wavelength of the photons and have discussed the diffraction of X-rays by crystals. All of this suggests the **wave** nature of electromagnetic radiation.

Electromagnetic radiation therefore displays what we call **duality**, both wave and particle properties are exhibited, even in the same experiment. For example, in X-ray diffraction by a crystal the diffraction process is a wave phenomenon, but the effect of the X-rays on the photographic plate is a particle phenomenon.

In 1930 De Broglie proposed that the wave-particle duality, accepted for electromagnetic radiation, also applied to matter. He gave theoretical arguments which predicted the equation,

$\lambda = h/p$

where λ is the wavelength associated with a particle of momentum p. If we consider a 1kg mass moving with a speed of 2m s^{-1} then $p = 1 \times 2 = 2$kg m s^{-1}, and $\lambda = 6.6 \times 10^{-34}/2 = 3.3 \times 10^{-34}$m. This wavelength is very much smaller than even the size of the nucleus ($\sim 10^{-14}$m), so such a particle would show no diffraction effects whatever.

Consider, however, the case of an electron accelerated by a p.d. of 200V.

$$\text{k.e.} = eV$$
$$= 1.6 \times 10^{-19} \times 200$$
$$= 3.2 \times 10^{-17} \text{J}$$

So $mv^2/2 = 3.2 \times 10^{-17}$

or $9.1 \times 10^{-3} v^2/2 = 3.2 \times 10^{-17}$

giving $v = 8.4 \times 10^6$m s^{-1}

The momentum of the electron, p, is given by

$$p = mv = 9.1 \times 10^{-31} \times 8.4 \times 10^6$$
$$= 7.7 \times 10^{-24} \text{kg m s}^{-1}$$

The wavelength λ associated with the electron is, by De Broglie's relation

$$\lambda = h/p = 6.6 \times 10^{-34}/7.7 \times 10^{-24}$$
$$= 8.6 \times 10^{-11} \text{m} = 0.86 \times 10^{-10} \text{m}$$

This is the same order of magnitude as the interatomic spacing in a crystal. This led De Broglie to predict that electrons should show diffraction effects when they fall on a crystal. In fact this prediction proved correct, including the relation $\lambda = h/p$. Electron diffraction is now a commonly used tool in research. Thus the wave-particle duality does extend to both matter and electromagnetic radiation. The following table summarises the relationships.

	Wave	Particle	Relation
E-M radiation	frequency f	Energy E	$E=hf$
Matter	wavelength λ	momentum p	$p=h/\lambda$

THE NUCLEAR ATOM AND RADIOACTIVITY

GEIGER AND MARSDEN'S EXPERIMENT

The present model of an atom, where most of the mass and all the positive charge of the atom is concentrated into a tiny nucleus, was confirmed by Geiger and Marsden in 1911. At the suggestion of Rutherford they fired α-particles at a very thin film of gold foil in an evacuated chamber as shown in Fig 9.6. The number of α-particles scattered by various angles θ, were measured by counting the number of scintillations on a movable zinc sulphide screen.

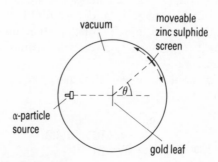

Fig 9.6 Geiger and Marsden's experiment

Previously Rutherford had calculated theoretically the variation of the number of particles scattered with an angle θ. He based his calculation on three ideas:

1 The nuclear model. Most of mass of the atom is concentrated into a tiny area which carries a positive charge.

2 The force causing the α-particles to deflect is the electrostatic repulsion between the positive charge and the α-particle and that on the nucleus.

3 Newtonian mechanics still applied at these very minute distances (10^{-14}m).

After the experiment the predicted variation of number scattered with θ was compared with the observed variation and very good agreement was found. This confirmed that the nuclear model was correct and that the α-particles were scattered by the nucleus due to electrostatic repulsion.

RADIATION DETECTORS AND TYPES OF RADIATION

Radioactivity produces ionization in its passage through matter. Radiation detectors rely on this property and it is assumed that the reader is already familiar with their action from a study of O level.

Ionization produced by radioactivity may be measured directly using an ionization chamber (Fig 9.7). The radioactive source S is positioned inside the metal ionization chamber C. A supply of several hundred volts is applied between C and the inner metal electrode E. Any positive ions will travel to E and negative ions and electrons to C. Provided the voltage is sufficiently high all the ions produced each second by the radioactive source will be collected. In this case the current in the circuit is equal to the total charge produced each second by ionization. So the current is proportional to the activity or intensity of the radiation.

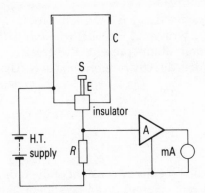

Fig 9.7 Ionization chamber

The current flows through the high resistor R and this sets up a p.d. which is measured on the high impedance voltmeter. Suppose the voltmeter produces full scale deflection for 1V input, and that $R = 10^9\Omega$. If the meter reads 0.3 of full scale deflection then the p.d. V across $R = 0.3$V. Hence the ionization current flowing $= V/R = 0.3/10^9 = 3 \times 10^{-10}A$.

By measuring the ionization current at various times the half-life of a radioactive source can be found using the graphical method described on page 313.

The three types of radiation are (a) α-particles consisting of two protons and two neutrons and which are therefore nuclei of helium; (b) β-particles, which are fast moving electrons; and (c) γ-rays, which are short wavelength ($\sim 10^{-10}$m) electromagnetic waves. α- and

β-decay cause the nucleus which emits them to become a different chemical element. If Z is the proton number (the number of protons in the nucleus) and A is the mass or nucleon number (the total number of protons and neutrons in the nucleus), the following changes occur in α- and β-decay:

$$^A_Z X \rightarrow {}^4_2\alpha + {}^{A-4}_{Z-2} Y$$

$$^A_Z X \rightarrow {}^0_{-1}\beta + {}^A_{Z+1}Y$$

The explanation of these changes of A and Z for α-decay is simply that the α-article takes away two protons, so Z decreases by two, and in total the α-particle removes four nucleons, so A decreases by four. In β-decay a neutron in the nucleus changes into a proton (which stays in the nucleus), and an electron (which is ejected as the β-particle). Hence A is unaltered but Z increases by one. γ-decay does not change the chemical element since γ-radiations are electromagnetic waves. Often, after α- or β-decay, the nucleus is left in an excited state and the excess energy is emitted as a quantum of energy in the very short wavelength region of the electromagnetic spectrum, that is, a γ-ray is emitted.

EXPONENTIAL DECAY

Radioactive decay is a random process. The rate of decay of a sample may be written as $-dN/dt$, where N is the number of atoms present. The more atoms present, the greater will be the decay rate and so $-dN/dt$ is directly proportional to N. That is,

$$-\frac{dN}{dt} \propto N$$

so

$$-\frac{dN}{dt} = \lambda N \tag{1}$$

λ is called the **decay constant** of the radioactive atoms. If λ is large the decay rate is large. It can be shown that λ can be interpreted as the probability per second of a particular atom decaying. Equation (1) can now be integrated to give

$$N = N_0 e^{-\lambda t}$$

where N_0 is the number of atoms of the radioactive substance present at $t=0$. This result shows that radioactive decay follows an *exponential* law. There will be statistical variations in N and in the count rate, see page 314.

The number of atoms of the original element thus decrease exponentially with time, and are replaced by atoms of another element. Fig 9.8 shows a typical graph. The time taken for half the atoms to decay is called the **half-life** T. In $N = N_0 e^{-\lambda t}$, $N = N_0/2$ when $t = T$. This gives

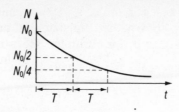

Fig 9.8 Radioactive decay
and half-life

$$\frac{N_0}{2}=N_0e^{-\lambda T}$$

or $2=e^{\lambda T}$

So $T=\ln2/\lambda=0.693/\lambda$

Hence the half-life and the decay constant have a simple relationship.

Since $-dN/dt \propto N$, the **decay rate** or **activity**, I also decreases exponentially with time, and this can be used as a basis for measuring short half-lives. The count rate near a source is measured at various times and then plotted graphically against time. The half-life can then be found from the graph as indicated in Fig 9.8.

A more accurate method is to use a logarithmic graph. Using I for dN/dt, then

$$I=I_0e^{-\lambda t},$$

where I_0 is the activity at $t=0$. Taking logs to the base e, then

$$\ln I=\ln I_0-\lambda t$$

A graph of $\ln I$ against t will therefore be a straight line. The slope of the graph can be determined and is equal to $-\lambda$. This method has the advantage of giving an average value of many values, as the best straight line is drawn passing as near as possible to most points.

For long half-lives a different method is necessary. This is indicated in worked example 9.5.

BACKGROUND COUNT   Whenever a count rate is measured, as in the last example, it is necessary to allow for the **background count** to get the true count. In the air radioactivity occurs naturally from radioactive minerals and cosmic rays, which gives an approximately constant count rate. This must be measured before any experiment and then subtracted from any count taken during the experiment.

ERRORS IN COUNTING EXPERIMENTS

It can be shown that the statistical error in any total count of N is equal to $\sqrt{N}$. If, for example, 100 counts is taken, the error is $\sqrt{100}$ $=10$. This means that is is possible to claim $\frac{10}{100} \times 100\% = 10\%$ accuracy for the result. If 400 counts were taken the error would be $\sqrt{400} = 20$. The percentage error in the answer would now be $\frac{20}{400} \times 100 = 5\%$. To get results of high accuracy it is therefore necessary to obtain a high total count.

RANGE OF PARTICLES; INVERSE-SQUARE LAW

If particles leave a radioactive source at a rate of N per second and none are absorbed, then N particles will pass per second through any sphere of radius r centred on the radioactive source. The number of particles per metre^2 at a distance r from the source will be $N/4\pi r^2$. Thus the count rate, measured on a geiger counter would be proportional to l/r^2. So the count rate from a source obeys an inverse-square law *provided that there is no absorption* of the radioactive particles.

In air, γ-radiation obeys an inverse-square law to a high degree of precision. β-particles obey the rule approximately, but at large distances from the source there is some deviation, showing that β-particles are slightly absorbed. α-particles do not obey the inverse-square law. α-particles have a definite *range* in air. As α-particles pass through the air they produce a great deal of ionization and after a certain distance (a few centimetres in air at atmospheric pressure) all their kinetic energy has been removed in producing ionization.

In passing through a material of thickness d, β- and γ-radiations obey a rule of the form $I = I_0 e^{-kd}$, where I_0 is the incident count rate and I the count rate on emerging from the material. k is a constant depending on the type of radiation and the density of the material used.

SAFETY IN RADIOACTIVE EXPERIMENTS

Radioactive sources can cause damage to cells in the body because of the ionization they produce. Small doses of radiation are not harmful but care must be taken to avoid severe doses. For safety then:

1   Sources are always kept in lead containers when not in use and kept in a safe.

2   The hands and body must be kept as far away from the sources as possible. Tongs are used for handling sources, and rubber gloves employed when using liquid sources.

3   Shields, e.g. of glass or perspex, help shield β-particles.

NUCLEAR STRUCTURE

ISOTOPES

The chemical properties of an atom are decided by the number of orbiting electrons. This is equal to the number of protons in the nucleus, that is, to Z. Atoms with the same number of protons in the nucleus but a different number of neutrons are called **isotopes**. For example, chlorine exists in two naturally occurring isotopes $^{37}_{17}Cl$ and $^{35}_{17}Cl$. These are identical chemically but the two isotopes have masses in the ratio 37:35 [approx]. The average mass number of chlorine is 35.5 and from this we can calculate the *relative abundance* of the two isotopes. Let there be x% of $^{35}_{17}Cl$ and hence $(100-x)$% of $^{37}_{17}Cl$. Since 100 atoms have an average mass of 35.5, we have

$$35x+(100-x)37=35.5\times100$$

So $\qquad 2x=100\times1.5$

or $\qquad x=75\%$

Isotopes which emit α- or β-particles or γ-radiation are described as **radioactive isotopes**. They can be used as tracer elements. For example, a radioactive isotope of iodine will be processed by the human body exactly as ordinary iodine, but the radioactivity which it emits allows the progress of the iodine around the body to be monitored.

MASS SPECTROMETER

For reasons which we shall see shortly it is important to be able to measure the mass of atoms and isotopes very precisely. Instruments for this purpose are called **mass spectrometers**.

Bainbridge's mass spectrometer has three important parts (Fig 9.9):

1 an *ion gun*. Here the sample chemicals are vaporized and, at very low pressure, bombarded with electrons (from an electron gun) to ionize them. They are then accelerated by a p.d. and fired into

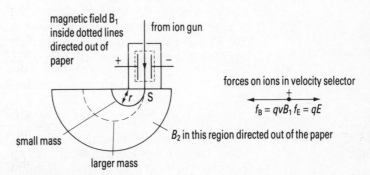

Fig 9.9 Mass spectrometer

2 a *velocity selector*. Here the ions pass through a magnetic field B_1 and an electric field E perpendicular to each other, and each perpen-

dicular to the ion beam. A slit, S, at the end of the velocity selectors allows *only those ions to pass which are not deflected by the fields.* For those ions the magnetic force B_1qv must be *equal and opposite* to the electric force qE. Hence $v=E/B_1$. Note that this result is not dependent on the mass of the ions – only those ions whose velocity is E/B_1 get through to the

3 *analyser.* Here a magnetic field, B_2, is applied perpendicular to the ion beam. Since all the ions have the same speed there will be the same magnetic force B_2qv on them (unless q is different, that is the ions were doubly rather than singly ionized). The curvature of the path therefore depends on the mass of the particles, the lighter particles being bent into the small radius orbit. We have

$$B_2qv=\text{centripetal force}=mv^2/r$$

So $$m=\frac{B_2qr}{v}=\frac{B_2B_1qr}{E}$$

Thus m is directly proportional to r. Using mass-spectrometers the masses of isotopes and atoms can be measured very precisely.

MASS UNITS

The **relative atomic mass** of an atom is defined as the mass of the atom relative to $\frac{1}{12}$th the mass of carbon-12 atom. The relative atomic mass of hydrogen is 1 very nearly, more accurately it is 1.008. Relative atomic mass has no units.

One **atomic mass unit** (u) is the actual mass of $\frac{1}{12}$th that of a carbon-12 atom. Now 1 atom of carbon has a mass $12g/(6\times10^{23})$.

Hence $$1\,u=\frac{1}{12}\times\frac{12}{6\times10^{23}}g=1.67\times10^{-24}g$$

$$=1.67\times10^{-27}kg.$$

The **mass number** A is the total number of protons and neutrons in the nucleus. It is now more usually called the **nucleon number**. Approximately, the mass number (which, by definition, must be an integer) is equal to the relative atomic mass. For example, uranium-238 has a nucleon number of 238 and a relative atomic mass of 238.051. The reasons for these slight differences will be examined next.

MASS DEFECT. BINDING ENERGY

Consider a helium nucleus. It contains two protons and two neutrons. Each proton has a mass of 1.0073u and each neutron a mass of 1.0087u. The total mass of the constituents is thus

$$2\times1.0073+2\times1.0087=4.0320\,u$$

The actual mass of a helium nucleus 4.0015 u. There is thus an apparent *mass-defect* of 4.0320−4.0015=0.0305 u.

In general, for an element $^A_Z X$ whose mass is M u the mass defect, δ, is given by

$$\delta = Zm_p + (A-Z)m_n - M,$$

since there are Z protons of mass m_p and $(A-Z)$ neutrons of mass m_n.

The significance of mass defect can be understood in terms of the famous Einstein relation

$$m = E/c^2$$

This states that *energy possesses inertia* (that is, *mass*). An amount of energy E joules possesses an inertia of m kg given by the above relation, where c is the velocity of light in a vacuum, $3 \times 10^8 \text{m s}^{-1}$.

When particles in a nucleus come together there are very strong short range forces (called the 'strong interactions') which come into play and hold the nucleus together, even though the protons repel electrostatically. In forming a stable nucleus there is thus a considerable loss of potential energy (and an equal release of electromagnetic energy). This enormous loss of energy means a corresponding loss of inertia, that is, mass. The assembled nucleus thus has a lower mass than the constituents, and there is therefore a 'mass-defect'. The changes in mass in nuclear processes are more noticeable than in other changes since the energies involved are very large.

The energy released when the nucleus is formed is called the **binding energy**. (Note that energy needs to be *removed* to create stability. Binding energy is *not* the energy needed to glue the nucleus together. Adding energy to the nucleus would tend to disrupt it.) Binding energy can also be thought of as the energy which must be supplied to a nucleus to split it up into individual components of protons and neutrons. Clearly from what we have said

$$\text{Mass defect (in kg)} = \frac{\text{Binding energy (in J)}}{c^2}$$

PROTON AND NEUTRON NUMBERS

In a stable nucleus the number of protons and the number of neutrons are approximately equal for light nuclei, but for heavier nuclei more neutrons than protons are found. This is because the protons have repulsive electrical forces between them and at higher proton number there has to be more neutrons to maintain stability. Figure 9.10(a) shows a graph of the number of neutrons plotted against the number of protons. Each isotope has one point on the graph and the line shows the mean of the various points. Not every stable nucleus lies on the line.

In β-decay, Z increases by one and the neutron number decreases by one. The movement in position on the graph is shown in Fig 9.10(b).

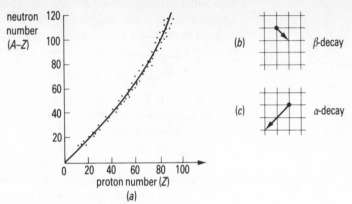

Fig 9.10 Proton–neutron curve for nuclei

In α-decay, Z decreases by two and the neutron number decreases by two. The movement in position on the graph is shown in Fig 9.10(c).

In any series of nuclear decays such as those starting from uranium, the α- and β-decays both take place so that the resulting isotopes do not deviate far from the average line shown in Fig 9.10(a).

BINDING ENERGY PER NUCLEON

From the example on page 316 the mass defect of helium is

$$0.0305\,u = 0.0305 \times 1.67 \times 10^{-27} kg$$

$$= 5.09 \times 10^{-29} kg$$

Hence   the binding energy $= 5.09 \times 10^{-29} \times (3 \times 10^8)^2 J$

$$= 4.58 \times 10^{-12} J$$

It is more usual to quote the binding energy in electron-volts (eV). Now $1.6 \times 10^{-19} J = 1 eV$ (see page 304). So

$$\text{binding energy of helium} = \frac{4.58 \times 10^{-12}}{1.6 \times 10^{-19}} eV$$

$$= 28.6 \times 10^6\, eV = 28.6\, MeV$$

In general, a similar method will show that $1 u$ is equivalent to $931\, MeV$.

The binding energy per nucleon is the total binding energy divided by the number of nucleons $= 28.6/4 = 7.15$ MeV/nucleon. The binding energy per nucleon is important in deciding the stability of a nucleus. The higher the value of the binding energy per nucleon, the more energy per nucleon would be needed to split the nucleus up into separate nucleons and so the nucleus is then *more stable*. The graph in Fig 9.11 shows the binding energy/nucleon plotted against A (the nucleon number). The peak occurs at the nucleon number for *iron* which is therefore the element with the most stable nucleus. The next section examines some further consequences of this graph.

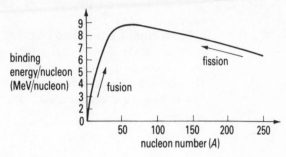

Fig 9.11 Binding energy per nucleon

NUCLEAR FISSION

Nuclear fission was discovered in 1938. Neutrons incident on uranium could cause the uranium nucleus to split into two. This process was called **fission** and the nuclei produced are called 'fission fragments'. The fission process releases large amounts of kinetic energy as we can see by considering the graph of Fig 9.11. In uranium-235 there are 235 nucleons each with a binding energy of approximately 7.5 MeV per nucleon. If the nucleus splits up, the elements produced will have a nucleon number of around 120 (in general the fragments do not have equal nucleon numbers although both fragments are usually quite large). At this point on the graph (where $A=120$) the binding energy is about 8.3 MeV/nucleon. Hence the change in energy of the particles is

$$(8.3-7.5)\times235=188\,\text{MeV}$$

This is a *release* of energy since the greater the binding energy the *more* energy is released in the initial formation of the nucleus.

The precise amount of energy released depends on the particular fission fragments produced and can be calculated as shown in worked example 9.8.

CHAIN REACTIONS

In the last example two neutrons were released, whereas one neutron was needed to initiate the fission. It is possible that these two neutrons may strike other nuclei and cause more fission. In this way a **chain reaction** can be set up. To achieve a chain reaction the neutrons must be used to produce fission before they escape from the material, and so it is not possible to achieve a chain reaction in small specimens.

A nuclear reactor has three main components: (i) rods of fissionable material which are surrounded by (ii) a moderator, which reduces the speed of the neutrons and increases their chances of causing fission and (iii) control rods of cadmium, which absorb neutrons and so can prevent the chain reaction from becoming too violent.

In an **atomic bomb** two pieces of fissionable material which are too small in size to allow a chain reaction (too many neutrons escape from the surface of the samples) are exploded together by a small charge. The resulting block of fissionable material is above the size needed to maintain a chain reaction and an uncontrolled release of energy then occurs.

NUCLEAR FUSION

In fission, a heavy atom near the falling right hand side of the graph of Fig 9.11 would split into light fragments which are situated nearer the centre of the graph, as shown by the arrow. The binding energy per nucleon increases in this process and so, as we have explained, energy is released.

It is also possible to move towards the peak of the graph by *fusing* two light nuclei together. Hence this process, called **nuclear fusion**, will also release energy.

For example, two isotopes of hydrogen, deuterium ^2_1H, and tritium, ^3_1H, can fuse together to form helium according to the reactions:

$$^2_1\text{H} + ^3_1\text{H} = ^4_2\text{He} + ^1_0\text{n}$$

By calculating the masses of each side of the equation, as shown for fission, it can be shown that 17.6 MeV of energy is released.

At the moment, considerable releases of fusion energy have only occurred in an uncontrolled manner in fusion (hydrogen) bombs. Here lithium deuteride is used. The neutrons released from a central fission bomb cause tritium to be formed from the lithium. The very high temperature then forces the deuterium and tritium nuclei to collide with very high speeds, so that the fusion can occur.

To control fusion is very difficult since very high temperatures (10^7 K) are needed for the colliding nuclei to get close enough together. Otherwise the repulsive electrostatic forces between nuclei prevent reactions occurring. At these temperatures the atoms are all ionized and a **plasma** of positive ions and electrons is produced. To maintain this high temperature plasma in existence, magnetic fields are used to make the charged particles move in circular orbits and so contain the plasma. Such an arrangement is called a 'magnetic bottle'. At the moment research is going on, financed by countries in Western Europe, to produce an experimental reactor. This is the JET project – the Joint European Torus.

WORKED EXAMPLES

**9.1 MOMENTA OF
CHARGED PARTICLES**

A proton and an α-particle are each accelerated through 1000V p.d. Find the ratio of their momenta after acceleration.

(**Overview** The α-particle will gain twice the energy as it has twice the charge. $mv^2/2$ is then used to find the ratio of the speeds, followed by mv to find the ratio of the momenta.)

The α-particle comprises two protons and two neutrons (see page 000). It thus has four times the mass m and twice the charge of the proton, that is, a mass $4m$ and a charge of $2e$. After acceleration the proton will have 1000 electron-volts of energy, the α-particle will have 2000 electron-volts of energy (as it has double the charge). If m=mass of a proton, v_p=proton velocity and v_α=α-particle velocity, then

$$\text{k.e. proton}=\tfrac{1}{2}mv_p{}^2=\tfrac{1}{2}(\text{k.e. }\alpha\text{-particle})=\tfrac{1}{2}(\tfrac{1}{2}.4mv_\alpha{}^2)$$

Hence $\quad v_p{}^2=2v_\alpha{}^2$, so $v_\alpha=v_p/\sqrt{2}$

Thus $\quad \dfrac{\text{momentum proton}}{\text{momentum }\alpha\text{-particle}}=\dfrac{mv_p}{4mv_\alpha}$

$$=\dfrac{v_p}{4v_p/\sqrt{2}}=\dfrac{\sqrt{2}}{4}$$

**9.2 PHOTOELECTRIC
EFFECT**

Light of wavelength 5×10^{-7}m emits electrons from a metal with a maximum kinetic energy of 2×10^{-19}J. Taking h as 6.6×10^{-34}J s, find (a) the work function of the metal, (b) the longest wavelength which will emit photoelectrons.

(**Overview** (a) is solved using photon energy=work function+maximum k.e. of electrons.
(b) is solved using least photon energy=work function, since the minimum k.e. is zero.)

(a) Using $hf=\omega_0+\text{max k.e.}$, and since $c=f\lambda$,

$$\frac{hc}{\lambda}=\omega_0+\text{max k.e.}$$

so $\quad \dfrac{6.6\times10^{-34}\times3\times10^8}{5\times10^{-7}}=\omega_0+2\times10^{-19}$

or $\quad\quad 3.96\times10^{-19}=\omega_0+2\times10^{-19}$

giving $\quad\quad \omega_0=1.96\times10^{-19}$J.

(b) The longest wavelength possible is obtained when the photons only just have enough energy to remove electrons, that is,

$$hf_{min} = \frac{hc}{\lambda_{max}} = \omega_0$$

giving $\lambda_{max} = \frac{hc}{\omega_0} = \frac{6.6 \times 10^{-34} \times 3 \times 10^8}{1.96 \times 10^{-19}} = 1.0 \times 10^{-6} m$

9.3 IONIZATION ENERGY

The ionization potential of the hydrogen atom is 13.6V. Use the data below to calculate (a) the speed of an electron which could just ionize the hydrogen atom, and (b) the minimum wavelength which the hydrogen atom can emit.

(**Overview** (a) The electron energy must be at least equal to the ionization energy.

(b) Minimum wavelength corresponds to the maximum energy – this is the ionization energy.)

Charge on an electron $= -1.6 \times 10^{-19} C$. Mass of an electron $9.11 \times 10^{-31} kg$. The Planck constant $h = 6.63 \times 10^{-34} J$ s. Speed of light $= 3 \times 10^8 m\ s^{-1}$.

(a)   The k.e. of the electron must be 13.6 electron-volts $= 13.6 \times 1.6 \times 10^{-19} J$.

Hence $\frac{1}{2}mv^2 = 13.6 \times 1.6 \times 10^{-19}$

So $v^2 = \frac{27.2 \times 1.6 \times 10^{-19}}{9.11 \times 10^{-31}}$

and $v = 2.19 \times 10^6 m\ s^{-1}$

(b)   The minimum wavelength corresponds to the highest frequency and so to the greatest photon energy. This occurs when a proton (a hydrogen ion) captures an electron and the atom falls to ground state. The energy emitted is then equal to the ionization energy of 13.6 electron-volts. The wavelength, λ, of the photon emitted is therefore given by

$$hf = \frac{hc}{\lambda} = 13.6 \times 1.6 \times 10^{-19}$$

or $\lambda = 6.6 \times 10^{-34} \times 3 \times 10^8 / 13.6 \times 1.6 \times 10^{-19}$

$$= 9.1 \times 10^{-8} m = 91nm$$

9.4 X-RAYS

(a)   Calculate the minimum wavelength which can be emitted by an X-ray tube operating at 10kV.
(b)   X-rays of this wavelength are used to study a crystal where one of the plane separations is 0.2nm. Find the angle which the

X-rays must make to these crystal planes and the angle of deviation of the diffracted beam. ($h=6.6\times10^{-34}$J s, $e=1.6\times10^{-19}$C, $c=3\times10^8$m s^{-1})

(**Overview** (a) The minimum wavelength corresponds to the maximum energy of the X-ray photon. This occurs when all the electron energy is transferred to the photon.

(b) can be solved using Bragg's Law.)

(a)   The highest frequency, f_0, which can be emitted from an X-ray tube is obtained where all the electron energy is transferred to an X-ray photon. In this case

$$hf_0=eV$$

So $$\frac{hc}{\lambda_0}=eV$$

So $$\lambda_0=\frac{hc}{eV}=\frac{6.6\times10^{-34}\times3\times10^8}{1.6\times10^{-19}\times10^4}=1.24\times10^{-10}\text{m}.$$

(b)   Using $2d\sin\theta=n\lambda$,

$$2\times0.2\times10^{-9}\sin\theta=1.24\times10^{-10}\text{ in the case where }n=1.$$

Simplyfying, we find $\theta=18°$.

From Fig 9.5 the angle of deviation is 2θ, which is $36°$.

9.5 CALCULATING LONG HALF-LIFE

A Geiger counter with an end window of area 5cm^2 is placed 10cm from 1g of a radioactive material. A count rate of 50 counts per second is observed. The relative atomic mass of the element of the sample is 238. Calculate the half-life of the material and the activity of the specimen in curies. (The Avogadro constant$=6\times10^{23}$mol^{-1}.)

(**Overview** The basis of the calculation is to use $-dN/dt=\lambda N$ directly. The number of atoms N in the specimen is found from its mass, relative atomic mass and the Avogadro Constant. dN/dt is calculated from the count rate. Allowance has to be made for the fact that particles are spreading in all directions from the source.)

The number of moles in the sample is 1/238.

Hence the number of atoms in the sample$=(1/238)\times6\times10^{23}$
$$=2.52\times10^{21}$$

The Geiger counter receives 50 counts/second in its 5cm^2 of area. But particles are radiated from the source in all directions, that is, over an area of a sphere of radius 10cm.

Thus the fraction of all the particles received by the GM tube is $5/(4\pi\times10^2)$.

So the total number of disintegrations per second$=50\times\dfrac{4\pi\times10^2}{5}$

$=12566\text{s}^{-1}$. Now this last figure is $-dN/dt$, when the number of atoms in the sample is N.

But
$$-\frac{dN}{dt}=\lambda N$$

So
$$12566=\lambda\times2.52\times10^{21}$$

From which
$$\lambda=4.99\times10^{-18}\,s^{-1}$$

But
$$T=\frac{\ln 2}{\lambda}=\frac{0.693}{4.99\times10^{-18}}=1.39\times10^{17}\text{ seconds}$$

or
$$T=\frac{1.39\times10^{17}}{60\times60\times24\times365}\text{ years}$$

$$=4.4\times10^{9}\text{ years}$$

This method of calculation is based on finding λ, the probability of decay, and then using this to calculate T. It would be clearly impossible to wait for the material to decay enough to measure any change in activity.

One curie is defined to be 3.7×10^{10} disintegrations per second. This source makes 12566 disintegrations per second and so has an activity of $12566/(3.7\times10^{10})=3.4\times10^{-7}$ curie

$$=0.34\ \mu\text{curie}$$

9.6 HALF-LIFE

A radioactive isotope of strontium, of half-life 28 years, providing a source of β-particles, has been in use for 14 years. If originally $5\mu g$ of the strontium were present show graphically, or otherwise, that the amount of this isotope has been reduced to approximately $3.5\mu g$.

(Overview Use $N=N_0e^{-\lambda t}$ and find λ from $\ln 2/$(half-life).)

We have
$$N=N_0e^{-\lambda t}$$

Hence for the mass m, we have

$$m=m_0e^{-\lambda t}$$

Now $\lambda=\ln 2/T$,

or
$$\lambda=(0.693/28)\ y^{-1}$$

So
$$m=5\times e^{-(0.693/28)\times14}=5\times e^{-0.347}$$

$$=3.53\ \mu g$$

9.7 HALF-LIFE AND INVERSE SQUARE LAW

A point source of γ-radiation has a half-life of 30 minutes. The initial count rate, recorded by a Geiger counter placed 2.0m from the source, is $360\,s^{-1}$. The distance between the counter and the source is altered.

After 1.5 hours the count rate recorded is $5\,s^{-1}$. What is the new distance from the source?

(**Overview** First calculate what is the new activity of the source after 1.5 hours. The inverse square law can then be used to relate this to the observed count rate.)

After 1.5 hours (3 half-lives) the count rate at 2.0m would be reduced to $360/2^3 = 45\,s^{-1}$.

Since γ-radiation obeys an inverse square law, the distance, r, at which the count rate is $5\,s^{-1}$ is given by $5 = \dfrac{k}{r^2}$.

But $45 = \dfrac{k}{2^2}$ and hence, by division,

$$\frac{45}{5} = \frac{r^2}{2^2}$$

So $r = \sqrt{2^2 \times \dfrac{45}{5}} = 6m$

9.8 ENERGY RELEASE IN FISSION

Possible fission reaction for $^{235}_{92}U$ is to split into molybdenum-95 and lanthanum-139. Two neutrons and seven electrons are produced, that is,

$$^{235}_{92}U + ^1_0n \rightarrow ^{95}_{42}Mo + ^{139}_{57}La + 2^1_0n$$

Calculate the energy released in MeV neglecting electron masses.

(**Overview** Find the change in mass and use $E = mc^2$. (Or remember the energy equivalent of 1 atomic mass unit.))

We first set out the masses of each side of the equation.

$^{235}_{92}U = 235.123$	$^{95}_{42}Mo = 94.945$
$^1_0n = \quad 1.009$	$^{139}_{57}La = 138.955$
	$2 \times ^1_0n = \quad 2.018$
$\overline{236.133\ u}$	$\overline{235.918\ u}$

Data for the masses in u is found from tables.

The increase in mass defect is thus $(236.133 - 235.918) = 0.215\,u$. Since $1\,u$ corresponds to 931 MeV (page 318) the energy release is $0.215 \times 931 = 198$ MeV.

9.9 VELOCITY SELECTORS IN MASS SPECTROMETERS

In the velocity selector of a mass-spectrometer, it is required to select singly-charged oxygen ions with energy 1000eV. The electric field in the velocity selector is provided by two parallel plates 1cm apart and having a p.d. of 200V applied between them. Calculate the magnetic flux density needed in the velocity selector.

(Overview Find the speed from $eV=mv^2/2$. Then use the velocity selector formula $v=E/B$.)

The kinetic energy of the particles is $1000\times1.6\times10^{-19}$J. The mass m of an oxygen atom is $16\times$mass of proton$=16\times1.7\times10^{-27}$kg.

So $\quad \frac{1}{2}mv^2=1000\times1.6\times10^{-19}$

and $\quad v^2=\dfrac{2\times1000\times1.6\times10^{-19}}{16\times1.7\times10^{-27}}$

Hence $\quad v=1.1\times10^5$m s^{-1}

The electric field intensity$=V/d=2000/0.01=2\times10^5$V m^{-1}. But using the velocity selector formula (page 316), $v=E/B$.

So $\quad B=\dfrac{E}{v}=\dfrac{2\times10^5}{1.1\times10^5}=1.8$T.

9.10 PARTICLE EMISSION

In the radioactive decay of uranium 238, the following reaction takes place when an α-particle is emitted.

$$^{238}_{92}U \rightarrow {}^{234}_{90}Th+{}^{4}_{2}\alpha$$

At some instant the centre of the emerging α-particle is 9×10^{-15} metre from the centre of the residual thorium nucleus. At this instant (a) what is the force on the α-particle, (b) what is the acceleration of the α-particle? (Take the mass of an α-particle to be 1.7×10^{-27}kg and $e=1.6\times10^{-19}$C.)

(Overview Use Coulomb's Law and then $F=ma$.)

(a) $\quad F=\dfrac{Q_1Q_2}{4\pi\varepsilon_0 r^2}=\dfrac{9\times10^9\times(2\times1.6\times10^{-19})\times(90\times1.6\times10^{-19})}{(9\times10^{-15})^2}$

$\quad=512$ N

(b) $\quad a=F/m=512/(1.7\times10^{-27})=3\times10^{29}$m s^{-2}

This is an enormous acceleration. The α-particle must be extremely stable to withstand these enormous forces and accelerations.

1 PHOTOELECTRIC EFFECT	Emission of electrons when light of suitable wavelength falls on metals.

Explained in terms of particles called *photons*.

Einstein equation: $hf = \omega_0 + \frac{1}{2}mv^2_{max}$, where ω_0 is called the *work function* of the metal.

2 ENERGY LEVELS

Atoms exist only in one of a definite series of energy levels.

When an atom goes from one energy level E_1 to a lower level E_2, electromagnetic radiation is emitted of frequency f, where $hf = E_1 - E_2$.

Energy just to remove an electron from an atom is called the *ionization energy*.

3 X-RAYS

Produced when fast moving electrons strike a metal target.

Two parts to spectrum, (*a*) background – minimum wavelength given by $\lambda_0 = hc/eV$, (*b*) characteristic – dependent on target metal.

X-rays used to study crystal structure. Bragg's law, $2d\sin\theta = n\lambda$.

4 WAVE-PARTICLE DUALITY

Electromagnetic radiation and matter both exhibit wave and particle behaviour, called wave-particle duality. The energy of an electromagnetic radiation photon is given by $E = hf$. The wavelength associated with a particle of momentum p is h/p.

5 NUCLEAR ATOM AND RADIOACTIVITY

Rutherford model confirmed by Geiger and Marsden's experiment.

Radioactivity produces ionization. Detected in ionization chamber.

GM tubes rely on ionizing effect of radiations in a gas.

In α-decay, helium nucleus emitted

$$^A_Z X \rightarrow {}^4_2\alpha + {}^{A-4}_{Z-2} Y$$

In β-decay, fast moving electron emitted

$$^A_Z X \rightarrow {}^0_{-1}\beta + {}^A_{Z+1} Y$$

Exponential decay, $-dN/dt = \lambda N$, $N = N_0 e^{-\lambda t}$. Half-life $= 0.693/\lambda$.

Experiments with radioactivity must take into account (a) background count, (b) errors due to statistical nature of decay – error in N counts $= \pm \sqrt{N}$, (c) safety precautions.

Range. γ- (and, approximately, β-) obey inverse square law in air showing no absorption. α-particles have a definite range in air.

---

6 NUCLEAR STRUCTURE　　Isotopes. Atoms with same proton number but different number of neutrons in nucleus.

Mass of atoms measured by mass-spectrometer – consists of ion gun, velocity selector, analyser. Radius of orbit in analyser is proportional to mass of ion.

Mass defect $\delta = $ (total mass of individual nucleons) $-$ (nuclear mass).

Caused by reduction in energy when nucleus formed. $\delta \times c^2$ is a measure of the *binding energy* = energy needed just to break up a nucleus into its component parts.

Nuclear fission = splitting up of large nucleus. Binding energy per nucleon increases, so energy released in process.

Nuclear fusion = joining together of two light nuclei. Binding energy per nucleon increases so energy released.

DIAGRAM SUMMARY

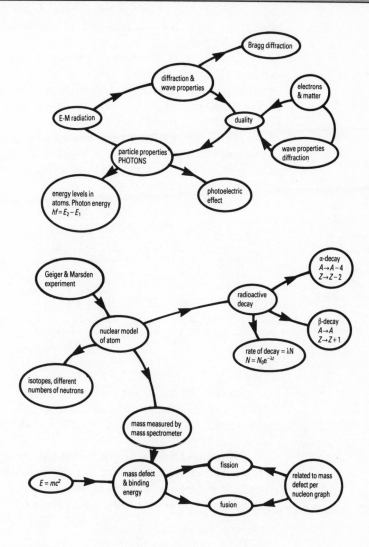

(a)   Write down the Einstein photoelectric equation and explain how it accounts for the emission of electrons from metal surfaces illuminated by light.

(b)   In Fig 9.12 P is a vacuum photocell with anode, A, and cathode, K, made from the same metal of work function 2.0eV. The cathode is illuminated by monochromatic light of constant intensity and of wavelength 4.4×10^{-7}m. Describe and explain in general terms how the current shown by the microammeter, M, will vary as the slider of the potential divider is moved from B to C. What will be the reading of the high-resistance voltmeter, V, when photoelectric emission just ceases? Assume that no secondary emission of electrons occurs from the anode.

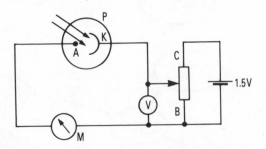

Fig 9.12

(c)   With the slider set midway between B and C, describe and explain how the reading of M would change if, in turn (i) the intensity of the light was increased, (ii) the wavelength of the light was changed to 5.5×10^{-7}m.

(d)   Outline *briefly* how you could use the arrangement described in (b), together with an additional light source of a different wavelength, to determine a value for the Planck constant assuming that the value of the work function was unknown.

(Charge of the electron=1.6×10^{-19}C.
Planck constant=6.6×10^{-34}J s.
Speed of light in vacuo=3.0×10^{8}m s^{-1}.)   (JMB)

2   (a)   Explain briefly what is meant by an emission spectrum. Describe a suitable source for use in observing the emission spectrum of hydrogen.

(b)   The diagram (Fig 9.13), which is to scale, shows some of the possible energy levels of the hydrogen atom.

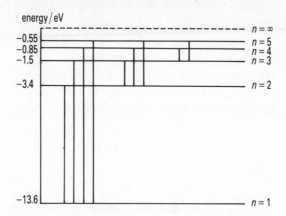

Fig 9.13

(i) Explain briefly how such a diagram can be used to account for the emission spectrum of hydrogen.

(ii) When the hydrogen atom is in its ground state, 13.6eV of energy are needed to ionize it. Calculate the highest possible frequency in the line spectrum of hydrogen. In what region of the electromagnetic spectrum does it lie?

(The frequency range of the visible spectrum is from 4×10^{14}Hz to 7.5×10^{14}Hz.)

(iii) A number of transitions are marked on the energy level diagram. Identify which of these transitions corresponds to the lowest frequency that *would be visible*.

(c)   The wavelengths of the lines in the emission spectrum of hydrogen can be measured using a spectrometer and a diffraction grating. If the grating has 5000 lines per cm, through what angle would light of frequency 4.6×10^{14}Hz be diffracted in the first order spectrum?

($1\,eV=1.60\times10^{-19}$J.

Planck constant$=6.6\times10^{-34}$J s.

Speed of light$=3.00\times10^{8}$m s^{-1}.)   (L)

3   In a modern X-ray tube electrons are accelerated through a large potential difference and the X-rays are produced when the electrons strike a tungsten target embedded in a large piece of copper.

(a)   What is the purpose of the large piece of copper?

(b)   Describe the energy changes taking place in the tube.

(c)   If the accelerating voltage is 40kV, calculate the kinetic energy of the electrons arriving at the target, and determine the minimum wavelength of the emitted X-rays.

(Electronic charge$=-1.6\times10^{-19}$C.

Planck constant$=6.6\times10^{-34}$J s.

Velocity of electromagnetic waves$=3.0\times10^{8}$m s^{-1}.)   (*AEB* 1984)

4   Molybdenum Kα X-rays have wavelength 7×10^{-11}m. Find (i) the minimum X-ray tube potential difference that can produce these X-rays, and (ii) their photon energy in electron volts.

(Electronic charge, $e=-1.6\times10^{-19}$C.
Planck constant, $h=6.6\times10^{-34}$J s.
Speed of light, $c=3\times10^8$m s^{-1}.)   (W)

5   The diagram (Fig 9.14) shows a gold foil mounted across the path of a narrow, parallel beam of alpha particles. The fraction of incident alpha particles reflected back through more than 90° is very small.

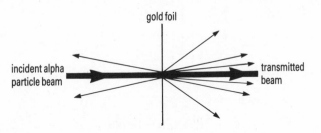

gold foil

incident alpha particle beam

transmitted beam

Fig 9.14

How does this result lead to the idea that an atom has a nucleus
(a)   whose diameter is small compared with the atomic diameter, and
(b)   which contains most of the atom's mass?   (L)

6   $^{32}_{15}$P is a β-emitter with a decay constant of 5.6×10^{-7}s^{-1}. For a particular application the initial rate of disintegration must yield 4.0×10^7 β-particles every second. What mass of pure $^{32}_{15}$P will give this decay rate?   (C)

7   Describe how you could detect and distinguish between β-radiation and γ-radiation from a radioactive source emitting both radiations.
     A source of radioactive potassium is known to contain two isotopes, $^{42}_{19}$K and $^{44}_{19}$K, both of which decay by emission of β-radiation to stable isotopes of calcium. Write a nuclear transformation equation for one of these decays.
     The source is placed in front of a β-radiation counter and the following count rates, corrected for background, are recorded.

time/hours	0	0.5	1.0	1.5	2.0	2.5	3.0
count rate/min^{-1}	10000	3980	2125	1260	955	890	832

time/hours	4.0	5.0	6.0	7.0	8.0	9.0	10.0
count rate/min^{-1}	790	750	710	670	630	600	575

Plot the data on a graph of lg (count rate/min^{-1}) against time/hours.

From your graph estimate values for

(a)   the half-life of $^{42}_{19}$K which is the longer-lived isotope,
(b)   the half-life of $^{44}_{19}$K which is the shorter-lived isotope,
(c)   the initial count rates due to $^{42}_{19}$K and $^{44}_{19}$K,
(d)   the ratio of the amounts of $^{42}_{19}$K and $^{44}_{19}$K present in the source
at the start of the measurements.   (O&C)

8   In the fusion reaction $^2_1$H+$^3_1$H=$^4_2$He+$^1_0$n, how much energy, in joules, is released?

$$(\text{Mass of } ^2_1\text{H}=3.345\times10^{-27}\text{kg,}$$
$$^3_1\text{H}=5.008\times10^{-27}\text{kg,}$$
$$^4_2\text{He}=6.647\times10^{-27}\text{kg,}$$
$$^1_0\text{n}=1.675\times10^{-27}\text{kg.}$$
$$\text{Speed of light}=3.0\times10^8\text{m s}^{-1}.)$$ (L)

9   Given

$$^{235}_{92}\text{U}+^1_0\text{n} \rightarrow ^x_{45}\text{Rh}+^{113}_y\text{Ag}+2^1_0\text{n}$$

and $^2_1$H+$^3_1$H $\rightarrow$ $^4_2$He+A

(i) explain what is meant by the 235 and 92 in $^{235}_{92}$U;
(ii) determine x, y, and A;
(iii) describe the importance of the reactions;
(iv) write down a similar equation for the fusion of two atoms of deuterium to form helium of atomic mass number 3.

Given the mass of the deuterium nucleus is 2.015u, that of one of the isotopes of helium is 3.017u and that of the neutron is 1.009u, calculate the energy released by the fusion of 1kg of deuterium. If 50% of this energy were used to produce 1MW of electricity continuously, for how many days would the station be able to function?

(Speed of light $c=3.00\times10^8$m s^{-1}.)   (W)

HEAT ENERGY: GASES; HEAT CAPACITIES OF GASES; THERMODYNAMICS

CONTENTS

Contents

GAS LAWS

Gases are used in engines of cars, aeroplanes and motor-cycles. The properties of gases and the laws they obey are therefore of particular importance to heat engineers.

Gases can vary in pressure p, volume V and temperature. We shall use the symbol T for temperature in kelvin and θ for temperature in °C.

BOYLE'S LAW

In 1660 Boyle investigated the relationship between the pressure p and volume V of a gas *at constant temperature*. His result, known as **Boyle's Law**, states:

The pressure of a given mass of gas at constant temperature is inversely proportional to its volume. So $p \propto 1/V$, or $pV = constant$.

It should be carefully noted that Boyle's law can only be applied to a given or fixed mass of gas. If some of the gas escapes while a change in pressure is made, we can *not* say that $p_1 V_1 = p_2 V_2$ where p_1, V_1 refer to the gas before some escaped and p_2, V_2 refer to the gas after some has escaped. Figure 10.1 shows how

(a) p and V vary, and
(b) p and $1/V$ vary, when Boyle's law is obeyed.

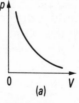
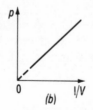

Fig 10.1 Boyle's law

Boyle's law (temperature constant): $pV = constant$ or $p \propto 1/V$

VOLUME AND TEMPERATURE (CONSTANT PRESSURE)

Fig 10.2 shows how the volume V of a gas varies with temperature in °C when the gas is kept at *constant pressure*. So gases expand uniformly with temperature rise and decrease uniformly with temperature fall. At about -273°C (more accurately, -273.15°C), the volume of a gas would *theoretically* become zero. The temperature of $-273°C$ is therefore called the **absolute zero** of temperature. The kelvin scale of temperature is measured from the absolute zero. So

$$0K = -273°C$$
and $$TK = (273 + \theta°C)$$

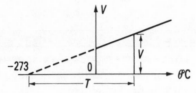

Fig 10.2 Variation of gas
volume and temperature –
constant pressure

Using the kelvin temperature scale, we see from Fig 10.2 that, when a given mass of gas is at constant pressure, its volume V is related to its temperature by

$$V \propto T \tag{1}$$

Suppose 1000cm³ of a given mass of oxygen in a cylinder is warmed from 27°C to 100°C while its pressure remained constant. Then, since $V \propto T$, the new volume V is given by $V/V_1 = T/T_1$. So

$$\frac{V}{1000} = \frac{273 + 100}{273 + 27} = \frac{373}{300}$$

So $$V = \frac{373}{300} \times 1000 = 1243cm^3$$

PRESSURE AND TEMPERATURE (CONSTANT VOLUME)

Figure 10.3 shows the result for the pressure-temperature changes of a given mass of gas when its volume is kept constant. The relationship is similar to that in (1): $p \propto T$ (see Fig 10.3).

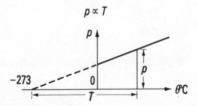

Fig 10.3 Variation of gas
pressure and temperature –
constant volume

•Consider the air in a car tyre which is initially at a pressure of 2.50×10^5 Pa (Nm^{-2}) at a temperature of 10°C. If the temperature increases to 18°C, then, assuming the tyre has practically a constant volume, the new pressure p is given by

$$\frac{p}{2.5 \times 10^5} = \frac{273 + 18}{273 + 10} = \frac{291}{283}$$

So
$$p = \frac{291 \times 2.5 \times 10^5}{283} = 2.57 \times 10^5 \text{ Pa}$$

IDEAL GAS EQUATION

From our previous relationships, we can see that when p, V and T of a fixed mass of gas all vary, we can write

$$\frac{pV}{T} = \text{constant}$$

The value of the constant depends on the amount of gas. If 1 mole of gas is used, the symbol for the constant is R. If n moles of gas are used, the volume V of the gas under the same conditions of pressure p and temperature T is n times that for 1 mole. So, from above,

$$pV = RT \text{ (1 mole)}$$
and $\quad pV = nRT \text{ ($n$ moles)}$

The equation $pV = nRT$ is called the *ideal gas equation* for n moles. It holds for real gases such as oxygen and nitrogen at near-normal pressures and for vapours at temperatures well above their 'critical temperatures', when the vapours begin to condense to liquids.

If M is the mass of 1 mole of a gas, then the number of moles in a mass m of the gas is m/M. So we can also write the ideal gas equation for a mass m as

$$pV = \left(\frac{m}{M}\right)RT$$

The mass of 1 mole of helium is 4g. So a mass of 12g, for example, is 3 moles; and its gas equation, from $pV = nRT$, is $pV = 3RT$. The mass of 1 mole of hydrogen is 2g. So 1kg of hydrogen is 1000/2 or 500 moles. Its gas equation is therefore $pV = 500RT$. Note that the number of moles n of a gas is given by

$$n = \frac{pV}{RT}$$

which is a useful relation for n when n keeps constant. This is illustrated in Example 10.3 on p341 which shows how to find the new pressure when two gases are mixed.

MOLAR GAS CONSTANT

The volume occupied by 1 mole of *any* gas is the same, about $22.3 \times 10^{-3} m^3$ at 0°C and 1.013×10^5 Pa pressure. So

$$R = \frac{pV}{T} = \frac{1.013 \times 10^5 \times 22.3 \times 10^{-3}}{273}$$
$$= 8.3 \, J \, mol^{-1} K^{-1}$$

R is known as the *molar gas constant*. Its value is the same for 1 mole of any gas. For example, the mass of 1 mole of hydrogen is 2g and the mass of 1 mole of oxygen is 32g. For both masses of gas, their equation is written as $pV = RT$ because this is the equation for 1 mole.

WORKED EXAMPLES ON GAS LAWS

10.1 MASS OF GAS

Calculate the mass of 2 litres of hydrogen at 27°C and 2×10^5 Pa pressure, given that the molar mass of hydrogen is 2g and the molar gas constant is $8.3 \, J \, mol^{-1} K^{-1}$.

(**Overview** (i) First we need to find the number of moles in the gas from $n = pV/RT$.
(ii) Then we use mass of 1 mole is 2g.)

Number of moles in gas,

$$n = \frac{pV}{RT} = \frac{2 \times 10^5 \times 2 \times 10^{-3}}{8.3 \times 300}$$

since 2 litres $= 2 \times 10^{-3} m^3$ and $T = 300K$. Simplifying, $n = 0.16$.

So   mass of hydrogen $= 0.16 \times 2g = 0.32g$

10.2 APPLYING GAS LAWS

(a)   3.0kg of gas is contained in a cylinder at 27°C and 2.0×10^5 Pa pressure. The gas is warmed so that its pressure increases by 10%. Calculate the new temperature of the gas.
(b)   The cylinder tap is now opened in a room where the temperature is 77°C and the pressure is 1.5×10^5 Pa. Some of the gas escapes. Calculate the mass of gas remaining in the cylinder.

(**Overview** (a) Here we need to use $p \propto T$.
(b)   We can find the initial number of moles and the final number of moles left after gas escapes by using $n = pV/RT$.)

(a)   Since the volume of gas is constant, $p \propto T$. The new pressure $= 2.2 \times 10^5$ Pa, an increase of 10%.

So $\frac{T}{273 + 27} = \frac{2.2 \times 10^5}{2.0 \times 10^5} = \frac{22}{20}$

$$T = \frac{22}{20} = 300 = 330\text{K} \ (57°C)$$

(b)   The initial number of moles of gas, $n_1 = \frac{pV}{RT} = \frac{2.0 \times 10^5 \times V}{R \times 300}$,

where V is the volume of the cylinder and R is the molar gas constant.
    The final number of moles of gas in the cylinder, n_2, since $T = 273 + 77 = 350\text{K}$ and $p = 1.5 \times 10^5$ Pa, is

$$n_2 = \frac{1.5 \times 10^5 \times V}{R \times 350}$$

So, by division, $\dfrac{n_2}{n_1} = \dfrac{1.5 \times 300}{2.0 \times 350} = 0.64 = \dfrac{m_2}{3\text{kg}}$

since the mass of a gas is proportional to its number of moles. Hence the mass m_2 left in the cylinder is

$$m_2 = 0.64 \times 3\text{kg} = 1.92\text{kg}$$

10.3 GAS MIXTURES

A vessel of volume 1 litre $(1 \times 10^{-3}\text{m}^3)$ contains a gas at 0°C and pressure 1.5×10^5 Pa. Another vessel of volume 2 litres contains the same gas at the same temperature and pressure. The two vessels are now connected by a tube of negligible volume and the 1 litre vessel is maintained at 0°C while the 2 litre vessel is heated to 100°C. Calculate the steady pressure produced in the vessels, assuming the volume change of the heated vessel is negligible.

(Overview Although some gas passes from one vessel to another, the *total* number of moles in the two gases remains constant. So we apply $n = pV/RT$ to the total number of moles of the two gases.)

From $n = pV/RT$,
initial number of moles in both vessels

$$= \frac{1.5 \times 10^5 \times 1 \times 10^{-3}}{R \times 273} + \frac{1.5 \times 10^5 \times 2 \times 10^{-3}}{R \times 273}$$

$$= \frac{150}{R \times 273} + \frac{300}{R \times 273}$$

$$= \frac{450}{R \times 273}$$

Suppose p is the final steady pressure in the vessels. Then number of moles in both vessels $= \dfrac{p \times 1 \times 10^{-3}}{R \times 273} + \dfrac{p \times 2 \times 10^{-3}}{R \times 373}$, since the respective temperatures are now 0°C and 100°C. But the *total* number of moles in both vessels is the same before and after one vessel is heated. So

$$\frac{p \times 1 \times 10^{-3}}{R \times 273} + \frac{p \times 2 \times 10^{-3}}{R \times 373} = \frac{450}{R \times 273}$$

Cancelling R throughout and simplifying, we find that

$$p = 1.8 \times 10^5 \text{ Pa}$$

KINETIC THEORY OF IDEAL GASES

ASSUMPTIONS FOR IDEAL GAS

In the kinetic theory of gases we aim to explain the gas laws and other behaviour of gases by considering the motion of their molecules. To simplify the calculation, we make some assumptions:

1 The attraction between molecules is negligible.
2 The volume occupied by the molecules is negligible compared with the volume occupied by the gas.
3 The molecules are perfectly elastic spheres, in which case any collisions are elastic ones and no energy is lost.
4 The time of a collision is negligible compared with the time between collisions.

PRESSURE OF GAS

1 Consider a cube of side l containing N molecules of gas each of mass m (Fig 10.4). The molecules all move in different directions with various velocities $c_1, c_2, \ldots, c_N$. A particular velocity, say c_1, has components in three directions Ox, Oy and Oz parallel to the three edges of the cube passing through a corner O. If the components are u, v and w respectively, then $c_1^2 = u^2 + v^2 + w^2$, since u, v, w are at 90° to each other.

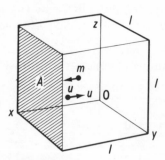

Fig 10.4 Kinetic theory of gases

Consider a molecule incident in a direction Ox normally to one face A of the cube, as shown. The incident velocity is u and the rebound velocity after impact is $-u$. So

$$\text{momentum change} = mu - (-mu) = 2mu$$

The time taken for the molecule to travel across to the opposite face and back is $2l/u$ and it then makes an impact again. So the number of impacts *per second* on a given face $= 1/\text{time} = 1/(2l/u) = u/2l$. Hence

$$\text{momentum change per second} = 2mu \times \frac{u}{2l} = \frac{mu^2}{l}$$

Now by definition, *force is the momentum change per second*. So

$$\text{force on face of cube} = \frac{mu^2}{l} \tag{1}$$

2   This is the force due to the component u of the velocity c_1. Adding the similar components of the velocities of all the other molecules, the total force on this face is given by

$$F = \frac{m}{l}(u_1^2 + u_2^2 + \ldots + u_N^2) \tag{2}$$

The *mean square* of the u components is the average of all the square values u_1^2, u_2^2, u_3^2, ... and is written $\overline{u^2}$, that is

$$\overline{u^2} = \frac{u_1^2 + u_2^2 + \ldots + u_N^2}{N}$$

or $N\overline{u^2} = u_1^2 + u_2^2 + \ldots + u_N^2$

So $F = \frac{Nm\overline{u^2}}{l}$ (3)

3   For a large number of molecules in the cube, the mean square of the components along Ox, Oy and Oz respectively must be equal. So

$$\overline{u^2} = \overline{v^2} = \overline{w^2}$$

But $\overline{u^2} + \overline{v^2} + \overline{w^2} = \overline{c^2}$, where $\overline{c^2}$ is the mean square speed of all the molecules. So $\overline{u^2} = \overline{c^2}/3$. From (3), it follows that

$$F = \frac{Nm\overline{c^2}}{3l}$$

4   The *pressure*, $p = F/A = F/l^2$, since pressure = force/area. So

$$p = \frac{Nm\overline{c^2}}{3l^3}$$

Since the volume V of the cube $= l^3$, then

$$p = \frac{Nm\overline{c^2}}{3V}$$

The mass M of the N molecules $= Nm$, and $M/V = \rho$, the density of the gas. So

$$p = \frac{1}{3}\rho\overline{c^2} \tag{4}$$

Note also that $\quad pV = \dfrac{1}{3}Nm\overline{c^2}$ (5)

The equations in (4) and (5) are the main results of the kinetic theory of gases, and should be memorized.

STEPS IN KINETIC THEORY

We can summarise the main steps in kinetic theory for finding the pressure:

(a)   Take one molecule in the Ox direction and find the force on the side of the cube using 'force = change in momentum per second = change in momentum × number of impacts per second'.

(b)   To take into account all the molecules, which move with different speed, use the 'mean square speed' because u^2 comes into the expression for the force due to one molecule (the 'mean square speed' is *not* the same as the 'mean speed').

(c)   The components of the mean square speed are the same in the three different directions, Ox, Oy, Oz, so $\overline{u^2} = \overline{c^2}/3$.

(d)   Finally, pressure = force per unit area or F/l^2.

DIFFERENCE BETWEEN R.M.S. SPEED AND MEAN SPEED

As we discuss later, the speeds of gas molecules vary from a very low to a very high value. The *mean speed* $\bar{c}$ of all the molecules is their arithmetical average. So for N molecules,

$$\bar{c} = \frac{c_1 + c_2 + \ldots + c_N}{N}$$

The mean speed determines how fast the gas molecules pass through or diffuse through a small hole in the side of a vessel containing the gas. It also determines how heat is transferred along a gas from one molecule to another or its thermal conductivity.

The *root-mean-square speed*, $\sqrt{\overline{c^2}}$, of all the gas molecules is different from the mean speed. From previous,

$$\text{r.m.s. speed} = \sqrt{c_1^2 + c_2^2 + \ldots + c_N^2 / N}$$

Calculations show that the r.m.s. speed is about 10% greater than the mean speed. The mean square speed determines the value of the gas pressure, since $p = \rho \overline{c^2}/3$, as we have shown.

ROOT-MEAN-SQUARE (R.M.S.) SPEED VALUES

From (4), the **root-mean-square** speed of all the molecules which is defined as $\sqrt{\overline{c^2}}$, is given by

$$\text{r.m.s. speed} = \sqrt{\overline{c^2}} = \sqrt{\frac{3p}{\rho}}$$

So if we know the pressure and density of a gas, we can calculate its root-mean-square velocity. For hydrogen, the density is 0.09kg m^{-3} at a pressure of 1.013×10^5 Pa (N m^{-2}) and temperature 0°C. So

$$\text{r.m.s. speed} = \sqrt{\frac{3 \times 1.013 \times 10^5}{0.09}} = 1840 \text{m s}^{-1} \text{ (approx)}$$

$$= 1.84 \text{km s}^{-1}$$

This is a high speed but it is reasonable because it is the same order as the speed of sound in hydrogen under these conditions. A sound wave is due to the movement of molecules in transferring energy from one molecule to another.

Other gases under the same conditions of pressure and temperature have a greater density than hydrogen, since their relative molecular mass is greater. So oxygen and nitrogen, for example, have a *lower* root-mean-square speed than hydrogen. Oxygen, nitrogen and carbon dioxide are gases found in the atmosphere close to the Earth's surface because the speed of many of their molecules is less than the so-called 'escape velocity' from the Earth; or to put it another way, gravitational attraction keeps these molecules near to the Earth's surface (p52).

TEMPERATURE ON KINETIC THEORY; BOYLE'S LAW

With N molecules each of mass m, the mass M of a gas $= Nm$. So, from (5)

$$pV = \frac{1}{3}M\overline{c^2} = \frac{2}{3}\left(\frac{1}{2}M\overline{c^2}\right) \qquad (6)$$

The quantity $\frac{1}{2}M\overline{c^2}$ is the **kinetic energy of translation** of the gas molecules. Now heat is a form of energy and it is reasonable to assume that the gas molecules have greater energy of motion or translation when its temperature is higher. We therefore make the *assumption* that

kinetic energy of translation, $\frac{1}{2}m\overline{c^2} \propto T$, kelvin temperature

So, from (6), with this assumption, it follows that, at constant temperature, pV = constant, which is *Boyle's law*. Further, at a kelvin temperature T, it also follows from (6) that

$$pV = \text{constant} \times T, \text{ or } pV = RT$$

where R is the constant for 1 mole of the gas. This is the ideal gas equation. See p339.

We can now extend our results. From (6), for one mole of a gas,

$$pV = \frac{1}{3}M\overline{c^2} = RT$$

so kinetic energy of translation, K.E. $= \frac{3}{2}RT$

The temperature T (kelvin) of a gas is a measure of the average translation energy of the molecules.

HOW R.M.S. SPEED VARIES WITH M AND T

Since $Mc^2/3 = RT$, it follows that

$$\text{r.m.s. velocity, } \sqrt{\overline{c^2}} = \sqrt{\frac{3RT}{M}} \qquad (1)$$

Now R is a constant. So, at a given temperature,

$$\text{r.m.s. velocity} \propto \frac{1}{\sqrt{M}}, \qquad (2)$$

where M is the molar mass of the gas. Suppose the r.m.s. velocity of hydrogen is 1800m s^{-1} at a given temperature, and the r.m.s. velocity of oxygen is required at the same temperature. The relative molecular masses of hydrogen and oxygen are 2 and 32 respectively. So $M_H/M_O = 2/32 = 1/16$, and, from above,

$$\frac{\text{r.m.s. velocity, oxygen}}{1800\text{m s}^{-1}} = \sqrt{\frac{M_H}{M_O}} = \sqrt{\frac{1}{16}} = \frac{1}{4}$$

So   r.m.s. velocity, oxygen $= \frac{1}{4} \times 1800 = 450\text{m s}^{-1}$

From (1), we also see that for the same gas at different temperatures,

$$\text{r.m.s. velocity, } \sqrt{\overline{c^2}} \propto \sqrt{T} \qquad (3)$$

Suppose the r.m.s. velocity of oxygen molecules is 400m s^{-1} at 0°C. Then at 100°C,

$$\frac{\text{r.m.s. velocity}}{400\text{m s}^{-1}} = \sqrt{\frac{273 + 100}{273}} = \sqrt{\frac{373}{273}}$$

So   r.m.s. velocity $= 400 \times \sqrt{\frac{373}{273}} = 468\text{m s}^{-1}$

DALTON'S LAW OF PARTIAL PRESSURES

In a mixture of gases, the *total pressure* is the sum of the pressures due to all the gases *individually*, assuming they occupy the volume of the mixture. This is known as *Dalton's law of partial pressures*.

To show how Dalton's law can be deduced from the kinetic theory, suppose we have a mixture of two gases A and B which occupy a volume V of a vessel at a kelvin temperature T.

At the same temperature T, there can be no exchange of energy between the individual molecules of A and B on average, otherwise the temperature would change. So the average kinetic energy of translation of the individual molecules of A

$$= \tfrac{1}{2}m_A\overline{c_A^2} = \tfrac{1}{2}m_B\overline{c_B^2}, \tag{1}$$

the average kinetic energy of translation of the individual molecules of B.

From our result for each gas of the kinetic theory, if V is the volume of the vessel, their individual pressures p_A and p_B will be given by

$$p_A V = \tfrac{1}{3}N_A m_A\overline{c_A^2} \text{ and } p_B V = \tfrac{1}{3}N_B m_B\overline{c_B^2} \tag{2}$$

where N_A and N_B are the respective number of molecules of the two gases. Adding the results in (2), we have

$$(p_A + p_B)V = \tfrac{1}{3}(N_A + N_B)m\overline{c^2}, \tag{3}$$

where $m\overline{c^2} = m_A\overline{c_A^2} = m_B\overline{c_B^2}$, from (1). If the *total* pressure is p, then

$$pV = \tfrac{1}{3}Nm\overline{c^2} \tag{4}$$

where N = total number of molecules = $N_A + N_B$: From (3) and (4), we see that

$$p = p_A + p_B$$

so the total pressure = sum of pressures of A and B.

Example 10.4 on p348 shows how Dalton's law is applied to a mixture of oxygen and nitrogen.

DISTRIBUTION OF MOLECULAR SPEEDS

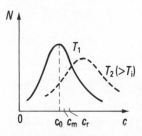

Fig 10.5 Velocity distribution of gas molecules

So far we have referred to the 'root-mean-square' and the 'mean' speed of molecules in a given gas. The actual speeds vary from low to high values. At a given temperature T_1, the variation follows a so-called Maxwellian distribution, shown roughly by the curve in Fig 10.5. The horizontal axis represents the speed c from 0 to a high value and the vertical axis represents the number N which have speeds within a given small range about a particular value c.

The maximum of the curve is the 'most probable' speed c_0 because more molecules have speeds round this value than any others. The mean speed c_m is greater than c_0 by about 13%, and the root-mean-square speed c_r is greater than c_0 by about 23%. At a higher temperature T_2 the speed of the molecules increases (Fig 10.5). The curve becomes less sharp at the maximum, showing a greater spread of speeds round the most probable value.

'Diffusion' of a gas is a phenomenon which concerns the mean (average) speed $\bar{c}$ of the individual molecules. The greater the value of $\bar{c}$, the faster will be the rate of diffusion. As we have seen, however, 'pressure' of a gas involves the mean square speed, $\overline{c^2}$.

BOLTZMANN CONSTANT AND IDEAL GAS EQUATION

Consider 1 mole of a monatomic gas. Its translational energy is related to temperature by (see p345).

$$\text{K.E.} = \frac{1}{2}M\overline{c^2} = \frac{3}{2}RT \tag{1}$$

The number of molecules in 1 mole is N_A, the Avogadro constant, which is about 6×10^{23}. So

$$\text{average energy per molecule} = \frac{3RT}{2N_A} = \frac{3}{2}kT \tag{2}$$

where $k = R/N_A$. k is known as the **Boltzmann constant**. Since $R = 8.3$ numerically (p340), $k(= R/N_A)$ is given by

$$k = 8.3/(6 \times 10^{23}) = 1.4 \times 10^{-23}\, \text{J K}^{-1}$$

$$k = \frac{R}{N_A} \tag{3}$$

If m is the mass of a molecule and $\overline{c^2}$ is the mean-square speed of the molecules, then

$$\tfrac{1}{2}m\overline{c^2} = \text{average kinetic energy per molecule} = \frac{3}{2}kT$$

Since $R = N_A k$ from (3) and $pV = RT$ for 1 mole, we can write the ideal gas equation as

$$pV = N_A kT$$

The Boltzmann constant k, and the average translational energy per molecule, $3kT/2$, is widely used where large numbers of particles are present and their thermal energy is needed. For example, electrons in metals at a temperature T have average thermal energy of the order $3kT/2$

WORKED EXAMPLE ON KINETIC THEORY

Assuming air has 20% oxygen molecules and 80% nitrogen molecules, of relative molecular mass 32 and 28 respectively, calculate
(a)   the ratio of the r.m.s. velocity of oxygen to that of nitrogen,
(b)   the ratio of the partial pressure of oxygen to that of nitrogen
in air.

(a)   The ratio of the r.m.s. velocities is *inversely* proportional to the square root of the ratio of the molar masses. So

$$\frac{\text{r.m.s. velocity, oxygen}}{\text{r.m.s. velocity, nitrogen}} = \sqrt{\frac{28}{32}} = 0.94$$

(b)   To use the number N of molecules, we have

$$pV = \frac{1}{3}Nm\overline{c^2}$$

The oxygen and nitrogen molecules *both* occupy the volume V of the air. So

for oxygen $p_O V = \frac{1}{3}N_O m_O \overline{c_O^2}$ (1)

and for nitrogen $p_N V = \frac{1}{3}N_N m_N \overline{c_N^2}$ (2)

Now at the same air temperature, there is no exchange of energy between the oxygen and nitrogen molecules otherwise the temperature would change. So the average kinetic energy of translation of oxygen molecules, $\frac{1}{2}m_O\overline{c_O^2}$ = average kinetic energy of translation of nitrogen molecules, $\frac{1}{2}m_N\overline{c_N^2}$. Dividing (1) and (2) and using this result, we obtain

$$\frac{p_O}{p_N} = \frac{N_O}{N_N} = \frac{20}{80} = \frac{1}{4} = \text{ratio of partial pressures}$$

HEAT CAPACITIES OF GASES. THERMODYNAMICS

INTERNAL ENERGY OF GAS

As we saw in the 'kinetic theory of gases', the molecules of a gas are moving about in different directions with different speeds at any instant. This is described as **random motion**. The *internal energy* of a gas is the kinetic energy of its random motion. This depends on the temperature of the gas. Generally, the higher its temperature, the greater is the internal energy of the gas. An **ideal gas** may be defined as one which obeys Boyle's law and whose internal energy depends only on the gas temperature and is *independent of its volume*.

> Internal energy of an ideal gas is the kinetic energy of random motion of its molecules

In the case of a solid, its internal energy is the kinetic energy of vibration of all its molecules about their mean positions. Throwing a metal in the air does not change its internal energy. But hitting the metal with a hammer increases the kinetic energy of vibration of all the molecules and so the internal energy is increased.

The symbol U will be used for internal energy. We shall be mainly concerned with *changes* in internal energy, written $\triangle U$, as nobody knows the actual value of U.

WORK DONE BY A GAS

Suppose we transfer a small amount of heat $\triangle Q$ to a gas in a vessel so that the internal energy of the gas increases (Fig 10.6). At the same time, suppose the piston in Fig 10.6 is pushed down steadily by a small amount to compress the gas. Some work $\triangle W$ is then done *on* the gas and this energy is also transferred to the gas to increase its internal energy. So the total increase in internal energy of the gas, $\triangle U$, is now given by

$$\triangle U = \triangle Q + \triangle W \tag{1}$$

Note carefully that the heat $\triangle Q$ is given *to* the gas and that the work $\triangle W$, often called the 'external work' because it is done by an outside or external force, is done *on* the gas.

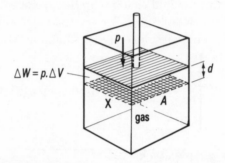

Fig 10.6 Work done on
compressing gas

The relation in (1) is often called the *First Law of Thermodynamics* – 'thermodynamics' is the science relating heat and energy. It is easily recognisable as a statement of the law of conservation of energy, that is, the gain in internal energy of the gas is equal to the heat energy given to it plus the work or mechanical energy transferred to it.

**SOME INTERNAL
ENERGY CHANGES
OF A GAS**

1   As stated above, the internal energy of an ideal gas depends only on its temperature. So when a gas is heated slowly in a metal container so that its temperature keeps *constant* at the value of the surroundings, there is no change in internal energy. So $U = 0$. From (1), $\triangle Q + \triangle W = 0$, or $\triangle Q = -\triangle W$. The minus means that work is done *by* the gas in expanding. So the heat $\triangle Q$ was used entirely for doing external work. When a gas is warmed under constant temperature conditions as here, we call this an *isothermal* change.

2   Suppose the vessel containing a gas is completely lagged together with the piston. No heat can then enter or leave the gas. Any change in volume or pressure or temperature of the gas is then said to be under *adiabatic* conditions. Compressing or expanding a gas swiftly approaches adiabatic conditions even without insulation or lagging.
     Suppose the gas in the vessel is compressed under adiabatic

conditions. Then $\triangle Q = 0$. From $\triangle U = \triangle Q + \triangle W$, then $\triangle U = \triangle W$. This means that the increase in internal energy = the external work done on the gas in compressing it.

Since the internal energy has increased, the temperature of the gas has risen. The molecules are moving about faster because they rebound with greater speed from the piston when it moves down.

If the piston is raised so that the gas expands under adiabatic conditions, then work is done by the gas. The energy to do this work comes from the internal energy of the gas because no heat enters or leaves the vessel. So the internal energy is *less* than before. The temperature of the gas therefore *falls*.

WORK DONE BY GAS; USE OF p-V GRAPH

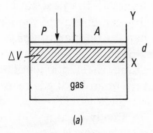 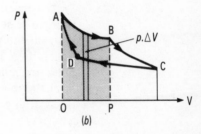

Fig 10.7 Work done by gas

For design of engines, the work done by a gas is needed. In Fig 10.7(*a*), suppose the external pressure is constant at a value of p while the volume of the gas expands by an amount $\triangle V$. Then if a piston, X, of area A moves a distance d,

$$\triangle W = \text{force} \times \text{distance} = p.A \times d = p \times A.d. = p.\triangle V,$$

since the volume increase = $A.d$. So

$$\text{work done by gas} = p.\triangle V \text{ (pressure} \times \text{volume change)}$$

We can find the work done by a gas from its p-V graph. Suppose the gas expands from a particular pressure, volume and temperature value to another value at B (Fig 10.7(*b*)). The $p.\triangle V$ represents a *strip of area* as shown. When we add all the strips in the expansion between A and B, we see that

work done from A to B = AREA ABPQ between graph and volume-axis

In Fig 10.7(*b*), suppose the gas expands isothermally along AB and then expands adiabatically along BC. Then the work done *by* the gas = area between ABC and the volume-axis. At C, suppose the gas is now compressed isothermically along CD and then compressed adiabatically along DA to return to A. Then work done *on* gas = area between CDA and volume-axis.

To find the *net* amount of work done by the gas we *subtract* the two areas. We see that in taking the gas round the cycle ABCDA,

net work done = AREA enclosed by the loop ABCD

MOLAR HEAT CAPACITY AT CONSTANT VOLUME

The heat capacity of a gas depends on the conditions under which it is heated. The **molar heat capacity at constant volume**, C_V, is the heat required to raise the temperature of 1 mole of the gas by 1K when its volume is constant (Fig 10.8(a)). Its unit is J mol^{-1} K^{-1}. The *specific heat capacity* at constant volume, c_V, is the heat required to raise the temperature of 1kg by 1K. C_V and c_V are simply related. For example, the molar mass of hydrogen is 2g and so 1kg, 1000g, is 500 times the mass of 1 mole. Hence $c_V = 500C_V$ for hydrogen. The unit of c_V is J kg^{-1} k^{-1}.

At constant volume, *all* the heat supplied to 1 mole of the gas is used to raise the internal energy of the gas. So C_V is the increase in internal energy when the temperature rises by 1K. For an ideal gas, in which there is no attraction between the molecules and the molecules have negligible volume, C_V is independent of the volume of the gas. So if the temperature of an ideal gas rises from T_1 to T_2, the gain in internal energy of 1 mole $= C_V(T_2 - T_1)$, no matter what volume the gas may initially or finally occupy. So the internal energy change depends *only on the temperature change*.

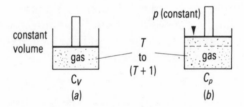

Fig 10.8 Molar heat capacities of gas

MOLAR HEAT CAPACITY AT CONSTANT PRESSURE

The **molar heat capacity at constant pressure**, C_p, is the heat required to raise the temperature of 1 mole at constant pressure by 1K (Fig 10.8(b)). Its unit is J mol^{-1} K^{-1}, The specific heat capacity at constant pressure, c_p, is the heat required to raise the temperature of 1kg at constant pressure by 1K.

Unlike the case of constant volume, the gas expands at constant pressure when its temperature rises. Fig 10.8(b) shows the volume increase, ΔV, after the temperature has risen 1K from T to $(T+1)$. So this time the heat supplied, C_p, is used not only for increasing the internal energy of the gas but also for doing external work. Now we have already emphasized that the gain in internal energy of 1 mole of a gas is C_V when its temperature rises by 1K and is independent of the volume. So, from the first law of thermodynamics,

$$\triangle Q = C_p = C_V + \triangle W$$

where $\triangle W$ is the external work done by the gas. On p351, we showed that $\triangle W = p.\triangle V$. So

$$C_p = C_V + p.\triangle V \tag{1}$$

For an ideal gas, $pV = RT$; and after warming so that the temperature increases to $(T+1)$ and the volume increases to $(V + \triangle V)$, then $p(V + \triangle V) = R(T+1)$ from the gas equation. So $pV + p.\triangle V = RT + R$, or $p.\triangle V = R$. From (1), it follows that

$$C_p = C_V + R, \text{ or } C_p - C_V = R \tag{2}$$

So, for an ideal gas, C_p is always greater than C_V. The difference R is the external work done by the gas when it is warmed at constant pressure so that its temperature changes by 1K.

The worked example 10.5 on p355 shows how the molar heat capacity of a gas can be calculated and how numerical values of external work done and of changes in internal energy can be found.

ISOTHERMAL AND ADIABATIC CHANGES

ISOTHERMAL CHANGE

When an ideal gas is allowed to expand under *constant temperature*, this is called an **isothermal** expansion. No change occurs in its internal energy because this depends only on temperature change. The heat supplied to maintain constant temperature is then equal to the external work done by the gas, from the first law of thermodynamics (see p350).

Since the temperature is constant, an ideal gas obeys Boyle's law $pV = $ constant during an isothermal change.

1 Isothermal change: $pV = $ constant (Boyle's law). For 1 mole of gas, value of constant $= RT$
2 Work done by gas $= RT \ln(V_2/V_1)$, where V_1 is initial gas volume and V_2 is final volume

ADIABATIC CHANGE

If a gas is allowed to expand *without any heat entering or leaving the gas*, for example, by insulating its cylinder and piston, the energy needed for the external work is taken from the internal energy of the gas. So the temperature of the gas falls. If the gas is compressed under these conditions, the work done *on* the gas produces a rise in internal energy equal to the work done and so the gas temperature rises. A gas is said to undergo an **adiabatic** change when no heat enters or leaves it.

Analysis shows that if p, V, T are the pressure, volume and kelvin temperature in an adiabatic change, then
(a) pV^γ = constant, where γ is the ratio C_p/C_v for the particular gas (see p352),
(b) $TV^{\gamma-1}$ = constant.

> 1 Adiabatic change: $pV\gamma$ = constant
>   and $T \times V^{\gamma-1}$ = constant
>   ($\gamma = C_p/C_V$; for air, $\gamma = 1.4$)
> 2 Work done = $(p_2V_2 - p_1V_1)/(\gamma - 1)$
>   where p_2V_2 and p_1, V_1 are final and initial values

The isothermal and adiabatic formulae given above only apply under changes in p, V and T which take place under *reversible conditions*. In this ideal case one assumes that there is no frictional force when a piston moves in gas expansion or contraction and that no heat is produced in the gas by eddies or swirls of gas in expansion or contraction.

SECOND LAW OF THERMODYNAMICS

When we do work against friction, the *whole* of the work can be transferred to heat. We can apply the first law of thermodynamics (the law of conservation of energy) to this change.

The reverse change, however, is not true. If an engine, for example, takes in an amount of heat, all the heat can *not* be transferred to work. There are several forms of the *second law of thermodynamics*. One form states that when a gas is taken through a complete cycle of changes so that it returns to its original condition, the amount of work done by the gas is always *less* than the total amount of heat taken in.

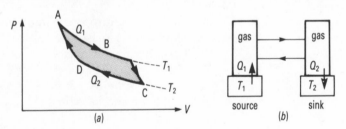

Fig 10.9 Heat and work in Carnot cycle

Engines contain a gas which goes through complete cycles of changes and does work in moving a piston continuously. Fig 10.9 shows diagrammatically a gas going through a so-called Carnot cycle ABCD. The pressure (p)-volume (V) changes along AB (expansion) and CD (compression) are isothermals and those along BC (expansion) and DA (compression) are adiabatics.

Along AB, an amount of heat Q_1 is taken in by the gas from a source at a constant temperature T_1 (Fig 10.9(b)). Along BC no heat is taken in or given up as this is an adiabatic change. Along DA, an amount of heat Q_2 is given up by the compressed gas to a sink at a temperature T_2 lower than T_1.

After the gas has gone through one cycle ABCDA, the net effect is:

(a)   the work done by the gas = area ABCD

(b)   the net amount of heat taken in by the gas = $Q_1 - Q_2$.

From the first law of thermodynamics or law of conservation of energy, the work done by the gas in its cycle = $Q_1 - Q_2$. So as stated in the second law of thermodynamics, not all the heat taken in, Q_1, is transferred to work.

So engines can never be 100% efficient from the second law. The efficiency is given by

$$\text{efficiency} = \frac{\text{work out}}{\text{heat in}} = \frac{Q_1 - Q_2}{Q_1}$$

The Carnot cycle gives the maximum possible efficiency. If the isothermal and adiabatic equations are applied to Fig 10.9(a), the final result gives a simple formula for the efficiency, assuming an ideal gas. See *Nelkon: Scholarship Physics* (Heinemann). This is:

$$\text{efficiency} = \frac{Q_1 - Q_2}{Q_1} = \frac{T_1 - T_2}{T_1} = 1 - \frac{T_2}{T_1}$$

So the greater the temperature T_1 of the source relative to that T_2 of the sink, the greater will be the efficiency.

WORKED EXAMPLE ON MOLAR HEAT CAPACITY, EXTERNAL WORK, INTERNAL ENERGY CHANGE

10.5 1 mole of a gas has a volume of $2.23 \times 10^{-2} m^3$ at a pressure of 1.01×10^5 Pa (N m^{-2}) at 0°C. If the molar capacity at constant pressure is 28.5J mol^{-1} K^{-1}, calculate the molar heat capacity at constant volume.

20g of this gas, initially at 27°C, is heated at constant pressure of 1.0×10^5 Pa so that its volume increases from 0.250m^3 to 0.375m^3. Calculate

(a)   the external work done,

(b)   the increase in internal energy,

(c)   the heat supplied.

(Relative molecular mass of gas = 2g.)

The molar gas constant $R = \dfrac{pV}{T} = \dfrac{1.01 \times 10^5 \times 2.23 \times 10^{-2}}{273}$

$$= 8.3J \text{ mol}^{-1} K^{-1}$$

So, from

$C_p = C_V + R$

$C_V = C_p - R = 28.5 - 8.3 = 20.2J \text{ mol}^{-1} k^{-1}$

(a)   External work done $= p . \Delta V = 1.0 \times 10^5 \times (0.375 - 0.250)$
$$= 1.25 \times 10^4 J$$

(b)   Number of moles in 20g $= 20/2 = 10$

So   internal energy rise $= 10 \times C_V \times (T_2 - T_1)$ (1)

where T_1 is the initial temperature $(273 + 27 = 300K)$ and T_2 is the final temperature. Now at constant pressure, volume $V \propto T$. So $V_2/V_1 = T_2/T_1$, or

$$\frac{T_2}{300} = \frac{0.375}{0.250} = 1.5$$

So $T_2 = 1.5 \times 300 = 450K$

From (1), internal energy rise $= 10 \times 20.2 \times (450 - 300) = 30\,300J$

(c)   Heat supplied $= 10 \times C_p \times (T_2 - T_1)$
$$= 10 \times 28.5 \times (450 - 300) = 42\,750J$$

VERBAL SUMMARY

GAS LAWS

For a given mass of gas,

(a) $p \propto 1/V$ or $pV = $ constant (Boyle)
(b) $V \propto T$ (kelvin) at constant pressure
(c) $p \propto T$ at constant volume

Ideal gas equation of state:

$$pV = RT \text{ (1 mole)}, \quad pV = nRT \text{ (} n \text{ moles)}$$

R = molar gas constant = 8.3J $mol^{-1}\,k^{-1}$ (approx) for all gases

When two bulbs containing the same gas are connected, total number of moles (pV/RT) is constant when one bulb is warmed.

2 KINETIC THEORY OF GASES

$$pV = \frac{1}{3}Nm\overline{c^2} = RT. \quad \text{K.E. of translation} = \frac{1}{2}M\overline{c^2} = \frac{3}{2}RT$$

Note that the ideal gas law $pV = RT$ only follows from kinetic theory by *assuming* that K.E. of translation $\propto T$ (kelvin).

$$\text{r.m.s. velocity } \sqrt{\overline{c^2}} = \sqrt{3p/\rho}$$

r.m.s. velocity:

(a) independent of pressure,
(b) proportional to $\sqrt{T}$,
(c) proportional to $1/\sqrt{M}$ where M is mass of 1 mole.

3 GAS MIXTURES

Dalton's law: In a gas mixture, the total pressure = sum of pressures of individual gases, each gas occupying the volume of the mixture. In problems with air and water vapour, apply the gas laws to the air.

4 GAS HEAT CAPACITIES; THERMODYNAMICS

Work done by gas $= pressure \times volume\ change = $ AREA between p-V graph and V-axis

C_V (constant volume) $=$ gain in internal energy of 1 mole for 1K temperature rise

C_p (constant pressure) $= C_V$ (internal energy rise) $+ R$ (external work done)

So C_p is always greater than C_V. In a real gas there are attractive forces between the molecules, so $C_p = C_V + R + W$, where W is the work done against the attractive forces.

Isothermal (constant temperature) change: $pV = $ constant.

Adiabatic (constant heat) change: $pV^\gamma = $ constant, $TV^{\gamma-1} = $ constant ($\gamma = C_p/C_V$)

First Law of Thermodynamics: principle of conservation of energy

Second Law of Thermodynamics: heat energy can never be transferred completely to mechanical energy when a gas is taken through a cycle

A Carnot cycle provides maximum efficiency of value $(T_1 - T_2)/T_1$

DIAGRAM SUMMARY

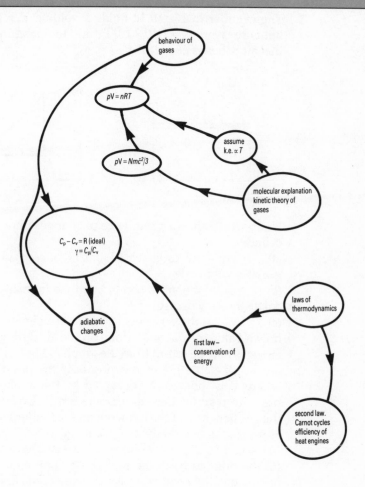

1   The cylinder in Fig 10.10 holds a volume $V_1 = 1000 cm^3$ of air at an initial pressure $p_1 = 1.10 \times 10^5$ Pa and temperature $T_1 = 300K$. Assume that air behaves like ideal gas.

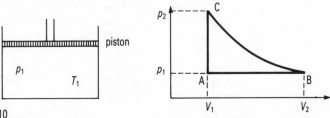

Fig 10.10

Figure 10.10 shows a sequence of changes imposed on the air in the cylinder.

(a)   AB – the air is heated to 375K at constant pressure. Calculate the new volume V_2.

(b)   BC – the air is compressed isothermally to volume V_1. Calculate the new pressure p_2.

(c)   CA – the air cools at constant volume to pressure p_1. State how a value for the work done on the air during the full sequence of changes may be found from the graph in Fig 10.10.   (L)

2   (a)   Sketch a graph showing how the product pV varies with θ, where V is the volume occupied by one mole of an ideal gas at a pressure p and a Celsius temperature θ. Explain the significance of the gradient and of the intercept on the temperature axis.

How, if at all, would the graph change if

(i) a second mole of the same gas were added to the first,

(ii) the original gas were replaced by one mole of another ideal gas having half the relative molecular mass of the first?

(b)   An ideal gas has a relative molecular mass of 4.00. The total translational kinetic energy of the molecules of a certain mass of this gas is 374J at a temperature of 27°C. Calculate

(i) the total translational kinetic energy of the molecules at a temperature of 127°C,

(ii) the mass of gas present. Molar gas constant $= 8.32J\ mol^{-1}\ K^{-1}$.

                                                            (JMB)

3   (a)   Write down an equation for the pressure, p, exerted by an

ideal gas in terms of its density, ρ, and the mean square speed of its molecules, $\overline{c^2}$.

(b)   Show that, for different gases at the same temperature, the mean square speeds of the molecules are inversely proportional to their relative molecular masses.

(c)   Explain why the Earth retains an atmosphere but the Moon does not.   (JMB)

4   (a)   (i) Explain how the molecules of a gas exert a pressure.

(ii) Give two reasons why the pressure exerted by the molecules of a gas, maintained at constant volume, increases as the temperature increases.

(iii) In an ideal gas pressure $p = \frac{1}{3}nm\overline{c^2}$, where $n =$ number of molecules per unit volume, $m =$ mass of one molecule, $\overline{c^2} =$ mean square speed of the molecules. Show how this equation leads to the relationship between the pressure and volume of an ideal gas at constant temperature.

(b)   (i) Calculate the root mean square speed of four molecules moving with speeds, in m s^{-1}, of 250, 500, 575 and 600 respectively.

(ii) Figure 10.11 shows a single molecule A of mass 4.6×10^{-26}kg moving with speed of 500m s^{-1} in a rigid cubical box. Calculate the change in momentum of the molecule when it strikes a wall W elastically. If the box has side length 0.25m, calculate the number of times the molecule strikes wall W each second, and deduce the average pressure exerted by molecule A on the wall W.

(iii) The box in (ii) now contains 4.2×10^{23} such molecules. Assuming that all the molecules move with the same speed and in the same direction as A, calculate the pressure now exerted on wall W.

(iv) The actual pressure at this density and temperature is 1.03×10^5 N m^{-2}. Show that this observation is consistent with your answer to (iii).

(AEB)

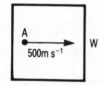

Fig 10.11

5   State the first law of thermodynamics and explain what is meant by the *internal energy* of the system. What constitutes the internal energy of an ideal gas? Starting from the expression $p = \frac{1}{3}\rho\overline{c^2}$, show that the internal energy of an ideal *monatomic* gas is $3pV/2$, and discuss the interpretation of temperature in the kinetic theory.

Explain the following observations:

(a)   When pumping up a bicycle tyre the pump barrel gets warm, and

(b)   when a gas at high pressure in a container is suddenly released, the container cools.

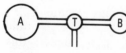

Fig 10.12

Two bulbs, A of volume 100cm^3 and B of volume 50cm^3, are connected to a three way tap T which enables them to be filled with gas or evacuated. The volume of the tubes may be neglected. Fig 10.12.

(c)   Initially bulb A is filled with an ideal gas at 10°C to a pressure of 3.0×10^5 Pa. Bulb B is filled with an ideal gas at 100°C to a pressure of 1.0×10^5 Pa. The two bulbs are connected with A maintained at 10°C and B at 100°C. Calculate the pressure at equilibrium.

(d)   Bulb A, filled at 10°C to a pressure of 3.0×10^5 Pa, is connected to a vacuum pump with a cylinder of 20cm³. Calculate the pressure in A after one inlet stroke of the pump. The air in the pump is now expelled into the atmosphere. Calculate the pressure in A after the second inlet stroke. Calculate the number of strokes of the pump to reduce the pressure in A to 1.0×10^5 Pa. The whole system is maintained at 10°C throughout the process. (O&C)

6 The graph in Fig 10.13 relates the pressure and volume of a fixed mass of an ideal gas as it is first allowed to expand under isothermal conditions and then compressed adiabatically.

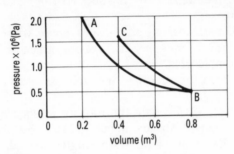

Fig 10.13

(a) What are meant by the phrases *expand under isothermal conditions* and *compressed adiabatically*?
(b) Use the graph to provide evidence that
(i) the expansion from A to B is isothermal;
(ii) the temperature at C is higher than that at A. In each case explain how you provide the evidence. (AEB 1985)

7 The kinetic theory leads to the equation $p = \frac{1}{3}\rho\overline{c^2}$, where p is the *pressure* ρ is the *density* and $\overline{c^2}$ is the *mean square molecular speed*. Explain the meaning of the terms in italics and list the simplifying assumptions necessary to derive this result. Discuss how this equation is related to Boyle's law.

Air may be taken to consist of 80% nitrogen molecules and 20% oxygen molecules of relative molecular masses 28 and 32 respectively. Calculate

(a) the ratio of the root mean square speed of nitrogen molecules to that of oxygen molecules in air,
(b) the ratio of the partial pressures of nitrogen and oxygen molecules in air, and
(c) the ratio of the root mean square speed of nitrogen molecules in air at 10°C to that at 100°C. (O&C)

8 A container holds 120cm³ of air which is just saturated with water vapour. The pressure in the container is 100kPa, the pressure exerted by the saturated water vapour at this temperature being 20kPa. The air in the container is now compressed at constant temperature until the pressure in the container is 150kPa. What is the new volume? (L)

9 A fixed mass of gas is trapped in a cylinder by a piston which is moved so that the gas is taken through the following cycle of volume

changes: A isothermal expansion; B adiabatic expansion; C isothermal compression; D adiabatic compression to original volume.

(*a*)   Sketch a graph of pressure against volume for these changes indicating which section of the graph corresponds to each volume change. What is the significance of the area enclosed?

(*b*)   Sketch a corresponding graph of pressure against temperature. *(AEB* 1984)

THERMOMETRY; HEAT ENERGY AND TRANSFER

CONTENTS

THERMOMETRY

TEMPERATURE AND HEAT; ZEROTH LAW

If a cold metal spoon is dipped into a wam liquid, energy or heat will be transferred from the liquid to the metal until 'thermal equilibrium' is reached between them. The direction in which energy or heat is transferred between two objects in contact depends on their respective *temperatures*. A warm metal spoon has a higher temperature than a cool liquid and heat will flow from the metal to the liquid when the spoon is placed in the liquid. 'Heat' and 'temperature' are different physical quantities. As we saw in the kinetic theory of gases, the temperature of a *gas* depends on the average kinetic energy of translation of all the molecules.

> Two objects in contact have the same temperature if they are each in thermal equilibrium with a third body. This is called the *Zeroth law*.

TEMPERATURE SCALES; THERMODYNAMIC SCALE

The temperature of an object is not a fixed number. As we discuss shortly, the temperature of the object depends on the type of thermometer used to measure it and on the temperature scale adopted. A *mercury-in-glass* thermometer uses the volume change of mercury relative to glass when its temperature changes. A *constant volume gas* thermometer uses the pressure change of a gas with temperature change when the volume of the gas is kept constant. A *platinum resistance* thermometer uses the resistance change of platinum with temperature change. The *thermoelectric* thermometer uses the change in e.m.f. thermocouple when the temperature difference between its junctions is varied. Each thermometer gives a different value to the temperature of the same liquid.

The **thermodynamic** scale is the standard temperature scale used in scientific measurement. The symbol for temperature on this scale is T and the temperature is measured in **kelvin**, symbol K, after Lord Kelvin who suggested the scale. Nowadays the thermodynamic scale is based on the temperature at which saturated water-vapour, pure water and melting ice are in equilibrium with each other, called the *triple point* of water. This is *defined* as the temperature 273.16K. The kelvin is 1/273.16 of this temperature. The *ice point*, the temperature when pure water and melting ice are in equilibrium, is 273.15K – the slight difference with the triple point value is due to the difference in pressure in the two cases. The Celsius temperature, $-273.15°C=0K$, $0°C=273.15K$ and a change of $1°C=$ a change of $1K$ (Fig 11.1).

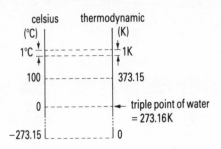

Fig 11.1

THERMODYNAMIC (KELVIN) AND CELSIUS SCALES

Suppose that the resistance of a platinum resistance thermometer is R_{tr} at the triple point of water and R_T at the temperature of a liquid when this is measured. Then, by definition, the thermodynamic temperature T of the liquid is calculated from

$$T=\frac{R_T}{R_{tr}}\times273.16K$$

A similar relation is used for measuring the temperature of the liquid using a constant volume gas thermometer, for example. Here p_T and p_{tr} are used in place of R_T and R_{tr} respectively, where p represents the pressure of the gas in this type of thermometer. The values of T differ slightly because the pressure of a gas at constant volume and the resistance of platinum have a different variation with temperature. In practice, *the gas thermometer is used as the standard thermometer*. Tables are drawn up which show the 'correction' to be made to the value of a temperature measured by a resistance thermometer, for example, to bring it to the value which would have been measured if the gas thermometer had been used.

The Celsius temperature, symbol t, is now defined by $t=T-273.15$, where T is the thermodynamic temperature. The ice point is $0°C$ and the temperature of steam at 760mmHg pressure is $100°C$. A temperature change of $1°C=1K$.

$$t(°C)=T(K)-273.15$$

HOW CELSIUS TEMPERATURE IS CALCULATED

On the Celsius scale, there are 100 divisions or degrees between the ice point, taken as $0°C$, and the steam point, taken as $100°C$. So if R_0, R_{100} and R_t represent the resistance of platinum at the ice point, steam point and the unknown temperature $t°C$ of a liquid, we calculate t from the relation.

$$t=\frac{R_t-R_0}{R_{100}-R_0}\times 100°C$$

A similar relation is used to find Celsius temperature when a gas thermometer is used. In this case, if p represents the pressure of the gas at constant volume,

$$t=\frac{p_t-p_0}{p_{100}-p_0}\times 100°C$$

If X is any property of a substance which changes with temperature (such as volume, pressure or resistance), then using this property the temperature t is calculated from

$$t=\frac{X_t-X_0}{X_{100}-X_0}\times 100°C$$

The worked examples 11.1 and 11.2 on page 371 show how to calculate a temperature using two different thermometers.

CONSTANT VOLUME GAS THERMOMETER
This is shown diagrammatically in Fig 11.2. A barometer is incorporated so that the pressure of the gas in the bulb B can be measured directly from the mercury height H. Only a very small correction is needed for the gas outside B in the 'dead space', which is not at the same temperature as the gas in B. At low temperatures hydrogen may be used to about −250°C and helium to about −260°C. Nitrogen is used at high temperatures as hydrogen diffuses through vessels at about 500°C.

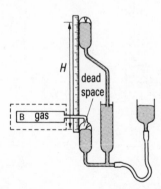

Fig 11.2 Gas thermometer

As previously stated, the gas thermometer is chosen as the standard thermometer. Firstly, the expansion of a gas is much higher than that of the containing vessel which has an irregular expansion. Secondly, and more importantly, at infinitely low pressures all gases obey the same gas laws. For this reason the temperature is corrected to the value it would have had if the gas pressure were infinitely low.

The temperature obtained with different gases should then all agree with each other.

Example 11.1 on page 371 shows how to calculate temperature on the constant gas thermometer.

RESISTANCE THERMOMETERS

The gas thermometer is not convenient for measuring temperatures in ordinary laboratory experiments on account of its size. The **platinum resistance thermometer** is much more convenient. Pure platinum is used and sealed into a tube; its resistance R_0 at 0°C is about 25 ohms (Fig 11.3(a)). A Wheatstone bridge circuit is used to measure the resistance at various temperatures, the resistance R_L of the leads being neutralized by the equal resistance R_D of 'dummy' leads on the other side of the bridge circuit. If t is the temperature in °C on the standard gas thermometer scale, the resistance at this temperature is found to vary according to the relation $R_t = R_0(1 + at + bt^2)$ over a wide temperature range, where R_0 is the resistance at 0°C and a and b are constants. Example 11.2 illustrates how the temperature measured by the platinum resistance thermometer differs from that measured by the gas thermometer. Tables of corrections are available so that the measured temperature on the platinum resistance thermometer scale can be converted to the gas thermometer scale. This resistance thermometer has a wide range of about −200°C to 1200°C. A disadvantage is the time needed to balance the Wheatstone bridge.

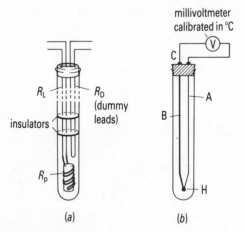

Fig 11.3 resistance and thermoelectric thermometers

Thermistors are semiconductors whose resistance is sensitive to temperature change. They are therefore used for temperature measurement. Some thermistors decrease in resistance when their temperature increases (negative temperature coefficient) and others increase in resistance when their temperature increases (positive temperature coefficient). In either case the temperature on the Celsius scale is found, if R represents resistance, by

$$t = \frac{R_t - R_0}{R_{100} - R_0} \times 100°C$$

Example 11.2 on page 372 shows how temperature is calculated using a resistance thermometer.

THERMOELECTRIC THERMOMETER

The **thermoelectric thermometer** (Fig 11.3(b)) has a wide range from about $-250°C$ to $1500°C$. At high temperatures above $1200°C$, metals with high melting-points such as platinum, A, and a platinum–rhodium alloy, B, must be used. An advantage of this type of thermometer is the low heat capacity of the thermocouple junction, so that it quickly reaches the temperature to be measured. It can therefore be used to measure varying temperatures. The small size of the junction also enables the temperature of metal surfaces or of a small metal block to be measured – the junction H can be inserted into a small hole bored into the metal. The variation of the e.m.f. E of a thermocouple with temperature $t°C$ of the hot junction, when the cold junction is kept at $0°C$, is the parabolic relation $E_t = E_0(1 + at + bt^2)$. So there are *two* temperatures which produce the same e.m.f. E. For this reason the range of the thermoelectric thermometer must have an upper temperature limit; it corresponds to the maximum of the parabolic relationship.

Using the thermoelectric thermometer, temperature t is calculated from

$$t = \frac{E_t - E_0}{E_{100} - E_0} \times 100°C$$

WORKED EXAMPLES ON TEMPERATURE USING GAS THERMOMETER AND RESISTANCE THERMOMETER

11.1 Using a constant volume gas thermometer, the pressure of the gas at $0°C$ is 300mmHg, at $100°C$ it is 580mmHg and at the temperature of a liquid, $t°C$, it is 350mmHg. Calculate the liquid temperature.

Here $p_0 = 300$mmHg, $p_{100} = 580$mmHg, and $p_t = 350$mmHg.

So $\quad t = \dfrac{p_t - p_0}{p_{100} - p_0} \times 100°C$

$\qquad = \dfrac{350 - 300}{580 - 300} \times 100°C$

$\qquad = \dfrac{50}{280} \times 100°C = 17.9°C$

11.2 At a temperature $t°C$, where t is measured on the gas thermometer scale, the resistance of a pure metal is given by $R_t = R_0(1+4\times10^{-3}t+10^{-5}t^2)$, where R_0 is the resistance at 0°C. Calculate the temperature on the resistance scale which corresponds to the temperature of 80.0°C measured on the gas thermometer scale.

From the formula for R_t,

at $t=0°C,$
      $R_t=R_0,$
at $t=100°C,$
      $R_t=R_{100}=R_0(1+4\times10^{-3}\times100+10^{-5}\times100^2)=R_0\times1.5$
at $t=80°C$
      $R_t=R_{80}=R_0(1+4\times10^{-3}\times80+10^{-5}\times80^2)=R_0\times1.384$

So $t_{res}=\dfrac{R_{80}-R_0}{R_{100}-R_0}\times100°C=\dfrac{1.384R_0-R_0}{1.5R_0-R_0}$

      $=\dfrac{0.384}{0.5}\times100°C=76.8°C$

HEAT ENERGY: HEAT CAPACITY AND LATENT HEAT CAPACITY

HEAT CAPACITY AND SPECIFIC HEAT CAPACITY

The **heat capacity** of an object is the heat required to raise its temperature by 1K (1K=1°C). The unit for heat capacity is therefore J K^{-1} (or kJ K^{-1} in larger heat units) and its symbol is C. A metal container of heat capacity 200J K^{-1}, heated from 15°C to 45°C, will therefore require a quantity of heat Q given by

$$Q=200\times(45-15)=6000J$$

Generally, if C is the thermal capacity and θ is the temperature *rise* or *fall*, the heat gained or lost would be

$$Q=C\times\theta$$

The **specific heat capacity** of a substance is the heat required to raise the temperature of *unit mass* (such as 1kg) by 1K. The unit for specific heat capacity is therefore J kg^{-1} K^{-1} (or kJ kg^{-1} K^{-1} in larger heat units) and its symbol is c. The specific heat capacity of water, c_w, is about 4200J kg^{-1} K^{-1}. Note that if m is the mass of a substance, its specific heat capacity $c=C/m$. So heat capacity $C=m\times c$, or mass×specific heat capacity. Generally, the quantity of heat Q needed to raise the temperature of an object of mass m and specific heat capacity c by θK (or by $\theta°C$) is

$$Q=mc\theta$$

Example 11.3 on page 375 shows how to use $Q=mc\theta$ to calculate the amount of heat needed to raise the temperature of water in a vessel.

SPECIFIC HEAT CAPACITY OF METAL

In experiments to measure specific heat capacity, electrical heating is often preferred as the amount of heat supplied can then be accurately measured. If a 50W heater is used, then 50J is supplied per second. If direct measurement of the current I and the potential difference V across the heating coil is preferred for greater accuracy, the energy Q supplied is calculated from IVt joules, where I is in amperes, V in volts and t is the time in seconds. (If the resistance R of the coil is constant, the heat supplied can also be calculated from the formulae I^2Rt or V^2t/R.)

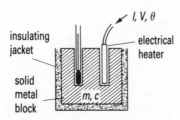

Fig 11.4 Electrical method for specific heat capacity of metal

Figure 11.4 shows a simple laboratory method for measuring the specific heat capacity c of a metal in the form of a cylindrical block. If θ is the temperature rise after a time t, then assuming negligible heat losses,

$$Q=IVt=mc\theta$$

where IVt is the heat supplied in joules and m is the mass of the block in kg. Here IVt is $\triangle Q$, the heat supplied to the block; $mc\theta$ is $\triangle U$, the gain in the internal energy of the block; and $\triangle W$, the work done when the metal expands and pushes back the atmosphere, and the heat lost to the atmosphere, are assumed negligible. From this relation,

$$c=IVt/m\theta.$$

Example 11.4 on page 375 shows how to calculate the specific heat capacity of a metal using electrical heating and making allowance for heat loss.

SPECIFIC HEAT CAPACITY OF LIQUID BY CONSTANT FLOW TUBE

The electrical method for measuring the specific heat capacity of a *liquid* such as water is illustrated in Fig 11.5. Under a constant pressure head K, the liquid flows steadily through a horizontal tube and is heated electrically by a coil R. The tube is surrounded by a vacuum and then by another tube containing water at the temperature of the surroundings. In this way the heat lost from the warmed liquid is only that due to radiation across the vacuum and, further, it can be eliminated from the calculation for c as seen shortly.

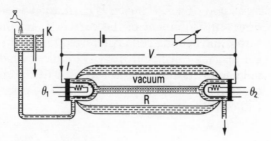

Fig 11.5 Flow tube method
for liquid specific heat
capacity

When the liquid outlet temperature is *steady*, the inlet (θ_1) and outlet (θ_2) temperatures are noted, as are the mass per second (m) of liquid flowing and the current I and potential difference V. The liquid flow is now increased to m_1 and the current and potential differences are increased to I_1 and V_1 respectively so that the liquid is again warmed to the *same* temperature θ_2 as before. Then

$$IV = \text{heat per second supplied} = mc(\theta_2 - \theta_1) + h$$

Also, $I_1 V_1 = m_1 c(\theta_2 - \theta_1) + h$

since the heat lost per second h in the second experiment is the same as in the first because the temperatures θ_1, θ_2 are the same. Subtracting to eliminate h,

$$c = \frac{I_1 V_1 - IV}{(m_1 - m)(\theta_2 - \theta_1)}$$

In accurate laboratory work, the current and p.d. are measured by a potentiometer, and θ_1 and θ_2 by platinum resistance thermometers. The advantages of the method are: (1) the unknown heat capacity of the flow tube is not required (when the temperatures are steady all the heat supplied is used only for warming the liquid from θ_1 to θ_2), (2) the heat lost can be eliminated, (3) current, p.d. and temperature can be measured accurately and at leisure with electrical meters, (4) the variation of the specific heat capacity with temperature can be measured by having a small temperature rise, for example, from 14°C to 16°C. In this case the value of c is that at 15°C. From 30°C to 32°C, the value of c would be that at 31°C.

In this way the variation of the specific heat capacity of water with temperature can be found.

Examples 11.5 and 11.6 on page 376 show how the rate of flow affects the voltage in a constant flow tube experiment and how specific heat capacity is calculated.

WORKED EXAMPLES ON HEAT CAPACITY, CONSTANT FLOW TUBE EXPERIMENT

11.3 A copper vessel of mass 0.2kg has a value of $c=400$J kg^{-1} K^{-1}. What is its heat capacity?

If a mass of water of 0.4kg is placed inside it and $c_w=4200$J kg^{-1} K^{-1}, calculate the heat needed to raise the temperature of the water and vessel from 10°C to 60°C.

Heat capacity $\quad C=mc=0.2\times400=80$J K^{-1}

Heat for water $\quad Q=mc\theta=0.4\times4200\times(60-10)=84\,000$J

Heat for vessel $\quad Q=mc\theta=0.2\times400\times(60-10)=4000$J

So $\quad$ total heat$=88\,000$J

11.4 A metal block of mass 0.5kg is heated electrically by a 24W heater in a room at 15°C. The temperature rises approximately uniformly to 35°C in 5 min and then becomes steady at 55°C. Assuming the rate of heat loss is proportional to the excess temperature over the surroundings, calculate (a) the rate of loss of heat at 25°C, (b) the specific heat capacity of the metal allowing for the heat loss while it was heated from 15°C to 35°C.

(a) $\quad$ At *steady* temperature of 55°C, an excess of 40°C over the surroundings temperature,

$$\text{rate of heat loss}=\text{rate of heat supplied}=24\text{W}$$

So at 25°C, an excess temperature of 10°C,

$$\text{rate of heat loss}=\frac{10}{40}\times24\text{W}=6\text{W}$$

(b) $\quad$ From 15°C to 35°C, average temperature$=(15+35)/2=25$°C.
So rate of loss of heat to surroundings while warming from 15°C to 35°C$=6$W. Hence, for 5 min or 300 s, since heat supplied$=$heat gained$+$heat lost,

$$24\times300=0.5\times c\times(35-15)+6\times300$$

So $\qquad c=\dfrac{18\times300}{0.5\times20}=540$J kg^{-1} K^{-1}

11.5 In a constant flow tube experiment, a p.d. of 8.0V was applied to the heating coil. When the rate of flow of liquid was halved, the p.d. was altered to a value V to produce the same inlet and outlet temperatures. Calculate V.

Heat per second supplied$=V^2/R$, where R is the resistance of the heating coil.

When the rate of flow is halved, only half the heat per second is required to produce the same outlet temperature. So

$$\frac{V^2}{R}=\frac{1}{2}\times\frac{8^2}{R}=\frac{32}{R}$$

Hence $V^2=32$ and $V=\sqrt{32}=5.7\text{V}$

11.6 In a constant flow tube experiment, the flow of liquid is $2\times10^{-3}\text{kg}$ s^{-1}, the heat supply is 24W and the temperature rise is 3K. Neglecting heat loss, calculate a value for the specific heat capacity of the liquid.

The true specific heat capacity is 3500J kg^{-1} K^{-1}. Calculate the percentage of heat lost in the experiment.

If c is the specific heat capacity of the liquid, neglecting the heat lost,

$$24=\text{heat per second}=2\times10^{-3}\times c\times3$$

So
$$c=\frac{24}{2\times10^{-3}\times3}=4000\text{J kg}^{-1}\text{K}^{-1}$$

If 3500J kg^{-1} K^{-1} is the true specific heat capacity of the liquid, then

$$\text{heat gained per second}=mc\theta=2\times10^{-3}\times3500\times3=21\text{W}$$

So
$$\text{heat lost per second}=24-21=3\text{W}$$

Hence
$$\text{percentage lost}=\frac{3\text{W}}{24\text{W}}\times100\%=12.5\%$$

LATENT HEAT

SPECIFIC LATENT HEAT OF FUSION

A solid consists of atoms or molecules held in a fixed structure by forces or bonds of attraction between them. The atoms or molecules vibrate about their mean positions. When heat is given to the solid the kinetic energy of vibration increases, thus increasing the temperature of the solid. At the melting point the heat given to the solid is used to overcome the forces of attraction between the atoms or molecules which keep the solid in its rigid form and the solid then melts. The **specific latent heat of fusion** is the heat required to change 1kg of the solid at the melting point to liquid at the same temperature.

Unlike a solid, a liquid has no definite form. Its molecules move in a random way inside the liquid, although the molecules are close enough to attract each other. Some of the molecules which have the greatest kinetic energy are able to escape through the surface and exist as vapour outside the liquid. This is evaporation and occurs at all temperatures. *Boiling*, however, occurs at a definite temperature, the boiling point, which depends on the external pressure. Water, for example, boils at 100°C at a pressure of 760mmHg and at lower temperatures when the external pressure is lower. Unlike evaporation, which is a surface phenomenon, boiling occurs throughout the whole volume of the liquid.

SPECIFIC LATENT HEAT OF VAPORIZATION

The **specific latent heat of vaporization** l is the heat required to change 1kg of liquid at the boiling point to vapour at the boiling point. For water, roughly $l=2.4\times10^6$J kg^{-1} at 100°C boiling or atmospheric pressure 760mmHg.

If a mass m of steam at 100°C condenses to water at 40°C, the heat given up

$$=ml+mc_w(100-40)$$

where c_w is the specific heat capacity of water.

Figure 11.6 shows an electrical method of measuring the specific latent heat of vaporization of a liquid. The liquid is heated by the coil R and when it boils the vapour passes down a central tube and is condensed by a surrounding water jacket J. After a suitable time, when conditions are *steady*, the mass of liquid condensed in a given time is measured. Then, if m is the mass per second, l is the latent heat and h is the heat loss per second of the vessel,

$$IV=ml+h$$

To eliminate h, the experiment can be repeated with a new current I' and p.d. V' and a new rate of vaporization m'. The value of h is the same as before since the temperature of the vessel is again the boiling point. So

$$I'V'=m'l+h$$

By subtraction to eliminate h and simplifying, then

$$l=\frac{IV-I'V'}{(m-m')}$$

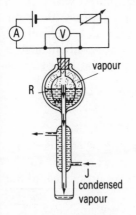

Fig 11.6 Electrical method for specific latent heat

Note that the vapour passing down the tube at the top is at the same temperature as the surrounding boiling liquid, which is an advantage. A double-walled vessel helps reduce heat losses.

LATENT HEAT AND INTERNAL ENERGY

When a liquid reaches its boiling point, the energy needed to change it to vapour is (1) the energy or work needed to separate the liquid molecules from their mutual attraction until they are relatively far apart in the gaseous state and (2) the energy or work needed to push back the external pressure so that the molecules can escape from the liquid. So the latent heat consists of (1) that needed to change the internal energy of the liquid and (2) that needed to do external work.

On page 351 we showed that the work done when a substance expands by a volume $\triangle V$ against a constant pressure p is calculated from $W=p.\triangle V$. Now 1g of water changes to about 1672cm^3 of steam at 100°C, so that $\triangle V=(1672-1)$cm$^3=1671\times10^{-6}$m^3. The external atmospheric pressure $p=1.013\times10^5$N m^{-2}. So

$$\text{external work}=1.013\times10^5\times1671\times10^{-6}=170\text{J}$$

The latent heat of vaporization per gram of water=2260J. So the internal part of the latent heat=2260−170=2090J, which is much greater than the external part. It can be noted here that the latent heat of fusion per gram of water is about 340J. So the energy needed to overcome the bonds between molecules in the solid state and form the liquid state is much less than the energy to form gaseous molecules from the liquid state.

HEAT TRANSFER: CONDUCTION

FUNDAMENTAL FORMULAE

Heat flows along a substance when a *temperature gradient* is set up in it. Figure 11.7 shows a small part of a metal of length l which has a high temperature θ_2 at one end and a lower temperature θ_1 at the other end. The temperature gradient is then $(\theta_2-\theta_1)/l$, by definition. The rate of heat flow Q/t depends on the temperature gradient and on the area of cross-section A of the metal normal to the temperature gradient. So we write:

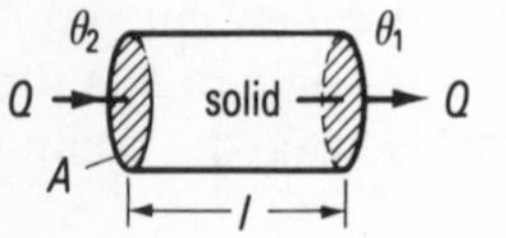

$$\frac{Q}{t}=kA\frac{\theta_2-\theta_1}{l} \tag{1}$$

Fig 11.7 Conduction through solid

where k is a constant for the metal known as its **thermal conductivity**. So k is numerically equal to the heat flow per second per unit area of cross-section when unit temperature gradient is set up normal to the area.

If we make the temperature difference $\theta_2-\theta_1$ very small and equal to $-\triangle\theta$ (the minus is necessary because the temperature θ *decreases* along the bar from the hot end) and the length l is made very small and equal to $\triangle x$, we can write the heat flow per second dQ/dt as

$$\frac{dQ}{dt}=-kA\frac{d\theta}{dx}$$

Here we have reached the limit of making the bar so short that dQ/dt is the heat flow per second through a *section* of the bar, and $d\theta/dx$ is the temperature gradient at that section. Note that $d\theta/dx$ is a negative value since the temperature θ decreases as x increases. So dQ/dt is a positive quantity from the formula.

LAGGED AND UNLAGGED BARS

Consider a long thick lagged metal bar with one end at a constant high temperature θ_1 and the other end at a constant lower temperature θ_2 (Fig 11.8(a)). From (1), the temperature gradient at any section of the bar is given by

$$\text{temperature gradient}=\frac{1}{kA}\times\frac{Q}{t}, \tag{2}$$

where Q/t is the quantity of heat per second flowing through that section. Now under steady conditions, the quantity of heat per second, Q/t, flowing through one section of the bar is equal to that flowing through the next section because no heat is lost from the sides when the bar is lagged. So, from (2), since k, A and Q/t is the same for all sections, the temperature gradient is *constant*. So the temperature variation is a straight line AB along the bar, as shown in Fig 11.8(a).

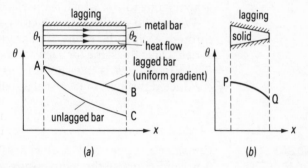

Fig 11.8 Temperature variation in lagged and unlagged bars

Suppose, however, that the bar is *unlagged*. Then, in the steady state, the heat per second Q/t flowing through one section is *less* than the section on its left at a higher temperature because some heat is lost from the sides of the bar. So, from (2), the gradient becomes smaller from the hot to the cold end of the bar. The temperature variation is thus a curve AC with diminishing gradient. See Fig 11.8(a).

Finally, suppose the cross-section of the bar decreases steadily from the hot to the cold end (Fig 11.8(b)) and the bar is *lagged*. Then, in the steady state, k and Q/t are the same for each section but this time A decreases. So, from (2), the temperature gradient *increases* from the hot to the cold end. The variation of temperature along the bar therefore follows the curve PQ shown in Fig 11.8(b), which has an increasing gradient from the hot to the cold end.

ANALOGY WITH ELECTRICAL CONDUCTION

In metals, electric current flow is due to moving electrons, which carry charge. The current flowing I is the quantity of charge per second, Q/t, flowing across a section of the metal. Now $I = V/R$, where V is the potential difference between the ends of the conductor and R is its electrical resistance. If the conductor is uniform in cross-section, then $R = \rho l/A$, where ρ is the resistivity of the metal (page 114). So

$$I=\frac{V}{R}=\frac{V}{\rho l/A}=\frac{1}{\rho}.A.\frac{V}{l}$$

Hence $\dfrac{Q}{t}=\dfrac{1}{\rho}A\dfrac{V}{l}$

So Q/t, the quantity of charge per second flowing, is proportional to the potential gradient across the conductor, V/l, and to the area A. In heat flow, $Q/t=kA\times$temperature gradient. So thermal conductivity is analogous to $1/\rho$, which is the electrical conductivity of the metal.

When two electrical conductors are in series, the electric current through each is the same. Similarly, when thermal conductors are in series, Q/t is the same for each conductor when conditions are steady. For example, Q/t is the same through brick and plaster when heat is conducted through the walls of a room at a steady rate.

GOOD AND BAD CONDUCTORS IN SERIES

Consider a copper tank of thickness 5mm lagged by felt of thickness 5cm, and suppose the temperature of water in the tank is constant at 40°C and the temperature of the outside of the felt is constant at 10°C (Fig 11.9). Copper, a good conductor, has a thermal conductivity of $400\text{W m}^{-1}\text{K}^{-1}$ and felt, a bad conductor, has a thermal conductivity of $0.04\text{W m}^{-1}\text{K}^{-1}$.

Under steady conditions, the heat flow Q/t is the same for the copper and felt. Now on page 380 we showed that

$$\text{temperature gradient}=\frac{1}{kA}\times\frac{Q}{t}$$

Since Q/t is the same for copper and felt, and assuming A is the same, it follows that, if g_c and g_f are the respective temperature gradients for copper and felt

$$\frac{g_c}{g_f}=\frac{k_f}{k_c}=\frac{0.04}{400}=\frac{1}{10\,000}$$

since the temperature gradient is *inversely* proportional to k. So the temperature gradient across the copper is 10 000 times *less* than the gradient across the felt. As illustrated in Fig 11.9, this means that the temperature of the outer surface of the copper tank is not much less than its inner surface at 40°C. Practically the whole of the temperature drop from 40°C to 10°C occurs across the felt, the bad conductor. This result is true whenever we have a good conductor in series with a bad conductor, since the temperature gradient g is inversely proportional to the thermal conductivity k.

In boilers, there is always a layer of gas between the flame and the underside of the boiler. The gas is a very bad conductor and the metal of the boiler is a good conductor. As we have just seen, the rate of heat flow is mainly determined by the gas, the bad conductor, since this has most of the temperature drop across it. It would not affect the

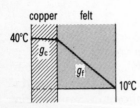

Fig 11.9 Good and bad conductors in series

rate of heat flow appreciably if the material of the boiler were steel or copper. A similar result is obtained in electricity. With a poor conductor in series with a good conductor, the current flowing is small due to the poor conductor, even if the good conductor is changed to an even better conductor.

Example 11.7 on page 383 shows how to calculate the thickness of air equivalent to 30cm of brick. Example 11.8 shows how to calculate heat flow in double glazing and the inside temperatures of the two pieces of glass used.

MEASURING THERMAL CONDUCTIVITY OF GOOD CONDUCTOR

To measure the thermal conductivity of a metal, we need to have (*a*) a thick bar so that appreciable heat flows through a cross-section, (*b*) a reasonably long bar so that the temperature gradient can be measured accurately, (*c*) a lagged bar so that no heat escapes and the heat flow is linear along the bar when conditions are steady.

Searle designed the apparatus required to measure *k*. In Searle's method, the lagged thick metal bar is heated at one end by a heating coil connected to an electrical supply and the final steady temperature can be measured by means of two thermometers P, Q placed in deep holes bored into the bar at a known distance *l* apart (Fig 11.10). A coiled tube R in thermal contact with the bar near one end enables the heat per second flowing to be measured. Water from a constant pressure head flows steadily through the tube R and the inlet and outlet temperatures, θ_3 and θ_4, are measured. After switching on the heat supply, the apparatus is left until all the temperatures are *steady*.

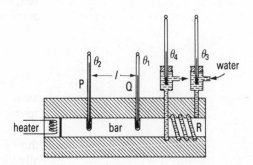

Fig 11.10 Thermal conductivity of good conductor (Searle)

The rate of flow of the water *m* is measured using a cylinder and stop-clock. The area *A* of cross-section of the bar is found using calipers to measure the diameter. From the steady temperature readings shown in Fig 11.10, *k* is found from

$$\frac{Q}{t}=mc_w(\theta_4-\theta_3)=kA\frac{\theta_2-\theta_1}{d}$$

Alternatively, Q/t could be measured electrically from IV for the heater without using R.

Lees designed the apparatus widely used for measuring the thermal conductivity of a bad conductor which can be made in the form of a slab, such as cardboard or glass. To make the heat flow measurable, the area A is made large – so a disc of large diameter is used, and the temperature gradient is made high – so a thin disc is used. Further, to keep the whole of each surface of the disc at the same temperature, thick metal slabs are pressed against each surface. To make good thermal contact with the disc the faces of the slabs must be flat and clean; a little Vaseline smeared on each face improves thermal contact.

Figure 11.11(a) shows one form of apparatus. A steam chamber D has a bottom made of thick brass plate A and the material, such as cardboard, is sandwiched between A and a lower thick brass plate B. Thermometers are placed inside holes drilled in A and B. The disc A is heated by passing steam through the chamber and after a time the temperatures θ_1 and θ_2 are read when they are *steady*. If $\theta_1=82°C$, for example, the heat per second Q/t flowing through the cardboard in the steady state is equal to the heat per second lost by B by radiation and convection at 82°C.

A second experiment is necessary to find this rate of loss of heat by B. As shown in Fig 11.11(b), the steam chamber is removed and with

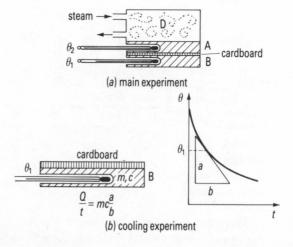

(a) main experiment

(b) cooling experiment

$$\frac{Q}{t}=mc\frac{a}{b}$$

Fig 11.11 Thermal
conductivity of bad
conductor (Lees)

the cardboard on top of B, this disc is warmed gently by a burner until its temperature is a few degrees above 82°C (θ_1). The burner is then removed, B is allowed to cool and readings of its temperature fall with time are taken. From the temperature-time graph, Fig 11.11(b), the rate of cooling a/b, is found at 82°C (θ_1). Then

$$Q/t = \text{heat lost by cooling at } 82°C = mc\,a/b,$$

where m is the mass of B and c is its specific heat capacity. From the main experiment, $Q/t = kA(\theta_2 - \theta_1)/d$, where A is the area and d is the thickness of the cardboard. So k can now be calculated.

WORKED EXAMPLES ON CONDUCTION IN SERIES CONDUCTORS AND IN DOUBLE GLAZING

11.7 Assuming the thermal conductivity of air and brick is 0.02 and $0.6\,\text{W m}^{-1}\text{K}^{-1}$ respectively, calculate the thickness of air equivalent to a thickness of 30cm of brick.

If two such brick walls are separated by an air gap of 3cm, how much heat per minute would flow through them in the steady state when the outside temperatures of the brick are 60°C and 10°C respectively, and the area of cross-section of each is 2m²?

Suppose the thickness of the air is d and the same temperature difference $(\theta_2 - \theta_1)$ is across the air and the brick. If they are equivalent, the same heat per second, Q/t, would flow. So since 30cm=0.3m,

for air
$$\frac{Q}{t} = kA\frac{\theta_2 - \theta_1}{d} = 0.02A\frac{\theta_2 - \theta_1}{d}$$

for brick
$$\frac{Q}{t} = kA\frac{\theta_2 - \theta_1}{0.3} = 0.6A\frac{\theta_2 - \theta_1}{0.3}$$

Therefore
$$\frac{0.02}{d} = \frac{0.6}{0.3}$$

So
$$d = \frac{0.02 \times 0.3}{0.6} = 0.01\text{m} = 1\text{cm air}$$

If 3cm of air is in series with 60cm of brick, the total equivalent thickness of air would be 3cm+2cm, because we can replace the 30cm of brick by 1cm of air, as we have just shown. So the arrangement is equivalent to a thickness of 5cm of air or 0.05m. So

$$Q \text{ per minute} = 60 \times \frac{Q}{t} = 60 \times 0.02 \times 2 \times \frac{60 - 10}{0.05}$$

since Q/t is the heat per second and the temperature gradient is $(60-10)\text{K}/0.05\text{m}$.

So Q per minute $= 2400\text{J min}^{-1}$

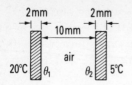

Fig 11.12 Calculation on
series conductors
(double glazing)

11.8 In double glazing, two sheets of glass 2mm thick are separated by 10mm of air (Fig 11.12). The temperatures of the outside glass surfaces are 20°C and 5°C respectively. Calculate the heat per second per unit area flowing by conduction, and the temperatures of the interior glass surfaces, assuming a steady state. (Thermal conductivities of glass and air=0.06 and 0.02W m^{-1}K^{-1} respectively.)

We can replace the 10mm of air by a thermally equivalent layer of $x$mm of glass. In this case, with the usual notation, if k_a=0.02 (air) and k_g=0.06 (glass),

$$\frac{Q}{t}=k_aA\frac{\theta_2-\theta_1}{10}=k_gA\frac{\theta_2-\theta_1}{x}$$

So $\quad \dfrac{k_a}{10}=\dfrac{k_g}{x}$, and $x=\dfrac{k_g}{k_a}\times10=\dfrac{0.06}{0.02}\times10=30$mm

Hence total equivalent glass thickness=2+2+30=34mm.
Since outside temperatures are 20°C and 5°C,

$$\text{heat per second per unit area}=\frac{1}{A}\frac{Q}{t}=0.06\times\frac{20-5}{34\times10^{-3}}$$

$$=26.5\text{J} \tag{1}$$

We can now find the temperatures θ_1 and θ_2 shown in Fig 11.12. For the 2mm glass on the left,

$$\frac{1}{A}\frac{Q}{t}=26.5=0.06\times\frac{20-\theta_1}{2\times10^{-3}}$$

Solving, $\quad\quad\quad \theta_1=19.1°C \text{ (approx.)} \tag{2}$

For θ_2, we take 2mm of glass on the right. In this case,

$$\frac{1}{A}\frac{Q}{t}=26.5=0.06\times\frac{\theta_2-5}{2\times10^{-3}}$$

Solving, $\quad\quad\quad \theta_2=5.9°C \text{ (approx.)} \tag{3}$

1 THERMOMETRY

On the *thermodynamic* temperature scale, $T=(P_T/P_{tr})\times273.16K$, where P_T and P_{tr} are the respective magnitudes of the property used (gas pressure, resistance or volume) at the unknown temperature T and at the triple point.

On the *Celsius scale*, $t=(P_t-P_0)\times100°C/(P_{100}-P_0)$, where P_t, P_{100}, and P_0 are the respective magnitudes of the property used at the unknown temperature t, at 100°C (steam temperature at 760mmHg) and at 0°C (melting ice temperature).

Thermometers: The *gas* thermometer is the 'standard' thermometer. It is too bulky for general use. So the *platinum resistance* thermometer and *thermoelectric* thermometer are used. The resistance thermometer may be used, for example, in the flow tube method for specific heat capacity of liquids. The thermoelectric thermometer is useful for measuring changing temperatures or the temperatures of metal surfaces.

2 HEAT CAPACITY AND SPECIFIC HEAT CAPACITY

Heat gain or loss:

$$Q=mc\theta \text{ where } \theta \text{ is the temperature change}$$

In electrical methods,

$$\text{heat supplied}=IVt=I^2Rt=V^2t/R$$

For a *metal*, $IVt=mc\theta+h$, where h is heat lost to surroundings.

For a *flow tube method* for a liquid, if m_1 is the mass per second of flowing liquid,

$$I_1V_1=m_1c(\theta_2-\theta_1)+h$$

To eliminate h, reduce I_1V_1 to IV, and m_1 to m, so that θ_2 and θ_1 have same values as before, and subtract the heat equations as h is the same in each.

3 LATENT HEAT AND SPECIFIC LATENT HEAT (l)

l is the heat needed to change 1kg of solid at the melting point to liquid at the same temperature (fusion) or 1kg of liquid at the boiling point to vapour at the same temperature (vaporization).

At *fusion* or *vaporization*, bonds are broken with surrounding molecules. This produces an internal energy change (change of state).

l for water to vapour$= l_{int} + l_{ext}$

where $l_{ext} =$ work done in expansion from liquid to vapour

$=$ atmospheric pressure (N m^{-2})$\times$volume change (m^3)

4 HEAT TRANSFER: CONDUCTION

Basic formula: $\dfrac{Q}{t} = kA\dfrac{\theta_2 - \theta_1}{l}$ or $\dfrac{dQ}{dt} = -kA\dfrac{dQ}{dx}$

Bar of uniform cross-section: (*a*) lagged – straight-line temperature variation; (*b*) unlagged – falling curve of temperature.

Series good and bad conductors: Temperature gradient very low across good conductor and high across bad conductor.

Problems: With two or more series conductors, it is often useful to find the equivalent thermal lengths and then use Q/t for one conductor.

In Searle's method for measuring k for a *good conductor*, a thick lagged bar is used and Q/t is measured by a coil with water passing through or by the electrical heater, in steady state conditions.

In Lees' disc method for a *bad conductor*, a thin disc of large area sandwiched between two thick metal slabs is used and Q/t is found by a cooling experiment with one slab.

DIAGRAM SUMMARY

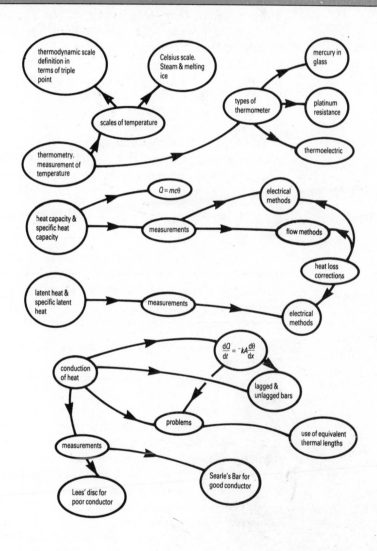

1   The resistance of the element of a platinum resistance thermometer is 2.00Ω at the ice point and 2.73Ω at the steam point. What temperature on the platinum resistance scale would correspond to a resistance value of 8.34Ω?

Measured on the gas scale, the same temperature corresponded to a value of 1200°C. Explain the discrepancy.   (L)

2   What are the main factors controlling the rate of heat conduction between the opposite faces of a disc of a given material? Explain why, in the experimental determination of the thermal conductivity of a poor conductor, a thin disc of the material is used?   (L)

3   The table below gives data for two thermometers at three different temperatures (the ice-point, steam-point and room temperature).

Type of thermometer	Property	Value of property		
		ice-point	steam-point	room temp
Gas	Pressure in mmHg	760	1040	795
Thermistor	Current in mA	12.0	54.0	15.0

(a)   Calculate the temperature of the room according to each thermometer.

(b)   State why the thermometers disagree in their value for room temperature.

(c)   Explain why a gas thermometer is seldom used for temperature measurement in a laboratory.   (AEB 1985)

4   A material used for thermal lagging, which can be cast or moulded into any shape, is claimed to have a thermal conductivity of $10^{-2}W$ $m^{-1} K^{-1}$. Describe an experiment you would make to test this claim, giving a labelled diagram of the apparatus and explaining the method of supplying and measuring the heat flow and making the necessary temperature measurements.

An unlagged hot-water tank is maintained at 70°C by an electrical heater and thermostat at times when no water is being drawn off. The tank is lagged with the above material. What is the average power needed to maintain the temperature? The lagging is 30mm thick, has an external surface area of $6m^2$, and loses heat to the surroundings at the rate of 3W per square metre per degree Centigrade difference in temperature between the surface and the surroundings. Assume the surrounding temperature to be 20°C.   (O&C)

5   In a simple continuous flow apparatus (i.e. one with no vacuum jacket) for determining the specific heat capacity of water the following measurements were made.

Inlet and outlet water temperature .... 11.0°C and 13.5°C
Ammeter and voltmeter readings .... 2.40A and 5.00V
Mass of water collected in seven minutes 450g

Using these results, calculate a value for the specific heat capacity of water.

Why is the value not likely to be very accurate? Is it likely to be high or low?

Explain how by repeating the experiment you would try to reduce the error?   (L)

6   (a)   When bodies are in thermal equilibrium, their temperatures are the same. Explain in energy terms the condition for two bodies to be in thermal equilibrium with one another.

(b)   The temperature of a beaker of water is to be measured using a mercury-in-glass thermometer.

(i) Why is it necessary to wait before taking the reading?

(ii) Explain briefly how you might estimate the heat capacity (energy required per unit temperature rise) of a mercury-in-glass thermometer.

(iii) If the beaker contains 120g of water at 60°C, what temperature would be recorded by the mercury-in-glass thermometer if it was initially at 18°C and had a heat capacity of 30J K^{-1}? (Assume the specific heat capacity of water to be 4200J kg^{-1} K^{-1} and ignore the heat losses to the beaker and surroundings while the temperature is being taken.)

(iv) Why, if a more accurate value of the temperature were required in this case, might you use a thermocouple?

(v) Describe briefly how you would calibrate a thermocouple and use it to measure the temperature of the water. Show how you would calculate the water temperature from your readings.   (L)

7   The average energy of an atom of a solid at a kelvin temperature T is $3kT$, where k is the Boltzmann constant. Calculate

(a)   the average energy of a copper atom in a mole of copper at 300K and

(b)   the total energy of all the atoms in a mole of copper at the same temperature.

If this total energy could be transformed into linear kinetic energy of the whole mass of copper, at what speed would the mass be travelling? (Boltzmann constant=1.38×10^{-23}J K^{-1}, Avogadro constant=6.02×10^{23}mol^{-1}, mass of 1 mole of copper atoms=0.064kg.)   (L)

8   A copper hot water cylinder of length 1.0m and radius 0.20m is lagged by 2.0cm of material of thermal conductivity 0.40W m^{-1} K^{-1}. Estimate the temperature of the outer surface of the lagging,

assuming heat loss is through the sides only, if heat has to be supplied at a rate of 0.25kW to maintain the water at a steady temperature of 60°C. Assume that the temperature of the inside surface of
the lagging is 60°C.   (L)

9   A student using a continuous flow calorimeter obtains the following
results:

Using water, which enters at 18.0°C and leaves at 22.0°C, the rate
of flow is 20g min^{-1}, the current in the heating element is 2.3A and
the potential difference is 3.3V. Using oil, which flows in and out at
the same temperature as the water, the rate of flow is 70g min^{-1}, the
current is 2.7A and the potential difference is 3.9V.

Taking the specific heat capacity of water to be 4200J kg^{-1} K^{-1},
calculate, explaining your method clearly,

(i) the rate of heat loss from the apparatus,
(ii) the specific heat capacity of the oil.

Explain carefully how, using this same method, the specific heat
capacity of the oil could be obtained without a knowledge of the
specific heat capacity of water.

Explain why readings should only be taken when a steady state
exists. How would you ensure that such a condition has been
attained?

Explain in principle what steps you would take to increase the
accuracy in the measurements of the values of

(i) the rate of flow of the liquid,
(ii) the temperature difference between the inflow and
outflow,
(iii) the potential difference across and the current in the
heating element.   (L)

ELASTICITY; YOUNG MODULUS; SOLID MATERIALS; MOLECULES

CONTENTS

The stretching of metals by forces, and their elasticity or ability to recover their original length when the forces are removed, must be investigated to decide if a particular metal is suitable for aeroplane struts or for lift cables, for example.

ELASTICITY AND YOUNG MODULUS

EXPERIMENT ON ELASTICITY

Figure 12.1(*a*) shows a laboratory apparatus to investigate the elasticity of metal in the form of wires. Two long similar wires X and Y are suspended from the same beam B. One wire X carries a millimetre scale M and the other Y a vernier V for reading small extensions. X is kept taut by a fixed weight A. Y is stretched by adding loads or weights F in steps of 10N from zero to 60N, for example.

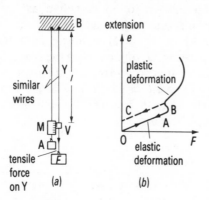

Fig 12.1 Young modulus
experiment and result

The purpose of using two *similar* wires is to eliminate error due to any sag of the beam and to any temperature changes when carrying out the experiment. These changes will affect both wires so the vernier measures only the extension of Y due to the load F.

The reading of V is taken for values of $F=0$, 10N, 20N, ..., 60N, and then again when F is 50N, 40N, ..., 0. If the readings of V on decreasing F is near to the readings of V when F was increased, we know that the wire has not been over-stretched or exceeded its elastic limit, discussed later. The average of the two readings is taken for a particular load F in this case.

HOOKE'S LAW, ELASTIC LIMIT AND YIELD POINT

Figure 12.1(*b*) shows the results when the load F, which is the stretching or tensile force in newtons, is varied and the extension e is measured. For moderate forces *the extension is directly proportional to the force*, as shown by the straight line OA. Also, the wire returns to its original length for any force corresponding to OA, so the wire is **elastic** if these forces are removed. The elastic limit is just beyond the end of the straight line, which is called the **limit of proportionality**.

As more loads are added to the wire to exceed the elastic limit, the graph curves upwards. The extension now increases considerably for small loads and the wire begins to thin or 'yield' at B. The **yield point** B is reached shortly after the elastic limit is passed. In this case the wire is permanently stretched after the load is removed, as at C, and is said to have undergone **plastic deformation**. In **elastic deformation**, when the metal recovers on removing the load, the wire gains molecular potential energy due to displacement of its molecules when it is stretched and recovers this energy when the molecules return to their original positions on removing the load. In plastic deformation, however, the energy of the displaced molecules is lost as heat in the wire when the load is removed and the molecules do not return to their original positions.

Materials are often classified as 'ductile' or 'brittle'. A ductile material is one which has a yield point. A brittle material, which has no yield point, breaks shortly after the elastic limit is exceeded. Compare Fig 12.1(*b*), which is the graph for ductile material.

Hooke investigated elasticity over 300 years ago. **Hooke's law** states: *Provided the elastic limit is not exceeded, the extension of a wire is proportional to its tension.* 'Tension' is the stretching force or tensile force in the wire.

STRESS, STRAIN AND YOUNG MODULUS

The tensile **stress** in a wire is defined as the *force per unit area*, or F/A where F is the force applied in newtons and A is the cross-section area in metre^2. So if a weight of 20N is attached to a wire of cross-sectional area 10^{-6} m^2 (this corresponds to a wire with a diameter about 1mm),

$$\text{stress} = \frac{F}{A} = \frac{20\text{N}}{10^{-6}\text{m}^2} = 2 \times 10^7 \text{N m}^{-2}$$

The tensile **strain** of a wire is defined as the *extension/original length*, or the ratio e/l where e is the extension and l is the original length. So if e is 0.2mm and l is 2m,

$$\text{strain} = \frac{e}{l} = \frac{0.2 \times 10^{-3}\text{m}}{2\text{m}} = 10^{-4}$$

Note that strain, the ratio of two lengths, has no unit, unlike stress.

In the region when the elastic limit is not exceeded, the **Young modulus** E is defined as the ratio *tensile stress/tensile strain*. So using the above values,

$$E = \frac{\text{stress}}{\text{strain}} = \frac{2 \times 10^7 \text{N m}^{-2}}{10^{-4}} = 2 \times 10^{11} \text{N m}^{-2}$$

The Young modulus E is a constant for the particular metal of the wire; its value does not depend on the length or the cross-section area of the wire considered.

E can be found from the experiment outlined in Fig 12.1(*a*). As shown in Fig 12.1(*b*), the force F in newtons is plotted against the extension e in metres when Hooke's law applies. A straight line such as OA is obtained and its gradient is found from the graph. Now $E = (F/A) \div (e/l) = (F/e) \times (l/A) = gradient \times (l/A)$. The length l is measured to the vernier V, and the area $A = \pi r^2$ where r is the radius of the wire, which is determined with a micrometer gauge at several places of the wire. Knowing the gradient from the graph, E can now be calculated.

FORMULAE

If E is the Young modulus, F is the force applied, A is the cross-section area of the wire, e is the extension and l is the original length, then

$$E = \frac{F/A}{e/l} \qquad\qquad (1)$$

and so $F = EAe/l$ $\qquad\qquad (2)$

F is in N when E is in N m^{-2} or Pa, A in m^2 and e and l in m.

Example 12.1 on page 397 shows how to calculate the force needed to extend a wire by a small length. Example 12.2 on page 397 deals with two different suspended wires which are connected at their ends with a weight on the connecting rod. It shows how the extension of each wire can be found.

ENERGY IN A STRETCHED WIRE

Suppose a weight of F newtons is placed on a wire and the extension produced is e metres. The molecules in the metal are then displaced from their normal positions against the force of molecular attraction and so the wire has gained molecular potential energy.

The molecular energy stored in the wire is equal to the work done in stretching the wire. The work done=average force in wire×extension. But the initial force or tension in the wire is zero before the weight is added and the final force after the weight has been placed on the wire is F newtons. Since the force in the wire increased steadily from zero to F as the extension increased to e,

$$\text{work done=average force} \times e = \frac{1}{2}(0+F) \times e$$

$$= \frac{1}{2}F.e \text{ joules} \tag{1}$$

$\frac{1}{2}F.e$ is the molecular potential energy stored in the wire. Since the weight F newtons falls a distance e after attaching it to the end of the wire, the loss in gravitational potential energy of the mass is $F.e$ joules. The wire has only gained $\frac{1}{2}F.e$ joules from this amount. The rest of the $F.e$ joules, which is $\frac{1}{2}F.e$ joules by subtraction, is lost finally as heat energy in the wire. This is produced when the load is released and falls, and vibrates for a short time at the end of the wire before coming to rest.

Since $A.l$ is the volume of the wire, we see that the molecular energy per unit volume of the stretched wire$=\frac{1}{2}F.e/A.l$. But F/A =stress and e/l=strain. So

$$\text{energy per unit volume} = \frac{1}{2}\text{ stress} \times \text{strain} \tag{2}$$

ENERGY STORED, AND WORK DONE, BY GRAPHICAL METHOD

The energy stored in stretching a wire can also be found from the graph of load or tension F plotted against the extension e.

In Fig 12.2(a), $F. \triangle e$, which is the small amount of work done in stretching the wire by an amount $\triangle e$, is represented by the *area* of the strip PABQ. By adding all the strips we see that the whole area below the straight line OA represents the total energy stored in the wire for an extension e equal to OB. The area of triangle OAB$=\frac{1}{2}F \times e$, as we previously obtained. The area represents joules when the axis of F is in newtons and that of e is in metres.

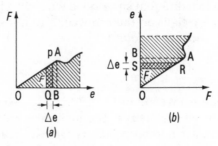

Fig 12.2 Energy stored in
stretched wire

Figure 12.2(b) shows e plotted against F. Here the area of the strip RS$=F. \triangle e$. So the energy stored in the wire is now represented by the area OAB shown shaded.

Even when the wire is stretched beyond its elastic limit so that the graph is no longer a straight line, the work done is still represented by the area between the graph and the e-axis. This time the work done is dissipated as heat in the wire, since the elastic limit is exceeded and plastic deformation occurs.

Example 12.3 on page 398 shows how the energy stored in a stretched wire is calculated.

WORKED EXAMPLES ON ELASTICITY AND YOUNG MODULUS

12.1 FORCE IN SINGLE WIRE

A wire is 2m long and its cross-section area is 0.8mm². Calculate the force to extend the wire by 1mm if the Young modulus is 2×10^{11}N m^{-2}.

$$F = EAe/l = 2 \times 10^{11} \times 0.8 \times 10^{-6} \times \frac{1 \times 10^{-3}}{2}$$

$$= 80\text{N}$$

12.2 FORCES IN PARALLEL WIRES

A steel wire and a brass wire, each 2m long, are suspended from a support so that each hangs vertically 90cm apart and are connected at their ends by a light rod 90cm long. A 60N weight is placed on the rod so that the same extension of 1mm is produced in each wire. If the cross-section area of the wires is the same, find (a) the position of the 60N weight on the rod, (b) the stress on each wire, (c) the cross-section area. ($E = 2 \times 10^{11}$N m^{-2} for steel and 10^{11}N m^{-2} for brass.)

(**Overview** (a) The extension e is $e = Fl/EA$, from $F = EAe/l$. If e is the same for both wires, the forces on each $F \propto E$.
(b) To find stress, use $E =$ stress/strain.
(c) The cross-section area A can be found from the stress value, since stress $= F/A$.)

(a)   With the usual notation, $F = EAe/l$. Since A, e and l are the same for each wire, it follows that $F \propto E$. So if F_1, F_2 are the forces on the steel and brass wires respectively, then

$$\frac{F_1}{F_2} = \frac{E_1}{E_2} = \frac{2 \times 10^{11}}{10^{11}} = 2$$

But $F_1 + F_2 = 60$N. So $F_1 = 40$N and $F_2 = 20$N, since $F_1 = 2F_2$. To produce a force F, at the steel wire, the 60N weight must be placed on the rod 30cm from this wire and 60cm from the other (brass) wire.

(b)   From $E =$ stress/strain,

$$\text{stress} = E \times \text{strain} = 2 \times 10^{11} \times (1 \times 10^{-3}/2)$$

$$= 10^8 \text{N m}^{-2} \text{ for steel}$$

and stress$=10^{11}\times(1\times10^{-3}/2)=5\times10^7\text{N m}^{-2}$ for brass

(c) For the steel wire,

$$\text{stress}=\frac{F}{A}=\frac{40\text{N}}{A}=10^8\text{N m}^{-2}, \text{ from above}$$

So $A=\frac{40}{10^8}=4\times10^{-7}\text{m}^2$

12.3 ENERGY STORED IN WIRE

A wire of length 3.0m and cross-sectional area $2\times10^{-6}\text{m}^2$ has a weight of 50N attached at one end. Calculate the energy in the wire if the Young modulus is $2\times10^{11}\text{N m}^{-2}$ and Hooke's law holds.

(**Overview** Energy stored$=\frac{1}{2}F\times e(=\frac{1}{2}\text{load}\times\text{extension})$. So e must be calculated.)

From $F=EAe/l$, the extension e is given by

$$e=\frac{F.l}{E.A}=\frac{50\times3}{2\times10^{11}\times2\times10^{-6}}$$

$$=3.75\times10^{-4}\text{m}$$

So energy stored$=\frac{1}{2}\times F\times e=\frac{1}{2}\times50\times3.75\times10^{-4}$

$$=9\times10^{-3}\text{J}$$

SOLID MATERIALS

We have just discussed the stress and strain of metals. We now describe briefly the microscopic arrangement of atoms and molecules inside metals and in some solid materials used in industry.

POLYCRYSTALLINE SOLIDS

All metal solids have a large number of very tiny crystals inside them pointing in different or random directions. These small crystals are

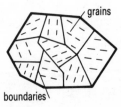

Fig 12.3 Polycrystalline solid-grains and boundaries

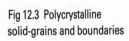

called *grains*. Under stress, the grain boundaries oppose plastic deformation (Fig 12.3).

GLASSY SOLIDS

In its solid state, silica is polycrystalline. When it is heated and melts, the liquid molecules have a disordered state. They remain in this state when the liquid is suddenly cooled and a glassy solid is then formed. Normally a liquid would change back to a polycrystalline solid when cooled to its freezing point, so the glassy solid is due to *supercooling*. Glass is a supercooled liquid.

AMORPHOUS SOLIDS

Glasses with a silica base such as porcelain have atoms packed in disordered directions. These solids are called amorphous solids. Unlike polycrystalline solids, amorphous solids have a completely irregular structure. Glass is practically amorphous.

POLYMER SOLIDS

Polymer solids, organic compounds, consist of long chains of molecules in units called *monomers* or *mers*. Polythene is an example of a polymer. Its molecular chain has a unit or mer ethylene, C_2H_4, shown diagrammatically in Fig 12.4(*a*). The long polythene molecule is shown partly in Fig 12.4(*b*).

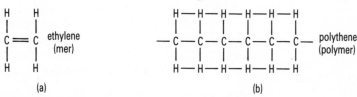

Fig 12.4 Polymer solid

Rubber is a polymer consisting of long coiled chains of molecules. When pulled, rubber may stretch by say eight times its original length as the coiled molecules unwind. When released, the rubber returns to its original length. So the strain can be as much as 800% under elastic deformation.

In contrast, metals and glass have a very low strain before breaking, such as 1% or less. In addition, Young's modulus is much less for rubber than for metals or glass. Figure 12.5 shows roughly the different stress-strain graphs for copper, glass and rubber. Glass extends very little (high stiffness) and is very brittle.

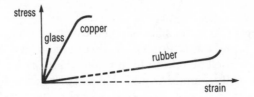

Fig 12.5 Stress-strain curves

OTHER PROPERTIES OF MATERIALS

GLASS

Glass fractures easily due to *cracks* in its surface. As experiment shows, there is a high stress concentration round any crack. Since the glass does not undergo plastic deformation, there is no movement or 'flow' inside the material to relieve the local stress and so the glass breaks. This explains why glass is brittle.

Toughened glass is made by rapid cooling of outer layers. So when the inside solidifies, the outer layers are in a state of compression. This tends to hold the 'cracks' together and so the glass is now toughened. Windscreens use toughened glass.

METALS

If cracks appear in metals, the material deforms plastically when the stress across the crack exceeds the yield stress. This 'flow' relieves the local stress and the metal is then non-brittle.

Work hardened metals are made by heating the metal and then quenching them in cold water. One type of imperfection in the crystal planes of metals, called a *dislocation*, enables the planes to move or *slip*. On quenching, however, the dislocations 'join up' and so their movement is very small. Cracks can then not be relieved. So the work hardened metal is brittle. Plastic flow in metals is due to slip.

RUBBER

Rubber shows elastic 'hysteresis', that is, it does not recover its original length immediately the stress is removed and absorbs energy, as Fig 12.6 shows. In oscillations, therefore, rubber absorbs energy. So it is used for damping oscillations in trailer suspensions and in a hi-fi stylus.

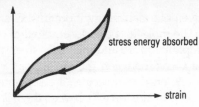

Fig 12.6 Rubber hysteresis

CONCRETE

Concrete material is relatively weak in tension but strong in compression since the particles are then forced together. The disadvantage of concrete is reduced by inserting stressed steel rods inside on laying the concrete. This holds the resulting concrete under compression after the concrete has set.

WOOD

Long fibres in wood give strength along the grain but the strength is relatively weak at right angles to the grain. This can be overcome by building up thin layers of wood into *plywood*, where the grains are alternately along and at right angles to each other and the layers are bonded by a suitable resin or glue.

MOLECULAR POTENTIAL ENERGY AND FORCE

The force between molecules is a complex matter. So is their potential energy. Analysis beyond the scope of this book shows that the *potential energy* V of a molecule varies with its separation r from another molecule along a curve PQRS similar to that shown in Fig 12.7. When the molecules are farther apart than a distance r_0, the

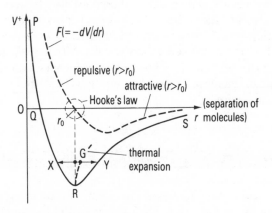

Fig 12.7 Molecular potential energy and force

negative potential energy V diminishes slowly along RS as r is increased. When the molecules are closer together than r_0, the negative potential V rises rapidly to zero and then rises steeply to positive values along QP.

The *force* F between the molecules can easily be found from the V-r graph. When the molecules are separated by a small distance $\triangle r$, the work done by the force F between them is $F.\triangle r$. This is equal to the loss in potential energy, $\triangle V$, of the molecules. So $-\triangle V = F.\triangle r$. Hence $F = -\triangle V/\triangle r = -dV/dr$ in the limit. So we can find F simply from the *gradient* to the V-r curve at any point.

INTERMOLECULAR FORCE

Figure 12.7 shows how F varies with r, when F is found from the gradient to the V-r graph. The minimum potential energy corresponds to a separation r_0; so this is the normal separation of molecules in a solid under given conditions. At distances greater than r_0, F is negative. Here the force F is *attractive*. At distances less than r_0, F is positive. Here the force is *repulsive*. At $r=r_0$, the attractive and repulsive forces balance each other and $F=0$. At large values of r the attractive force between the molecules is very small and molecules are then in the gaseous state.

PROPERTIES OF SOLIDS

We can deduce some of the common properties of metals from the F-r and V-r curves in Fig 12.7.

Near $r=r_0$, when the metal undergoes extension or compression by a force F, the curve is practically a straight line. So a small extension is proportional to the applied force, which is *Hooke's law* of elasticity.

The attractive force F increases from $r=r_0$ to the minimum of the F-r curve but then becomes smaller as r is increased further. So the minimum point is one beyond which the metal would break (page 394).

The minimum value R of the potential energy V of molecules in a metal occurs at the absolute zero of temperature. At higher temperatures the molecules gain energy and then oscillate between X and Y, for example. The average or mean separation is now G, half-way between X and Y. Since the V-r curve is not symmetrical. G is slightly to the right of R, so that the mean separation is greater than r_0. At higher temperatures the mean separation of the molecules increases more. So the metal expands with increasing temperature, which is the phenomenon of **thermal expansion**.

VISCOSITY OR FLUID FRICTION

FLUID FLOW AND VELOCITY GRADIENT

Liquids have frictional forces between their layers as they flow through pipes. So do gases. The name **viscosity** is given to the frictional forces in fluids. It is an important subject because it concerns the lubrication of moving parts in machines such as car engines, for example. Further, the viscosity of a liquid or gas affects the volume per second flowing along a pipe, as we see shortly.

It is not often realized that a liquid flowing along a pipe has regions moving with different velocities. The liquid flows fastest in the centre, along the axis. As we move towards the wall of the pipe the velocity decreases and very close to the solid wall of the pipe the velocity is practically zero. Figure 12.8(a) shows roughly, by vector lengths, the variation of the velocity from the axis to the boundary of the pipe for steady flow.

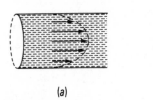

(a)

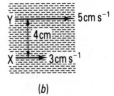

(b)

Fig 12.8 Steady or laminar liquid flow and velocity gradient

Figure 12.8(b) shows two layers X and Y in the liquid which are say 4cm (r) apart. Their velocities are respectively 5cm s^{-1} (v_1) and 3cm s^{-1} (v_2). So the average **velocity gradient** in the region of liquid between X and Y

$$=\frac{v_1-v_2}{r}=\frac{(5-3)\text{cm s}^{-1}}{4\text{cm}}$$

$$=0.5\,\text{s}^{-1}$$

Note that the unit of length cancelled.

COEFFICIENT OF VISCOSITY

A steady or non-turbulent flow of fluid is described as 'uniform or orderly flow' or 'laminar flow'. In this case Newton stated that the frictional force F in any region of liquid was proportional to (a) the velocity gradient g there and (b) the area A parallel to fluid flow of the liquid considered. So for so-called Newtonian liquids, $F \propto A \times g$, or

$$F = \eta A g, \tag{1}$$

where η is a constant for the given liquid called its **coefficient of viscosity**. From (1), η can be defined as the frictional force per unit area in a region of unit velocity gradient.

A unit of η, from (1), is N s m^{-2}, since g has a unit s^{-1}. The dimensions of η from (1) are ML^{-1}T^{-1}, so another unit is kg m^{-1} s^{-1}. In practice the name **poise** is adopted for the unit of η (after Poiseuille, a French scientist who first found the formula for the volume per second of fluid flowing along a pipe under steady conditions). The viscosity of one brand of motor oil is of the order 2 poises at 20°C; at 10°C the viscosity of water is of the order 1×10^{-3} poises. The viscosity usually diminishes with temperature but motor oils have been developed which can be used in summer or winter as their values of η are fairly independent of temperature.

From (1), it follows that an area A of $2\times10^{-3} m^2$ in a region of velocity gradient $0.5 s^{-1}$ inside an oil of viscosity coefficient $1 N\ s\ m^{-2}$ has a force F on it$=1\times(2\times10^{-3})\times0.5=10^{-3}N$.

STOKES' LAW AND TERMINAL VELOCITY

Stokes showed that when a sphere of radius a moves through an infinitely large fluid of viscosity coefficient η with a velocity v, then the frictional force F acting on the sphere is given by

$$F=6\pi\eta av \tag{1}$$

Example 12.4 on page 405 shows this formula is dimensionally correct, although the mathematical proof is beyond the scope of this book.

Consider an object such as a small steel ball-bearing X dropped gently into a very viscous liquid such as glycerine in a tall wide jar (Fig 12.9(a)). At first the frictional force F is small because the velocity v of X is small. The net force on X is then equal to its weight minus the liquid upthrust, say $m'g$, so X accelerates.

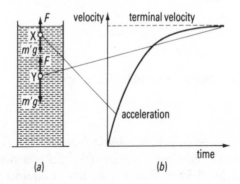

Fig 12.9 Stokes' law and terminal velocity

As X increases in velocity v, the frictional force F increases. So the net force becomes less and so does the acceleration. At a particular velocity, say that at Y, the net force becomes zero. So a maximum velocity, called the *terminal velocity*, is reached by X. Figure 12.9(b) shows how the velocity of X grows from a small value to its constant or terminal velocity.

An object released from an aeroplane at a great height, reaches a terminal velocity owing to the viscosity of air. A small oil drop released in air soon reaches a terminal velocity. As shown in Example 12.5 below, this can be used to find the radius of the small drop.

WORKED EXAMPLES ON VISCOSITY

12.4 DIMENSIONS AND STOKES' FORMULA

Show that Stokes' formula, $F=6\pi\eta av$, for frictional force is dimensionally correct.

The dimensions of $\eta=ML^{-1}T^{-1}$, of $a=L$, of $v=LT^{-1}$

So   dimensions of $\eta av=ML^{-1}T^{-1}\times L\times LT^{-1}$

$$=MLT^{-2}$$

But   dimensions of $F=MLT^{-2}$

So Stokes' law is dimensionally correct. (Note that the numerical quantities in a formula, such as 6π here, can never be found by dimensions as numbers have no dimensions.)

12.5 RADIUS OF DROP USING TERMINAL VELOCITY

A small oil drop falls in air with a terminal velocity of 2×10^{-2}m s^{-1}. The density of the oil is 900kg m^{-3}, the coefficient of viscosity of air is 2×10^{-5}N s m^{-2} and $g=10$m s^{-2}. Estimate the radius of the drop, neglecting the upthrust in air.

The volume of the spherical drop $=\frac{4}{3}\pi a^3$, where a is the radius.

So   weight of drop $=mg=\frac{4}{3}\pi a^3\rho g$,

where ρ is the density of oil. Neglecting the upthrust, which is reasonable since the density of air is very small, the weight of the drop=the frictional force at terminal velocity.

So $\frac{4}{3}\pi a^3\rho g=6\pi\eta av_0$

where v_0 is the terminal velocity. Simplifying,

$$a=\left(\frac{9\eta v_0}{2\rho g}\right)^{1/2}=\left(\frac{9\times2\times10^{-5}\times2\times10^{-2}}{2\times900\times10}\right)^{1/2}$$

$$=1.4\times10^{-5}m$$

So the order of the radius is 10^{-5}m or 10^{-2}mm, which corresponds to a tiny drop.

VERBAL SUMMARY

1 ELASTICITY

$$\text{Young modulus, } E = \frac{F/A \text{ (stress)}}{e/l \text{ (strain)}} \text{ (Unit: N m}^{-2})$$

$$F = EAe/l$$

During *elastic deformation,*
(a)   energy stored$=\frac{1}{2}F \times e=$area of triangle in graph of F against e
(b)   energy per unit volume$=\frac{1}{2}$ stress$\times$strain
(c)   energy recovered when strain removed

Generally, work done during deformation=area between graph of F against e and e-axis.

Heat (not recoverable) is produced during *plastic deformation.*

Plastic deformation is caused by the slip of crystal atomic planes as a result of movement of dislocations (irregularities) inside the crystal.

2 SOLID MATERIALS

(a)   Polycrystalline solids have a large number of very small crystals (grains) inside them, all pointing in different directions. Example: solid metals, silica.
(b)   Glassy solids have atoms in disordered states due to super-cooling of silica. Example: window glass, Pyrex.
(c)   Amorphous solids have their atoms packed in disordered directions. Example: porcelain, ceramics.
(d)   Polymer solids are organic compounds having a long chain of molecules in units called 'monomers' ('mers'). Example: rubber, polythene.
   Rubber molecules are coiled and unwind on stretching rubber. So rubber may have high strain such as 800% whereas metals have low strain such as $\frac{1}{2}$%. Young's modulus is much less for rubber than for metals or glass.

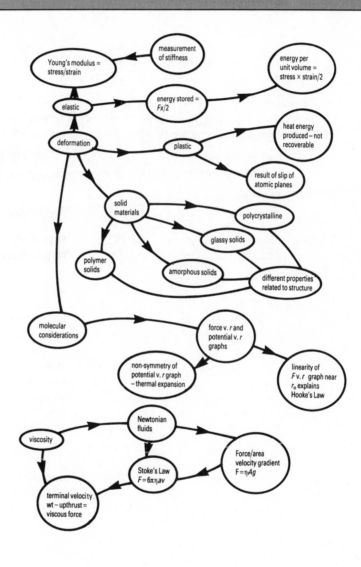

3 MOLECULAR POTENTIAL ENERGY (V) AND FORCE (F)

$F=-dV/dr=$ negative gradient of V against r graph.

At normal separation r_0, p.e. is a minimum and attractive force balances repulsive force ($F=0$).

Hooke's law is due to straight-line variation of F against r round $r=r_0$ value.

Thermal expansion is explained by non-symmetry of V against r graph as temperature rises.

4 VISCOSITY

In Newtonian fluids, frictional force $F=\eta Ag$, where g is the velocity gradient round the area A of the fluid and η is the coefficient of viscosity.

Stokes' law: The frictional force on a sphere of radius a moving with velocity v through a fluid of infinite extent is $F=6\pi\eta av$.

Terminal velocity: A sphere falling through a fluid reaches a terminal constant velocity v_0. In this case, weight − upthrust = frictional force.

12 QUESTIONS

1   In an experiment to measure the Young modulus for copper, a 3.0m length of copper wire of diameter 1.0mm is stretched by 2.0mm when a tensile load of 60N is applied.
(*a*)   Calculate the stress in the copper and the strain it produces.
(*b*)   Hence calculate the Young modulus for copper.
(*c*)   What is the advantage of using a long wire of small diameter in this experiment?   (*AEB* 1984)

2   Compare concisely the microscopic (molecular-scale) structure of glass with that of
(i) a crystalline solid, and
(ii) a liquid.
A freshly-drawn glass thread may be bent to a great extent; but if the thread is first touched with a feather, it shatters after only slight bending. Explain briefly why this is so.   (*W*)

3   A copper wire LM is fused at one end, M, to an iron wire MN (Fig 12.10). The copper wire has length 0.900m and cross-section $0.90 \times 10^{-6} m^2$. The iron wire has length 1.400m and cross-section $1.30 \times 10^{-6} m^2$. The compound wire is stretched; its total length increases by 0.0100m. Calculate
(*a*)   the ratio of the extensions of the two wires,
(*b*)   the extension of each wire,
(*c*)   the tension applied to the compound wire. (Young modulus of copper$=1.30 \times 10^{11} N\,m^{-2}$, of iron$=2.10 \times 10^{11} N\,m^{-2}$.)   (*L*)

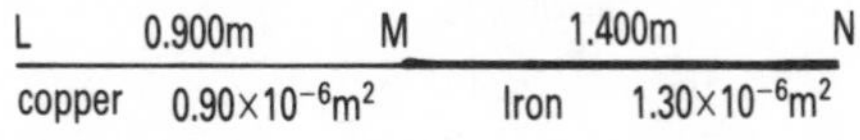

L	0.900m	M	1.400m	N
copper	$0.90 \times 10^{-6} m^2$	Iron	$1.30 \times 10^{-6} m^2$	

Fig 12.10

4   The graph in Fig 12.11 shows a much simplified model of the force between two atoms plotted against their distance of separation. Express
(i) the maximum restoring force between the atoms when they are pulled apart,
(ii) the equilibrium separation of the atoms,
(iii) the energy required to separate the two atoms, in terms of the forces and distances given on the graph.
With the aid of the graph, explain why solids show resistance to both stretching and compressing forces. Explain over what region you

expect Hooke's law to apply. What would this model predict about the elastic limit and the yield point for a material whose atoms followed the model?   (L)

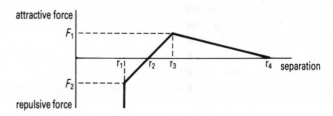

Fig 12.11

5   A long rod is stretched by the application of a tensile force, F; define
           (i) stress,
           (ii) strain, for such a case.
Prove that the work done in the stretching is the area under the (Force-Extension) curve.
     The figures in Fig 12.12 give force-extension curves for two identically shaped rods of a polymeric plastic and of a metal.
           (i) Compare the breaking stresses of the two materials.
           (ii) Compare the work done in breaking them.
           (iii) Explain your results in terms of the molecular structure of such materials.   (W)

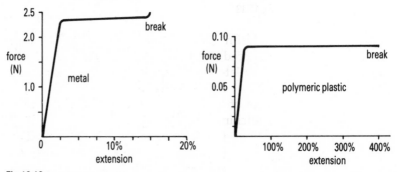

Fig 12.12

6   In the model of a crystalline solid the particles are assumed to exert both attractive and repulsive forces on each other. Sketch a graph of the potential energy between two particles as a function of the separation of the particles. Explain how the shape of the graph is related to the assumed properties of the particles.
     The force F in N, of attraction between two particles in a given solid varies with their separation, d, in m, according to the relation

$$F = \frac{7.8 \times 10^{-20}}{d^2} - \frac{3.0 \times 10^{-96}}{d^{10}}$$

State, giving a reason, the resultant force between the two particles at their equilibrium separation. Calculate a value for this equilibrium separation.

The graph in Fig 12.13 displays a load against extension plot for a metal wire of diameter 1.5mm and original length 1.0m. When the load reached the value at A the wire broke. From the graph deduce values of

(a)   the stress in the wire when it broke,
(b)   the work done in breaking the wire,
(c)   the Young modulus for the metal of the wire.

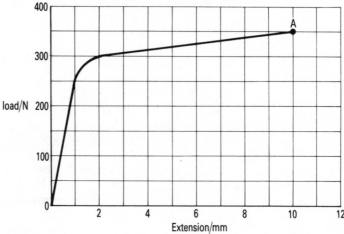

Fig 12.13

Define *elastic* deformation. A wire of the same metal as the above is required to support a load of 1.0kN without exceeding its elastic limit. Calculate the minimum diameter of such a wire.   (O&C)

7   (a)   Figure 12.14(i) is a graph showing the extension of a steel wire of length 1.2m and area of cross-section 0.012mm² alters as a stretching force is applied.

(i) Use the graph to calculate the Young modulus for steel.
(ii) Draw a labelled diagram of an experimental arrangement suitable for obtaining such a set of results.

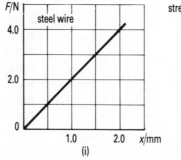
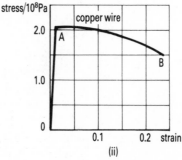

Fig 12.14

(b)   Figure 12.14(ii) shows the results of a similar experiment done with a copper wire. In this case the wire has been stretched until it breaks.

(i) The graph drawn in this instance is a stress-strain curve. Explain *one* advantage of representing the results in this way.

(ii) Account in molecular terms for the behaviour of the wire as it stretched from A to B.

(iii) The copper wire used was 2.0m long and 0.25mm² in cross-section. Calculate the tension of the wire at A and an approximate value for the work done in producing a strain of 0.1.

(c)   A length of rubber cord is suspended from a rigid support and stretched by means of weights attached to the lower end.

(i) Sketch a stress-strain curve to represent the behaviour of such a cord as it is first loaded and then unloaded.

(ii) Suppose the cord were continuously stretched and relaxed at a rapid rate. What might you notice? How would this be explained by the stress-strain curve?   (L)

8 (a)   Figure 12.15 shows the way in which the resultant force F between a pair of atoms in a metal depends on the distance r between them. State the physical significance of the following features of the graph:

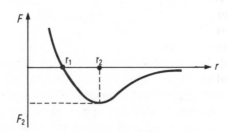

Fig 12.15

(i) positive values of F,
(ii) r_1, the value of r when $F=0$
(iii) the slope of the graph at $r=r_1$,
(iv) r_2, the value of r when F is a minimum,
(v) F_2, the minimum value of F.

(b)   A certain wire of diameter 0.91mm has an unstretched length of 1.80m. For extensions within the Hooke's law region, the constant of proportionality between tension and extension is 5.78×10^4N m^{-1}.

(i) With the aid of a sketch, describe how this constant might be measured.

(ii) Find the Young modulus of the wire.

(c)   Predictions of the values of the Young modulus from curves such as Fig 12.15 are oftenm in reasonable agreement with values obtained from experiments with wires. However, the breaking strains of such wires are normally very much smaller (by a factor of 100 or more) than values predicted from the curves. How do you account for agreement in the former case and disagreement in the latter when both are based on the same simple model of forces between atoms?

(C)

9   (a)   Sketch a graph which shows how the force between two atoms varies with the distance between their centres. With reference to this graph explain why

(i) any reversible change in volume of a solid is always a small fraction of the unstressed volume,

(ii) a metal wire obeys Hooke's law for small extensions.

(b)   (i) Sketch a graph which shows how the length of a rubber cord varies with the tension in the cord as the tension increases from zero until the cord breaks.

(ii) Account for the shape of the curve you have drawn in terms of the molecular structure of rubber.

(iii) In what important way does the structure of a metal at the molecular level differ from that of rubber? How do you account for the large extension of a rubber cord compared with the extension of a mild steel wire of the same dimensions and acted on by the same force?

(c)   Figure 12.16 shows a trolley of total mass 560kg which is used for testing seat belts. The trolley runs on rails and is attached to six identical, parallel rubber cords whose unstretched lengths are 40m each. When the trolley is pulled back far enough to extend the cords by 21m each and then released, it reaches a speed of 15m s^{-1} just as the cords begin to slacken. If the cords are assumed to obey Hooke's law over the full range of their extension, if the system is assumed free of friction and Young modulus for rubber is 2.2×10^7N m^{-2}, calculate

(i) the maximum force applied to each cord,

(ii) the area of cross-section of each cord when stretched.

(L)

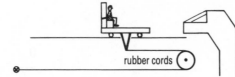

rubber cords

Fig 12.16

ANSWERS

2 (a) 4.0m s^{-2} (b) 7.1m s^{-1}
3 2.5N s
4 (i) 3m s^{-2} (ii) 45kJ (iii) 18kW
5 (i) $v=\sqrt{2gl(1\text{-cos }\theta)}$ (ii) $v/2,3v/2$
6 $15°$
8 $1.0\times10^5\text{J}$, 833kN
9 (i) 1m s^{-1}, 2m s^{-1}

1   (a) 0.25m s^{-1} (b) $1.58\times10^{-2}\text{J}$
2   (i) $45°$ (ii) 1.4m (iii) 5J
3   14.4N, 24.5h
4   (i) 5.2m s^{-1} (ii) 439N (iii) 912N
5   $35.5\times10^3\text{km}$
6   5%
7   (a)   A   (b)   C   (c)   E
8   $0.44\text{s}/0.037\text{m}$ from centre, 0.47m s^{-1}, 7.4m s^{-2}
9   $5.8\times10^{33}\text{kg m}^2\text{ s}^{-1}$, $2.1\times10^{29}\text{J}$
10 (i) 4 rad s^{-2} (ii) 16 rad s^{-1}, 28.8J (iii) 0.46N m

2 (i) $1.41\times10^3\text{N C}^{-1}$, $-3.38\times10^3\text{V}$ (ii) 0.3m from $-3Q$ towards $+Q$
3 (a) $1.45\times10^{-10}\text{N}$ downwards   (b) $1.45\times10^{-9}\text{J}$
4 (a)   (iii) $15\text{V}/16$, 7 times   (b)   (i) none (ii) $1/\epsilon_r$ (iii) ϵ_r
   (iv) $1/\epsilon_r$
5 (a) 0.1C (b) 10^{-2}F (c) $5\text{k}\Omega$
6 (a) 10^{-3}C (b) 1.35V (c) 6.9s
7 (a) 4V (b) $4\mu\text{C}$ (c) $8\mu\text{J}$
8 (c)   (i) $5.8\times10^{-10}\text{F}$ (ii) $1.7\times10^{-8}\text{C}$ (iii) $2.7\times10^{-7}\text{J}$
   (d)   (i) 100V (ii) $8.5\times10^{-7}\text{J}$

QUESTIONS 4 (PAGE 131)

2   (b)   (i) 0.9mA   (ii) 0.09, 0.81, 8.1, 81Ω
3   $R_1=16.7\Omega$, $R_2=12.5\Omega$
4   (b)   (i) 2×10^{-4}A   (ii) 4.99MΩ
5   (a)   360Ω (b)   0.96V
6   (a)   (i) 80cm (ii) 75cm
8   (c) 6.9×10^{-5}m s^{-1}

QUESTIONS 5 (PAGE 172)

1   4×10^{-6}N (about)
3   (a)   (ii) 1.4 rad   (iii) 2mm   (iv) 50μA
    (b)   (i) 0.100Ω
5   (b)   (i) $0.02\pi f \sin 2\pi ft$ (ii) 67.5Hz
6   (c)   (ii) 8m s^{-1}
8   (ii) 0.1T   (iii) 5×10^{-11}J
9   (a)   (i) 5V   (ii) 53°
10   (i) 240V   (ii) 2.1A   (iii) 0.5H, 50μF

QUESTIONS 6 (PAGE 198)

1   (b)   1.5   (c)   longer
2   −15cm
3   (a)   1.5, 3mm   (b)   0.004 rad
4   (i) 0.12 rad   (ii) 0.02   (iii) 6
5   2.15×10^3km
6   20.4mm from eyepiece
7   (ii) 442mm   (iii) 8.6×10^{-2} rad, 9.6   (iv) 32.4mm

QUESTIONS 7 (PAGE 243)

1   23.5 wavelengths
2   (b)   (ii) 1.2mm, 2.4mm from centre
4   (c)   (i) 0.06°   (ii) 0.2mm
5   97 000m^{-1}
7   (b)   (ii) 290.6K
8   (b)   (i) 0.8cm   (ii) 269m s^{-1} (c)   (i) 10/11   (ii) 440Hz
9   (b)   (iii) 5.3×10^{-3}kg m^{-1}
10   7×10^8m
11   5.1×10^9Hz

QUESTIONS 8 (PAGE 296)

2   (c)   (ii) 0.93m   (iii) 3.4×10^5Hz   (iv) 880mm
3   (b) 4.2×10^4V m^{-1}
6   34V
9   (b)   0.505V   (c)   2MΩ

QUESTIONS 9 (PAGE 330)

1   0.56V
2   (b)   (ii) 3.3×10^{15}Hz   (c)   19°
3   (c) 3.1×10^{-11}m
4   17.7kV, 17.7keV
6   3.4×10^{-9}g
7   (a)   13.3h   (b)   0.3h   (c)   960, 9040 min^{-1} (d)   4:1
8   2.8×10^{-12}J
9   (ii) $x=121$, $y=47$, $_0^1$n   (iv) 9×10^{13}J, 520 days

QUESTIONS 10 (PAGE 360)

1   (a)   1250cm^3 (b) 137.5×10^3Pa
2   (i) 499J   (ii) 0.4g
4   (i) 501m s^{-1} (ii) 4.6×10^{-23}kg m s^{-1}, 1000, 7.4×10^{-19}Pa
   (iii) 3.09×10^5Pa
5   (c) 2.45×10^5Pa   (d) 2.5×10^5Pa, 2.1×10^5Pa, 6
7   (a)   1.07   (b)   4/1   (c)   0.87
8   73.8cm^3
10   (iii) 40 atm

QUESTIONS 11 (PAGE 388)

1   881°C
3   12.5°C, 7.1°C
4   90W
5   4480J kg^{-1} K^{-1}
6   (iii) 57.6°C
7   1.24×10^{-20}, 7476J, 483m s^{-1}
8   51°C
9   1.99W, 1829J kg^{-1} K^{-1}

1 (a) $7.6 \times 10^7 \mathrm{N\,m^{-2}}$, 6.7×10^{-4} (b) $1.1 \times 10^{11} \mathrm{N\,m^{-2}}$

3 (a)   1.5   (b)   6mm, 4mm   (c)   780N

4 (i) F_1 (ii) r_2 (iii) $\frac{1}{2}F_1(r_4 - r_2)$

5 (i) 28:1 (ii) 1:1

6 $2.8 \times 10^{-10}\mathrm{m}$;   (a) $2.0 \times 10^8 \mathrm{N\,m^{-2}}$ (b)   3J
    (c) $1.4 \times 10^{11}\mathrm{N\,m^{-2}}$; 3mm

7 (a) $2 \times 10^{11}\mathrm{N\,m^{-2}}$ (b)   50N, 10J

8 $1.6 \times 10^{11}\mathrm{N\,m^{-2}}$

9 1000N, $8.7 \times 10^{-5}\mathrm{m^2}$

INDEX

Index compiled by Peva Keane

More advanced level exam help from Pan

BRODIE'S NOTES
on English Literature texts

This popular and respected series provides reliable guidance on texts commonly set for literature exams in the UK and the Republic of Ireland.

Brodie's Notes on texts set for A level, Highers and Leaving Certificate reflect the deeper critical appreciation and analysis required for study at such levels.

Some selected Brodie's Notes for advanced level are:

William Shakespeare
ANTONY AND CLEOPATRA
CORIOLANUS
HAMLET
KING LEAR
OTHELLO
THE TEMPEST

Geoffrey Chaucer
THE MILLER'S TALE†
THE NUN'S PRIEST'S
  TALE†
THE PARDONER'S TALE†
THE WIFE OF BATH'S TALE†
(With parallel texts)

Jane Austen	EMMA
	MANSFIELD PARK
	SENSE AND SENSIBILITY
Emily Brontë	WUTHERING HEIGHTS
Joseph Conrad	HEART OF DARKNESS
T. S. Eliot	MURDER IN THE CATHEDRAL
Thomas Hardy	RETURN OF THE NATIVE
	TESS OF THE D'URBERVILLES
James Joyce	DUBLINERS
Christopher Marlowe	DR FAUSTUS
Thomas Middleton	
& William Rowley	THE CHANGELING
John Milton	COMUS/SAMSON AGONISTES
Tom Stoppard	ROSENCRANTZ AND GUILDENSTERN ARE
	DEAD
Jonathan Swift	GULLIVER'S TRAVELS
John Webster	THE DUCHESS OF MALFI
	THE WHITE DEVIL
William Wordsworth	THE PRELUDE Books 1 and 2
William Wycherley	THE COUNTRY WIFE
Various	THE METAPHYSICAL POETS

*Written by experienced teachers and examiners who can give you effective advice

*Designed to increase your understanding, appreciation and enjoyment of a set work or author

*With textual notes, commentaries, critical analysis, background information and revision questions

£1.25 each except for †£1.75. On sale in bookshops.

For information write to:

Pan Study Aids & Brodie's Notes
Pan Books Ltd
18–21 Cavaye Place
London SW10 9PG

For business courses at school or college

BREAKTHROUGH BUSINESS BOOKS

This series covers a wide range of subjects for students following business and professional training syllabuses. The Breakthrough books make ideal texts for BTEC, SCOTVEC, LCCI and RSA exams.

Many business teachers and lecturers have praised the books for their *excellent value*, *clear presentation*, *practical, down-to-earth style* and *modern approach to learning*.

'An excellent range of texts at a price students can afford.'
'The self-study presentation and style make these ideal college books.'

The range of 30 includes the following major titles:

BACKGROUND TO BUSINESS	£3.50
BUSINESS ADMINISTRATION	
A fresh approach	£3.95
THE BUSINESS OF COMMUNICATING	£3.95
WHAT DO YOU MEAN 'COMMUNICATION'	£3.95
THE ECONOMICS OF BUSINESS	£2.95
EFFECTIVE ADVERTISING AND PR	£2.95
MANAGEMENT	
A fresh approach	£3.95
MARKETING	
A fresh approach	£3.50
PRACTICAL COST AND MANAGEMENT ACCOUNTING	£3.50
UNDERSTANDING COMPANY ACCOUNTS	£2.95
PRACTICAL BUSINESS LAW	£3.95

On sale in bookshops.

For information write to:

Business Books
Pan Books Ltd
18–21 Cavaye Place
London SW10 9PG